Heat and Thermodynamics

Heat and Thermodynamics

Pramila Shukla
Assistant Professor
Department of Applied Physics
Amity Institute of Applied Sciences
Amity University, Noida (U.P.)

Shivani
Assistant Professor
Department of Applied Physics
Amity Institute of Applied Sciences
Amity University, Noida (U.P.)

Shefali Kanwar
Assistant Professor
Department of Applied Physics
Amity Institute of Applied Sciences
Amity University, Noida (U.P.)

TECHSAR™
NEW DELHI

Published and Printed by

TechSar Pvt. Ltd.
4435-36/7, Ansari Road, Daryaganj
New Delhi–110 002 (India)
E-mail: tech_info@techsarworld.com
Website: www.techsarworld.com

ISBN: 978-93-86768-78-0

Reprint 2025

Preface

This book covers all the topics related to Heat and Thermodynamics as prescribed by most of the universities. It is mainly designed for those streams which have heat and thermodynamics as a single course. The most unique feature of the book is that it contains all the latest concepts. A sufficient number of solved and unsolved numericals are given to facilitate quick understanding of concepts, this is followed by a variety of thought-provoking practice exercises at the end of each chapter. Many examples and problems have been selected from recent papers of various engineering examinations.

Heat and Thermodynamics is the study of the relationships between heat (thermal) and work (dynamics). This book has nine chapters:

The first chapter covers the theory of ideal gases, it includes equation of ideal gases, Brownian motion, degree of freedom, equipartition theorem and specific heat of different gases. The second chapter presents real gases, van der Waals real gas equation and different constant of this equation has been discussed here. Joule-Thomson effect of cooling and its difference with adiabatic expansion have also been discussed in this chapter. Chapter three deals with various transport phenomena like, viscosity, thermal conductivity, diffusion of gas and mean free path. Chapter four covers basic laws of radiation such as black body radiation, Kirchhoff's law, Stefan-Boltzmann law, Wien's law, Rayleigh-Jeans law, and Planck's radiation law. In the fifth chapter, we discuss, zeroth and first law of thermodynamics, heat engine, Carnot's heat engine and second law of thermodynamics which deals with heat engines. Chapter six covers, concept of entropy and the third law of thermodynamics. Chapter seven deals with thermodynamical variables, Maxwell's thermodynamical relations, and their applications for different problems. Chapter eight presents state of matter, latent heat, freezing point, melting point, vaporization, condensation, cooling effect and refrigeration. Chapter nine covers transmission of heat in metals, glass, rubber, bad conductors, convection process and some important applications of convection.

We express our sincere thanks to respected Dr. Ashok K. Chauhan, Founder President, Amity University; Atul Chauhan, Chancellor, Amity University; Prof. Balvinder Shukla, V.C., Amity University, Noida; and Prof. Sunita Rattan, Director, AIAS, Amity University, Noida. We are also grateful to our respected teachers, colleagues, family members and students for their support and encouragement. We would like to express our appreciation to our husbands and our children for their patience, support and understanding throughout the preparation of this book.

We would be grateful to readers for bringing to notice any suggestions/errors for further improvement of this book.

Pramila Shukla
Shivani
Shefali Kanwar

Contents

1

Ideal Gases and Thermodynamics

1.1 IDEAL OR PERFECT GAS

At very low densities or at very low pressure, the mean distance between the molecules is large enough and the intermolecular potential energy may be neglected. In this case, the particles would be independent of one another and gas is referred to be an ideal gas.

An ideal gas follows the equation

$$Pv = RT \tag{1.1}$$

Here, R is the gas constant, and its value is different for different gases. The common gases such as oxygen, nitrogen, atmospheric air, etc., can be taken as ideal gases under specific conditions.

Equation (1.1) can also be written as

$$P(mv) = mRT$$

$$PV = mRT \tag{1.2}$$

Here, V is total volume and m is total mass of the gas. Equation (1.2) can also be written in terms of total number of moles (n) of the gas and molecular mass, M.

$$PV = n\bar{R}T \tag{1.3}$$

where $\bar{R} = MR$ is known as the universal gas constant and its value 8.3145 J/mol·K remains constant for all gases.

Equation (1.3) can also be written in terms of molar volume as

$$\bar{v} = \frac{V}{n} = Mv \tag{1.4}$$

Substituting this in equation (1.3), we obtain the equation for an ideal gas in the given form

$$p\bar{v} = \bar{R}T \tag{1.5}$$

Thus, equations (1.1), (1.2), (1.3) and (1.4) are four different forms of the ideal gas equation of state.

1.2 KINETIC THEORY OF IDEAL GASES

The kinetic theory of gases, which is a microscopic theory, predicts many macroscopically observed quantities. According to this theory, a gas contained in a vessel consists of a huge number of atoms and molecules moving randomly in all direction with all possible velocities. The assumptions made in kinetic theory of gases are as follows:

1. The gas consists of small individual particles which are known as molecules. The molecules of same gas are similar but are different from molecules of other gases.
2. The distance between any two molecules is large in comparison to liquid or solid and thus, the forces of attraction are negligible.
3. The size of the molecule is very small as compared to the mean distance traversed between any two consecutive collisions, i.e., the molecules are point masses. The distance between any two consecutive collisions is known as free path and the mean distance is called the mean free path. If pressure is low, mean free path is low and vice versa.
4. The collison between molecules does not affect the molecular density of the gas.
5. The time of collision is negligible in comparison to the time utilized to complete the free path.
6. The molecules of gases are perfectly hard elastic spheres. Collisions between molecules are perfectly elastic and are of very short duration, compared to time taken between collisions. Thus, there are no frictional forces and all energy is kinetic.
7. The molecular density and the randomness of motion are maintained.

1.3 PRESSURE EXERTED BY A PERFECT GAS

Let us consider a gas contained in a cubical container (Figure 1.1). Suppose that n is the number of molecules of gas in unit volume and m is the mass of single molecule. A molecule collides with the walls of the container with a velocity (u_x, u_y, u_z). This collision is elastic so molecule comes back with the same velocity. There will be no change in the velocity in Y and Z direction.

The change in momentum in x direction due to collision is

$$mu_x - (-mu_x) = 2mu_x$$

In Δt small interval of time, the distance traversed by the molecules is $u_x \Delta t$. Hence, the molecules which exist in the volume $Au_x \Delta t$ can only strike the wall of the container in time. So the number of molecules striking area A of the wall in per unit volume is given by

$$\frac{1}{2}(Au_x \Delta t)n$$

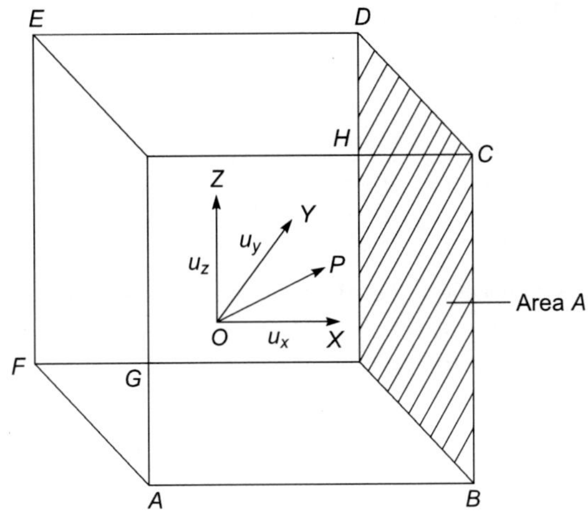

Figure 1.1 Pressure exerted by the molecule of a perfect gas

So the net momentum imparted to the wall in time interval Δt is

$$\Delta p_x = \frac{1}{2}(Au_x\Delta t)n \times 2mu_x = n(mu_x^2 A\Delta t)$$

Hence, force acted on the area A of wall $= \frac{\Delta p_x}{\Delta t} = n(mu_x^2 A)$

and the pressure on the wall of container, $P_x = \frac{n(mu_x^2 A)}{A} = nmu_x^2$

The different molecules of gas will move with different velocities, thus we can replace u_x^2 by its mean value $\overline{u}_x^2$. Hence,

$$\text{Pressure } P_x = nm\overline{u}_x^2$$

As the pressure of a gas is equal in all directions, the net pressure will be given by

$$P = \frac{P_x + P_y + P_z}{3} P_x = \frac{1}{3}nmu^2$$

or

$$P = \frac{1}{3}\rho u^2 \tag{1.6}$$

Here, $u^2 = \overline{u}_x^2 + \overline{u}_y^2 + \overline{u}_z^2$ is a mean square velocity of the molecules and ρ is the density of gas. Equation (1.1) gives the expression for pressure exerted by gas molecules on the wall of the container.

So mean square velocity is given by

$$u^2 = \frac{3P}{\rho}$$

or
$$u = \sqrt{\frac{3P}{\rho}} \quad (1.7)$$

Here, u is the square root of the average of squares of the velocities of different molecules, i.e., root mean square velocity.

Equation (1.6) can also be written as

$$P = \frac{1}{3}\rho u^2$$

$$= \frac{2}{3}\left(\frac{1}{2}\right)\rho u^2$$

$$= \frac{2}{3}E \quad (1.8)$$

where, $E = \left(\frac{1}{2}\right)\rho u^2$ is the translational kinetic energy per unit volume of the perfect gas and ρ is the density of the gas.

Again from equation (1.7), we can write

$$P = \frac{1}{3}\rho u^2 = \frac{1}{3}\frac{M}{V}u^2$$

or
$$PV = \frac{1}{3}Mu^2$$

We know that from perfect gas equation of 1 mole

$$PV = RT$$

or
$$\frac{1}{3}Mu^2 = RT$$

or
$$\frac{1}{2}Mu^2 = \frac{3}{2}RT$$

$$\frac{1}{2}mNu^2 = \frac{3}{2}RT$$

$$\frac{1}{2}mu^2 = \frac{3R}{2N}T = \frac{3}{2}kT$$

Mean kinetic energy per molecule,

$$\bar{E} = \frac{3}{2}kT \tag{1.9}$$

where, $M = mN$ and N is Avogadro's number and k is Boltzmann constant. Hence, from equation (1.9), it is clear that mean kinetic energy per molecule is directly proportional to absolute temperature of the perfect gas and independent of volume and pressure of the gas. As the temperature of gas increases, the mean kinetic energy also increases and as heat is taken from the gas, mean kinetic energy of the molecules of gas decreases. In the state of absolute zero temperature of gas, the mean kinetic energy will also be zero and the molecules will be in the state of rest. Before molecules reach in the absolute zero temperature state, all gases convert their states to solids and liquids.

1.4 BROWNIAN MOTION

If a solid particle of microscopic size is immersed into a liquid, highly irregular movement is observed under a microscope. As the surface area of the immersed particle is small, the number of molecules of liquid with which it collides at a time is relatively very small. The movement of particle is spontaneous and continuous. This non-stop and irregular motion was first observed in 1827 by botanist Robert Brown and is therefore called the Brownian motion.

This motion gets affected with the rise in temperature and viscosity of the liquid. Smaller size particles appear to be more agitated than the larger ones. The laws of kinetic theory of gases are also applicable to Brownian particles. There are some essential features of Brownian motion:

(i) The motion of the particles is continuous.
(ii) Each particle has irregular and random motion.
(iii) The bigger size particles appear to be less agitated than the smaller size particles.
(iv) On increasing the temperature, motion becomes more random.
(v) The postulates of kinetic theory of ideal gases are also applicable to Brownian motion.
(vi) The motion cannot be observed with bigger size particles.

1.5 DETERMINATION OF AVOGADRO'S NUMBER AND EXPERIMENTAL VERIFICATION OF BROWNIAN MOTION

Under the action of the earth's gravitation, the Brownian particles of a gas are in equilibrium. Consider two different layers of the particles at height h and $(h + dh)$, ρ is the density of substance at height h, P and $(P + dP)$, the pressure at height h and $(h + dh)$, respectively.

On considering unit area of the layer, the force due to the effect of gravity between two layers = $\rho g dh$.

Resultant net force on the layer = $(P+dP)-P+\rho gdh$

At equilibrium, the resultant net force between the layers must be zero, hence,

$$(P+dP)-P+\rho gdh=0$$

or

$$\rho gdh+dP=0$$

or

$$dP=-\rho gdh \tag{1.10}$$

We know that the gas equation for a perfect gas is

$$P=nkT$$

$$\rho=mn$$

and

$$dP=dnKT$$

where, m is the molecular mass and n is the number of molecules present in unit volume.

Substituting the values of dP and ρ in equation (1.10), we get

$$dnKT=-mngdh$$

$$\frac{1}{n}dn=-\frac{dh}{KT}mg \tag{1.11}$$

If at the bottom height h = 0 and concentration is $n_{0,}$ on integrating equation (1.11) between the limit n_0 and n, we get

$$\int_{n_0}^{n}\frac{1}{n}dn=-\int_{0}^{h}\frac{dh}{KT}mg$$

$$\log\frac{n}{n_0}=-\frac{1}{kT}mgh$$

$$n=n_0e^{-\frac{mgh}{kT}} \tag{1.12}$$

Equation (1.12) shows that the concentration of motion decreases exponentially with respect to height.

We can write equation (1.12) as

$$\frac{R}{mgh}T\log\left(\frac{n}{n_0}\right)=N$$

$$N=\frac{R}{V(\rho-\rho_0)gh}T\log\left(\frac{n}{n_0}\right) \tag{1.13}$$

where, $m=V(\rho-\rho_0)$ is the effective mass of the particle. We can calculate the value of Avogadro's number (N) using equation (1.13).

1.6 LANGEVIN'S THEORY OF BROWNIAN MOTION

A suspended particle experienced the following two forces:

(i) Frictional force which is proportional to the velocity of the molecule and is given by

$$f \cdot \frac{dy}{dt} = 6(\pi\eta av)$$

(ii) External force by means of other influences.

Let us consider the particle is moving in a particular direction y. The force experienced by the suspended particle is expressed by the equation,

$$m \cdot \frac{d^2y}{dt^2} = -f \cdot \frac{dy}{dt} + F \tag{1.14}$$

where, $-f \cdot \frac{dy}{dt}$ is the component of the frictional force in the y direction. F is the force due to external influences.

Multiplying equation (1.14) by y, we get

$$my \cdot \frac{d^2y}{dt^2} = -fy \cdot \frac{dy}{dt} + Fy \tag{1.15}$$

On differentiating y^2 we get,

$$\frac{dy^2}{dt} = 2y \cdot \frac{dy}{dt} \text{ and } y \cdot \frac{dy}{dt} = \left(\frac{1}{2}\right)\frac{dy^2}{dt}$$

$$\frac{d}{dt}\left(\frac{dy^2}{dt}\right) = 2\left[\frac{dy}{dt} \cdot \frac{dy}{dt} + y \cdot \left(\frac{d^2y}{dt^2}\right)\right]$$

$$\frac{d}{dt}\left(\frac{dy^2}{dt}\right) = 2\left[\left(\frac{dy}{dt}\right)^2 + y \cdot \left(\frac{d^2y}{dt^2}\right)\right]$$

$$y \cdot \left(\frac{d^2y}{dt^2}\right) = \frac{1}{2}\left[\frac{d}{dt}\left(\frac{dy^2}{dt}\right) - \left(\frac{dy}{dt}\right)^2\right]$$

Substituting the above values in equation (1.15) we obtained,

$$m\frac{1}{2}\left[\frac{d}{dt}\left(\frac{dy^2}{dt}\right) - \left(\frac{dy}{dt}\right)^2\right] = -f \cdot \frac{1}{2} \cdot \frac{dy^2}{dt} + Fy \tag{1.16}$$

Equation (1.16) is the motion of a single particle. Hence, mean value of the above equation can be taken for all particles and will be expressed as

$$\overline{m\frac{1}{2}\left[\frac{d}{dt}\left(\frac{dy^2}{dt}\right)-\left(\frac{dy}{dt}\right)^2\right]}=\overline{-f\cdot\frac{1}{2}\cdot\frac{dy^2}{dt}+Fy}$$

$$m\frac{1}{2}\frac{d}{dt}\overline{\left(\frac{dy^2}{dt}\right)}-m\overline{\left(\frac{dy}{dt}\right)^2}=-f\cdot\frac{1}{2}\cdot\overline{\frac{dy^2}{dt}}+\overline{Fy} \tag{1.17}$$

Since the force F varies in an irregular way, it can be taken as $\overline{Fy}=0$.
We know that from the law of equipartition of velocities,

$$m\overline{\left(\frac{dy}{dt}\right)^2}=kT$$

and
$$\overline{\frac{dy^2}{dt}}=\mathcal{P}$$

substituting the above values in equation (1.17), we get

$$m\frac{1}{2}\frac{d\mathcal{P}}{dT}+\frac{1}{2}f\mathcal{P}=kT$$

or
$$\frac{d\mathcal{P}}{dT}+\frac{f}{m}\mathcal{P}=\frac{2}{m}kT \tag{1.18}$$

The general solution of the above equation is

$$\mathcal{P}=\frac{2}{f}kT+Be^{-(f/m)t} \tag{1.19}$$

where B is constant.

In the above expression, the particle size is very small so f/m is very large and the value of $e^{-(f/m)t}$ is very small and thus we can neglect it.

Hence,
$$\mathcal{P}=\frac{2}{f}kT=\frac{\overline{dy^2}}{dt} \tag{1.20}$$

Integrating the above equation in the range of time interval between $t=0$ to $t=\tau$, we obtain

$$\overline{y^2}-\overline{y_0^2}=\frac{2}{f}kT\tau$$

At initial time, $t=0$ and $y_0=0$. For very small values of y^2, we can write it as Δy^2, hence, the above equation becomes

$$\overline{\Delta y^2}=\frac{2}{f}kT\tau$$

The mean square value of $\overline{\Delta y^2}$ can be calculated. In Brownian motion of the particle, the particle makes a number of collisions and travels in a zig-zag path. In such case, the value of $\overline{\Delta y^2}$ is less related to the original path.

As we know from the Stokes' formula of viscosity

$$f\frac{dy}{dt} = 6\pi\eta r v = 6\pi\eta r\frac{dy}{dt}$$

as

$$f = 6\pi\eta r$$

So,

$$\overline{\Delta y^2} = \frac{2}{6\pi\eta r}kT\tau = \frac{1}{3}\left(\frac{kT\tau}{\pi\eta r}\right)$$

or

$$(\overline{\Delta y^2})^{1/2} = \frac{2}{6\pi\eta r}kT\tau = \left(\frac{kT\tau}{3\pi\eta r}\right)^{1/2} \quad (1.21)$$

From the above expression, the following results are obtained:

(i) $\overline{\Delta y^2}$ is independent of mass of the particle

(ii) $(\overline{\Delta y^2})^{1/2} \propto \frac{1}{\eta^{1/2}}$

(iii) $(\overline{\Delta y^2})^{1/2} \propto T^{1/2}$

(iv) $(\overline{\Delta y^2})^{1/2} \propto \tau^{1/2}$

From the relation $(\overline{\Delta y^2})^{1/2} \propto T^{1/2}$, it is clear that effect of temperature is very small. However, viscosity depends on temperature and decreases with increase in temperature. Hence, the real temperature effect is very small in comparison to viscosity effect.

1.7 DEGREE OF FREEDOM

The degree of freedom of a system f is defined as the total number of independent variables which are required to describe the state of the system. To specify the state of a gas consisting of N particles enclosed in a box, we require $3N$ quantum numbers, so we can say that each possible state of a dynamical system can be described by some set of quantum numbers, f. This set of numbers (f) is called the degree of freedom of dynamical system. Mathematically, f is defined as $f = 3N–k$, where k is the number of independent ways between the particles.

1.8 DEGREE OF FREEDOM OF MONOATOMIC, DIATOMIC AND TRIATOMIC GASES

(i) Monoatomic Gases: The gases which consist of a single atom are called monoatomic gases, like Helium (He), Neon (Ne), Argon (Ar), etc. A monoatomic gas molecule has three degrees of freedom, which are due to translatory motion (Figure 1.2).

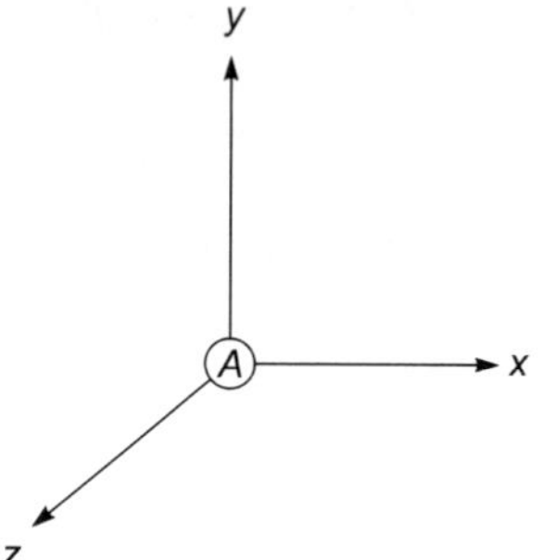

Figure 1.2 Monoatomic gas molecule

(ii) Diatomic gases: The gases which consist of two atoms at a fixed distance between them are called diatomic gases, e.g., like N2, HCl, O2, Cl2, CO, etc. A diatomic gas molecule has a dumbbell shape and 5 degrees of freedom. Out of five, three degrees of freedom are due to the translational motion and two degrees of freedom correspond to rotational motion (Figure 1.3).

The diatomic gases change their behavior at high temperature. At very high temperature, in addition to rotational and translational motion, a diatomic gas molecule has vibrational motion also. So at high temperature, the diatomic gas molecule has total 7 degrees of freedom (3 translational + 3 rotational + 2 vibrational).

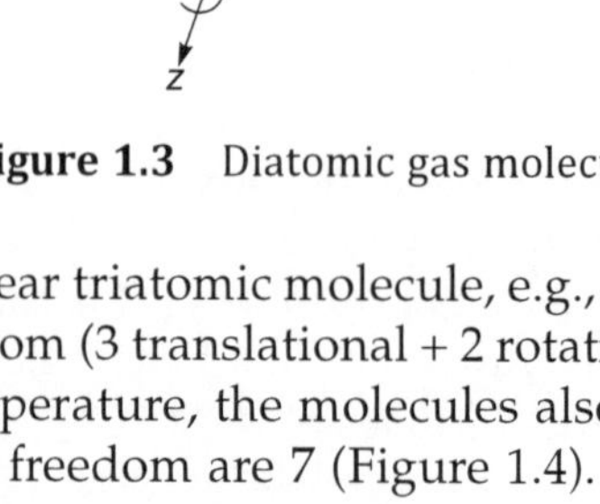

Figure 1.3 Diatomic gas molecule

(iii) Triatomic Gases: A linear triatomic molecule, e.g., CO_2, CH_2, CS_2, HCN etc., has total 5 degrees of freedom (3 translational + 2 rotational), similar to diatomic gas molecule. At high temperature, the molecules also possess vibrational motion, so the total degrees of freedom are 7 (Figure 1.4).

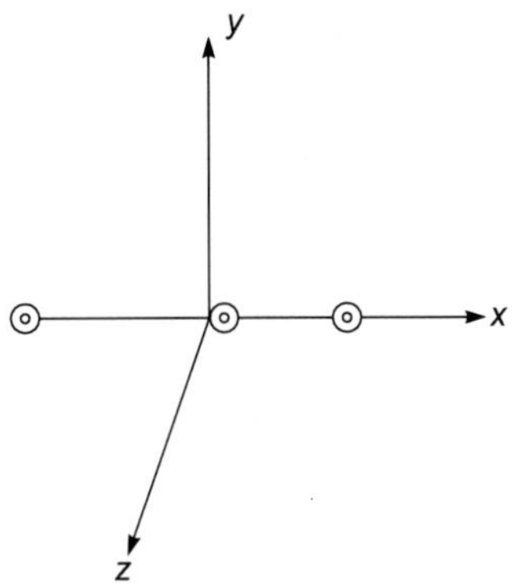

Figure 1.4 Linear triatomic gas molecule

A nonlinear triatomic gas molecule, e.g., SO_2, H_2O, etc., has total 6 degrees of freedom (3 translational + 3 rotational) (Figure 1.5).

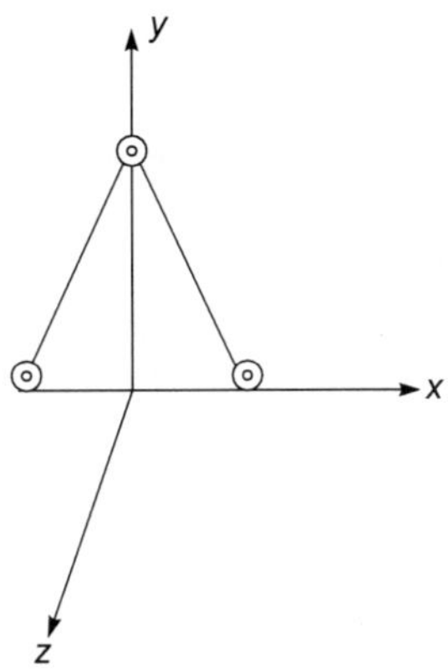

Figure 1.5 Nonlinear triatomic gas molecule

1.9 EQUIPARTITION THEOREM OF ENERGY

This theorem states that, in thermal equilibrium of a system at temperature T, the total energy is equally distributed among all the different degrees of freedom. This average energy associated with each different degree of freedom is $\frac{1}{2}kT$, where k is Boltzmann constant.

According to kinetic theory of gases, the average kinetic energy of a molecule at temperature T is given by

$$\bar{E} = \frac{1}{2}m\bar{u}^2 = \frac{3}{2}kT \tag{1.22}$$

We know that the mean square velocity $\bar{u}^2$ of three axes X, Y and Z is defined by

$$\bar{u}^2 = u_x^2 + u_y^2 + u_z^2 \tag{1.23}$$

We also know that the average velocities along all three axes are same.

So, $$u_x^2 = u_y^2 = u_z^2 \text{ or } u_x^2 = u_y^2 = u_z^2 = \frac{1}{3}\bar{u}^2$$

or $$\begin{bmatrix} \frac{1}{2}mu_x^2 = \frac{1}{2}kT \\ \frac{1}{2}mu_y^2 = \frac{1}{2}kT \\ \frac{1}{2}mu_z^2 = \frac{1}{2}kT \end{bmatrix} \tag{1.24}$$

Therefore, the total kinetic energy $\frac{3}{2}kT$ is equally distributed in each degree of freedom and the energy associated with each degree of freedom is $\frac{1}{2}kT$.

1.10 SPECIFIC HEATS, ATOMICITY AND DEGREE OF FREEDOM

The molecules of ideal gases are hard spheres and are not deformed due to collisions between the molecules. As there is no internal potential energy, so the entire internal energy is kinetic. This kinetic energy depends on temperature. When a gas is heated at a particular constant volume, a temperature change occurs due to supplied heat energy and if we increase this temperature by one degree kelvin, the molar specific heat at constant volume, $C_v = \left(\frac{dU}{dT}\right)_v$ is obtained. If we keep constant pressure in place of constant volume, then we obtain the specific heat at constant pressure.

The total kinetic energy of ideal gas molecules is wholly kinetic. For f degree of freedom, according to the equipartition theorem of energy, mean energy per molecule

$$= \frac{1}{2}(fkT)$$

Average total energy per gm molecule, $U = N\left(\frac{1}{2}fkT\right) = \frac{1}{2}(fRT)$

where $\frac{R}{N} = k$, N is Avogadro's number and R is universal gas constant.

The molar specific heat at constant volume is defined as

$$C_V = \frac{dU}{dT} = \frac{d}{dT}\left(\frac{fRT}{2}\right) = \frac{fR}{2}$$

As we know that molar-specific heat at constant pressure is

$$C_P = R + C_V = R + \frac{fR}{2} = R\left(1 + \frac{f}{2}\right)$$

So we can find the ratio of specific heat of a gas at constant pressure and at constant volume

$$\frac{C_P}{C_V} = \frac{R\left(1 + \frac{f}{2}\right)}{\frac{R}{2}f}$$

or $$\gamma = \frac{C_P}{C_V} = \left(1 + \frac{f}{2}\right) \tag{1.25}$$

1.11 SPECIFIC HEATS OF MONOATOMIC, DIATOMIC AND POLYATOMIC GASES

(i) Monoatomic Gas: As we know that a monoatomic gas has three degrees of freedom and energy associated with one degree of freedom is $\frac{kT}{2}$. So kinetic energy associated with all three degrees of freedom is $\frac{3kT}{2}$.

If we consider 1 gm molecule of a gas, then energy associated with it is internal energy and is defined as

$$= \frac{3NkT}{2}$$

$$U = \frac{3RT}{2} \tag{1.26}$$

Specific heat of a gas molecule at constant volume, C_V is

$$C_V = \frac{dU}{dT} = \frac{3R}{2}$$

and $$C_P = R + C_v = \frac{5R}{2}$$

Therefore, $$\gamma = \frac{C_P}{C_V} = \frac{5R/2}{3R/2} = \frac{5}{3} = 1.67 \tag{1.27}$$

This value of γ is experimentally true for monoatomic gases, e.g., Ar and He.

(ii) Diatomic Gas: As we know that at ordinary temperature, a diatomic gas has five degrees of freedom and energy associated with one degree of freedom is $\frac{kT}{2}$. So kinetic energy associated with all three degrees of freedom is $\frac{5kT}{2}$.

If we consider 1 gm molecule of a gas, then energy associated with it is internal energy and is defined as

$$U = \frac{5NkT}{2}$$

$$= \frac{5RT}{2} \tag{1.28}$$

Specific heat of a gas molecule at constant volume, C_V is

$$C_V = \frac{dU}{dT} = \frac{5R}{2}$$

and $$C_P = R + C_v = \frac{7R}{2}$$

Therefore, $$\gamma = \frac{C_P}{C_V} = \frac{7R/2}{5R/2} = \frac{7}{5} = 1.40 \tag{1.29}$$

This value of γ is experimentally true for diatomic gases, e.g., O_2, N_2, H_2, etc.

(iii) Triatomic Gas: As we know that at ordinary temperature, a triatomic gas has six degrees of freedom and energy associated with one degree of freedom is $\frac{kT}{2}$.

So kinetic energy associated with all three degrees of freedom is $\frac{6kT}{2} = 3kT$.

If we consider 1 gm molecule of a gas, then energy associated with it is internal energy and is defined as

$$U = 3NkT$$

$$= 3RT \tag{1.30}$$

Specific heat of a gas molecule at constant volume, C_V is

$$C_V = \frac{dU}{dT} = 3R$$

and $$C_P = R + C_v = 4R$$

Therefore, $$\gamma = \frac{C_P}{C_V} = \frac{4R}{3R} = \frac{4}{3} = 1.33 \tag{1.31}$$

At high temperature, the total degrees of freedom are seven so

$$\gamma = \frac{C_P}{C_V} = \frac{9R}{7R} = \frac{9}{7} = 1.29 \tag{1.32}$$

For H_2 molecule at low temperature of the order 70 K, only translatory motion is possible, like monoatomic gas molecule.
Therefore,

$$\gamma = \frac{C_P}{C_V} = \frac{5R/2}{3R/2} = \frac{5}{3} = 1.67 \tag{1.33}$$

1.12 INTERPRETATION OF TEMPERATURE BY KINETIC THEORY

In a system of N molecules of monatomic gas, if E is the net energy of the system, then total energy will be divided in translational, rotational and vibrational energies. If E_t is the translational energy of the system, then according to the theorem of equipartition, we can write

$$\frac{1}{N}E_t = \frac{3}{2}kT \tag{1.34}$$

where, T is temperature and k is Boltzmann constant. But we know that the mean translational energy per molecule,

$$\bar{E} = \frac{1}{N}E_t = \frac{3}{2}kT \tag{1.35}$$

Hence, $$\bar{E} \propto T \tag{1.36}$$

Therefore, "the mean translational kinetic energy per molecule of a monoatomic ideal gas is proportional to the absolute temperature of the gas".

So the mean translational kinetic energy of the gas is zero at $T = 0$.

Therefore, the temperature at which the mean translational kinetic energy of each molecule of the ideal gas is zero is known as the absolute zero temperature or we can say that the temperature at which molecular translational velocities of an ideal gas approach to zero is known as absolute temperature.

Maxwell gives an interpretation of temperature that in a mixture of gases, energy can be exchanged between different kinds of molecules until the mean translational kinetic energy of each molecule is equal. Therefore, if different gases having the same translational kinetic energy are mixed, their net exchange of energy will be zero. Therefore, the ideal gases having same translational kinetic energy will be at same temperature, or we can say that the mean translational kinetic energy is proportional to the absolute temperature.

1.13 MAXWELL-BOLTZMANN LAW OF SPEED DISTRIBUTION

Maxwell and Boltzmann made some assumptions to give their speed distribution law. They assumed that the molecules of a gas move randomly in all directions and they collide with each other as well as with the wall of the container. As a result of collision, they change their direction frequently. Due to exchange of kinetic energy between the molecules of a gas, the speed of gas molecules is not fixed and different molecules of a gas have different speeds, ranging from minimum to high value. The molecular distribution of a gas on the basis of their speed was studied by Maxwell-Boltzmann. According to Maxwell's distribution law of speed, the number of molecules having speed v is given by

$$dN = 4\pi N \times \left(\frac{m}{2\pi kT}\right)^{\frac{3}{2}} v^2 e^{-\left(\frac{m}{2kT}\right)v^2} dv \tag{1.37}$$

and the fraction of number of molecules is given by

$$\frac{dN}{N} = 4\pi v^2 \left(\frac{m}{2\pi kT}\right)^{\frac{3}{2}} e^{-\left(\frac{m}{2kT}\right)v^2} dv \tag{1.38}$$

They plotted a graph between the fraction of number of molecules in Y direction and speed v in X direction at different temperatures (Figure 1.6). This graph is known as Maxwell's velocity distribution curve. The curve first rises parabolically, gets a maximum value and then falls exponentially to zero.

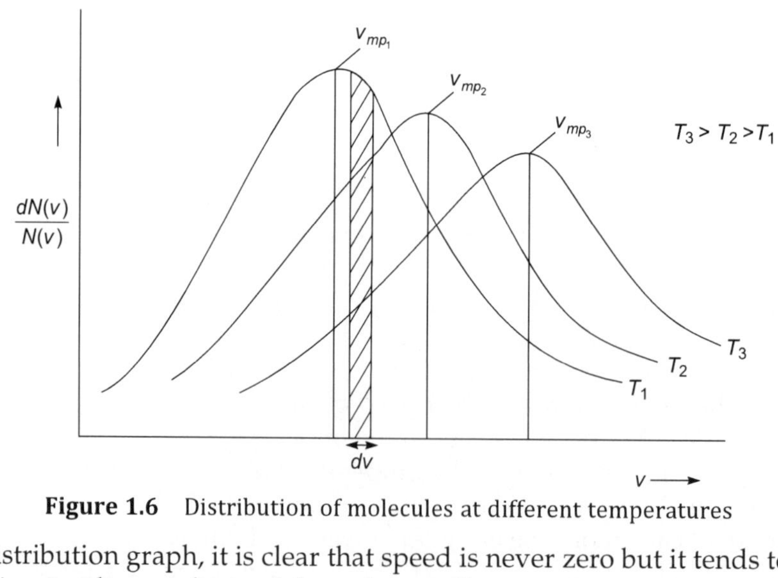

Figure 1.6 Distribution of molecules at different temperatures

From distribution graph, it is clear that speed is never zero but it tends to zero and the following results are obtained from the graph:

(i) The peak of the distribution curve corresponds to speed of the maximum number of fraction of molecules and is known as most probable speed and represented by v_{mp}.

(ii) The mean speed $\bar{v}$ is greater than v_{mp}.

(iii) The area under the shaded region of the curve provides the fraction of molecules lying in the velocity range v to $v+dv$.

(iv) The speed of the gas molecules is higher for higher temperatures.

(v) The whole area between the velocity curve and x-axis provides the total molecules n, having speed in the range 0 to ∞.

1.14 AVERAGE, ROOT MEAN SQUARE AND MOST PROBABLE SPEED OF GAS MOLECULES

1. Average or Mean Speed

Average or mean speed is defined as arithmetic mean of the molecular speed at a particular given temperature.

Let us suppose that there are N_r molecules having speed v_r, the mean speed $\bar{v}$ is defined by

$$\bar{v} = \frac{\sum_r N_r \times v_r}{\sum_r N_r} = \frac{\sum_r N_r \times v_r}{N}$$

Here, the molecular speed is approximately continuous and so, summation may be replaced by integration, we get

$$\bar{v} = \frac{1}{N}\int_0^\infty v \cdot dN$$

$$= \frac{1}{N}\int_0^\infty v \cdot 4\pi N \times \left(\frac{m}{2\pi kT}\right)^{\frac{3}{2}} v^2 e^{-\left(\frac{m}{2kT}\right)v^2} dv$$

$$= 4\pi \times \left(\frac{m}{2\pi kT}\right)^{\frac{3}{2}} \int_0^\infty v^3 e^{-\left(\frac{m}{2kT}\right)v^2} dv \tag{1.39}$$

We know that $\int_0^\infty v^3 e^{-\alpha v^2} dv = \frac{1}{2\alpha^2}$, here $\alpha = \frac{m}{2kT}$

Using above definite integral, equation (1.39) will be

$$\bar{v} = 4\pi \times \left(\frac{m}{2\pi kT}\right)^{\frac{3}{2}} \times \left[\frac{2(kT)^2}{m^2}\right]$$

$$= \sqrt{\frac{8kT}{\pi m}} = 1.596\sqrt{\frac{kT}{m}} \tag{1.40}$$

Equation (1.40) gives the expression for the average or mean speed of molecules of gas.

2. Root Mean Square Speed

The square root of the mean of squares of the speed of different gas molecules is defined as the root mean square speed.

The mean of the square of speed is given by,

$$\overline{v^2} = \frac{1}{N}\int_0^\infty v^2 \cdot dN$$

$$= \frac{1}{N}\int_0^\infty v^2 \cdot 4\pi N \times \left(\frac{m}{2\pi kT}\right)^{\frac{3}{2}} v^2 e^{-\left(\frac{m}{2kT}\right)v^2} dv$$

$$= 4\pi \times \left(\frac{m}{2\pi kT}\right)^{\frac{3}{2}} \int_0^\infty v^4 e^{-\left(\frac{m}{2kT}\right)v^2} dv \tag{1.41}$$

If $\alpha = \frac{m}{2kT}$, we know that $\int_0^\infty v^4 e^{-\alpha v^2} dv = \frac{3}{8}\left(\sqrt{\frac{\pi}{\alpha^5}}\right)$,

Using the above definite integral, equation (1.41) will be

$$\overline{v^2} = 4\pi \times \left(\frac{m}{2\pi kT}\right)^{\frac{3}{2}} \times \frac{3}{8} \times \sqrt{\left[\frac{\pi (2kT)^5}{m^5}\right]}$$

$$= 4\pi \times \left(\frac{m}{2\pi kT}\right)^{\frac{3}{2}} \times \frac{3}{8} \times \sqrt{\left[\frac{32\pi (kT)^5}{m^5}\right]}$$

$$= \frac{3kT}{m}$$

This is the expression for the mean of square of speed. Taking square root of the above expression will give the root mean square speed, i.e.,

$$v_{rms} = \sqrt{\overline{v^2}} = \sqrt{\frac{3kT}{m}} = 1.732\sqrt{\frac{kT}{m}} \tag{1.42}$$

The above expression is for root mean square speed of a gas molecule.

3. Most Probable Speed

The speed attained by maximum number of gas molecules in a given sample of gas at a particular temperature is defined as most probable speed of gas molecule.

To find out the most probable speed of gas molecules, the probability or fraction of molecules in the velocity range v to $v + dv$ will be maximum, i.e.,

$$\frac{d}{dv}\left(\frac{dN}{N}\right)=\frac{d}{dv}\left[4\pi v^2\left(\frac{m}{2\pi kT}\right)^{\frac{3}{2}} e^{-\left(\frac{m}{2kT}\right)v^2} dv\right]=0$$

or

$$\frac{d}{dv}\left[v^2 e^{-\left(\frac{m}{2kT}\right)v^2} dv\right]=0$$

or

$$\left[2ve^{-\left(\frac{m}{2kT}\right)v^2} - v^2 \times \frac{m}{2kT} \times 2v\right]=0$$

$$2v \times e^{-\left(\frac{m}{2kT}\right)v^2} = v^3 \times \frac{m}{kT} \times e^{-\left(\frac{m}{2kT}\right)v^2}$$

$$2v = v^3 \times \frac{m}{kT}$$

$$v^2 = \frac{2kT}{m}$$

This expression gives the most probable speed of gas molecules.

From the expression of mean speed, root mean square speed and most probable speed, it is observed that $v_{rms} > \overline{v} > v_{mp}$ and the ratio between the velocities is given by

$$v_{rms} : \overline{v} : v_{mp} = 1.73 : 1.59 : 1.41 \tag{1.43}$$

Numerical Examples

1. Determine the root mean square speed of the oxygen gas molecules at 300°K.

Solution: The r.m.s. velocity of oxygen molecules at N.T.P. is given by

$$(V_{rms})_1 = \sqrt{\frac{3P}{\rho}}$$

Where, $P = 76 \times 13.6 \times 980 = 1.012928 \times 10^6$ dynes/cm^2,

$\rho = 16 \times 89 \times 10^{-6} = 1.424 \times 10^{-3}$

Hence, $$(V_{rms})_1 = \sqrt{\frac{3 \times 1.0129 \times 10^6}{1.424 \times 10^{-3}}} = 46.2 \times 10^3 \text{ cm/sec.}$$

Let us assume that the root mean square speed of oxygen molecule at 300°K is $(V_{rms})_2$,

$$\frac{(V_{rms})_2}{(V_{rms})_1} = \sqrt{\frac{T_1}{T}}$$

or
$$(V_{rms})_2 = \sqrt{\frac{T_1}{T}} \times (V_{rms})_1$$

$$= \sqrt{\frac{300}{273}} \times 46.2 \times 10^3$$

$$= 47.99 \times 10^3 \text{ cm/sec.}$$

2. The oxygen gas molecules have the same root mean square speed as that of hydrogen gas molecules at 173°K. Find out the temperature of oxygen gas molecules.

Solution: We know that the energy of gas molecules is given by

$$\frac{3}{2}kT = \frac{1}{2}mv_{rms}^2$$

For oxygen gas molecules

$$\frac{3}{2}kT_1 = \frac{1}{2}m_1(v_{rms}^2)_{\text{oxygen}}$$

For hydrogen gas molecules

$$\frac{3}{2}kT_2 = \frac{1}{2}m_2(v_{rms}^2)_{\text{hydrogen}}$$

So from above two equations, we get

$$\frac{T_2}{T_1} = \frac{m_2(v_{rms}^2)_{\text{hydrogen}}}{m_1(v_{rms}^2)_{\text{oxygen}}}$$

Given that,

$$(v_{rms}^2)_{\text{hydrogen}} = (v_{rms}^2)_{\text{oxygen}}$$

So,
$$T_2 = \frac{m_2}{m_1} \times T_1 = 16 \times 173$$

$$= 27.68 \times 10^2 \text{ °K.}$$

3. Find out the number of molecules present in one meter cube of an perfect gas at N.T.P.

Solution: We know that for 1 gm gas molecule, gas equation is

$$PV = (Nk) \times T$$

If n is the number of molecules present in per cc volume, then the gas equation is

$$P = (nk) \times T$$

$$n = \frac{P \times N}{T \times R} \qquad \left(k = \frac{R}{N}\right)$$

Number of molecules present in one meter cube volume is

$$a = n \times 10^6$$

$$a = \frac{P \times N}{T \times R} \times 10^6$$

$$= \frac{(76 \times 13.6 \times 980) \times 6.02 \times 10^{23}}{273 \times 8.32 \times 10^7} \times 10^6$$

$$= 26.9 \times 10^{24}$$

4. Find out the temperature at which the root mean square velocity of a nitrogen molecule will be equal to the speed of earth's first satellite.

Solution: We know that the energy of 1 gm of nitrogen molecule is given by

$$\frac{3}{2}(RT) = \frac{1}{2} m v_{rms}^2$$

So,

$$T = \frac{m v_{rms}^2}{3R}$$

Here, $M = 28$, $v_{rms} = 8 \times 10^5$ cm/sec and $R = 8.32 \times 10^7$ ergs/mole/°K

or

$$T = \frac{28 \times 8 \times 10^5 \times 8 \times 10^5}{3 \times 8.32 \times 10^7}$$

$$= 1.43 \times 10^5 \text{ °K}$$

5. Calculate the volume engaged by 0.32 gm of oxygen at 300°K temperature and 76 cm of Hg pressure.

Solution:

Given that, $T = 300$°K

$P = 76 \times 13.6 \times 980$ dynes/cm^2

$R = 8.32 \times 10^7$ erg/gm mole°K

We know that the perfect gas equation is

$$PV = RT$$

So, $V = \frac{RT}{P} = \frac{8.32 \times 10^7 \times 300}{76 \times 13.6 \times 980} = 2.4612 \times 10^4 \text{ cm}^3$ per gm mole

Hence, volume for 3.2 gm of oxygen

$$= \frac{2.4612 \times 10^4 \times 0.32}{32} = 2.4612 \times 10^2 \text{ cc}$$

6. The r.m.s velocity of oxygen molecules at 0°C is 4.61×10^2 m/sec and mass of one molecule of oxygen is 52.80×10^{-27} kg. Calculate the molecules present in 1 cc of oxygen at N.T.P.

Solution:

Given that,

$$m = 52.80 \times 10^{-27} \times 10^3 \text{ g}$$

$$P = 76 \times 13.6 \times 980 \text{ dynes/cm}^2$$

$$v_{rms} = 4.61 \times 10^2 \times 10^2 \text{ cm/sec}$$

We know that the pressure exerted by gas molecules is given by

$$P = \frac{1}{3}\rho v_{rms}^2 = \frac{1}{3} mnv_{rms}^2$$

$$n = \frac{3P}{mv_{rms}^2} = \frac{3(76 \times 13.6 \times 980)}{52.80 \times 10^{-24} \times 4.61 \times 10^4 \times 4.61 \times 10^4}$$

$$= 26.99 \times 10^{18}$$

7. If the pressure remains constant, at what temperature the r.m.s velocity of a gas will be half of its value at 1°C.

Solution:

Given that,

$$(v_{rms})_2 = \frac{(v_{rms})_1}{2} \text{ and } T_1 = 274°\text{K}$$

We know that, $v_{rms} \propto T$

So, $$\frac{(v_{rms})_1}{(v_{rms})_2} = \sqrt{\frac{T_1}{T_2}}$$

or $$\frac{2}{1} = \sqrt{\frac{274}{T_2}}$$

or $$\frac{4}{1} = \frac{274}{T_2}$$

or $$T_2 = \frac{274}{4} = 68.5°\text{K} = -204.5°\text{C}$$

8. Calculate the net random kinetic energy of 4 gm of helium at 400°K.

Solution: The net random kinetic energy of 1 gm of He

$$= \frac{3}{2} \times \left(\frac{RT}{M}\right)$$

So the net random kinetic energy of 4 gm of helium

$$= \frac{4 \times 3}{2} \times \left(\frac{RT}{M}\right)$$

$$= 6 \times \left(\frac{RT}{M}\right)$$

$$= 6 \times \left(\frac{8.32 \times 10^7 \times 400}{4}\right)$$

$$= 12.44 \times 10^2 \text{ Joules}$$

9. Calculate the net random kinetic energy of 1 gm mole of oxygen at 200°K.

Solution: The net random kinetic energy of 1 gm mole of oxygen

$$= \frac{3}{2} \times RT$$

$$= \frac{3}{2} \times 8.32 \times 200$$

$$= \frac{3}{2} \times 8.32 \times 200 = 2.496 \times 10^3 \text{ joule}$$

10. The root mean square speed of hydrogen at N.T.P. is 1830 m/s. Calculate the root mean square speed of oxygen at N.T.P.

Solution: Given that,

$$(v_{rms})_1 = 1830 \text{ cm/sec}$$

$$M_1 = 2 \text{ and } M_2 = 32$$

We know that,

$$\frac{(v_{rms})_1}{(v_{rms})_2} = \sqrt{\frac{M_2}{M_1}}$$

or $$\frac{(v_{rms})_1}{1840} = \sqrt{\frac{32}{2}}$$

$$\frac{1830}{(v_{rms})_2} = \sqrt{\frac{32}{2}} = \sqrt{\frac{16}{1}} = 4$$

or $$(v_{rms})_2 = \frac{1830}{4} = 457.5 \text{ m/s}$$

11. Particles of mass 6.19×10^{-14} gm are slung in a liquid at 300°K. The root mean square speed of particles is 1.39 cm/sec. Calculate the Avogadro's number using equipartition theorem.

Solution:

We know from equipartition theorem that the average kinetic energy of particle

$$= \frac{3}{2}\frac{R}{N}T$$

According to kinetic theory of gases, average kinetic energy of the particle

$$= \frac{1}{2}mv_{rms}^2$$

So, $$\frac{3}{2}kT = \frac{1}{2}mv_{rms}^2$$

or $$\frac{3}{2}\frac{R}{N}T = \frac{1}{2}mv_{rms}^2$$

or

$$N = \frac{3RT}{mv_{rms}^2} = \frac{3 \times 8.32 \times 300}{6.19 \times 10^{-17} \times 1.39 \times 1.39 \times 10^{-4}} = 6.26 \times 10^{23} \text{/mol}$$

12. The ratio of specific heat at constant pressure and at constant volume is 1.32. Determine the values of the C_p and C_v.

Solution: We know that $\frac{C_P}{C_V} = \gamma$ and $C_P - C_V = R$

So, $$C_p = \gamma \cdot C_V \text{ and } C_V(\gamma - 1) = R$$

or $$C_V = \frac{R}{(\gamma - 1)}$$

Hence, $C_V = \dfrac{R}{(\gamma - 1)} = \dfrac{8.32}{1.32 - 1} = 26 \text{ J/mol°K}$

$$C_P = \gamma \cdot C_V = 1.32 \times 26 = 34.32 \text{ J/mol°K}$$

13. Calculate the temperature at which the root mean square velocity of helium molecule is triple of its value at N.T.P. and pressure remains unchanged.

Solution: We know that the root mean square velocity is given by

$$v_{rms} \propto \sqrt{T}$$

$$\frac{(v_{rms})_1}{(v_{rms})_2} = \sqrt{\frac{T_1}{T_2}}$$

$$\frac{(v_{rms})_1}{3(v_{rms})_1} = \sqrt{\frac{273}{T_2}}$$

$$\frac{1}{3} = \sqrt{\frac{273}{T_2}}$$

$$\frac{1}{9} = \frac{273}{T_2}$$

$$T_2 = 273 \times 9 = 2457°\text{K}$$

or $$T_2 = 2457 - 273 = 2184°\text{K}$$

14. Prove that the ratio of two molar-specific heats is given by

$$\gamma = \left(1 + \frac{2}{f}\right)$$

Solution: Let us assume that the degree of freedom of each molecule of 1 mole of a gas is *f*. According to the equipartition theorem, the kinetic energy associated with each degree of freedom of molecule is (½)*kT*.

Hence, the total kinetic energy associated with *f* degree of freedom

$$= f \times \left(\frac{1}{2}\right) kT$$

so the total internal energy of 1 mole the given gas

$$U = Nf \times \left(\frac{1}{2}\right)\frac{R}{N}T = \frac{1}{2}fRT$$

Molar specific heat at constant volume $C_V = \left(\frac{\partial U}{\partial T}\right)_V = \frac{\partial}{\partial T}\left(\frac{1}{2}f \times R \times T\right) = \frac{1}{2}fR$

We know that gas equation for perfect gas

$$C_P - C_V = R$$

or $$C_P = C_V + R = \frac{1}{2}fR + R$$

or $$= \left(\frac{1}{2}f + 1\right)R$$

So, $$= \frac{\left(\frac{1}{2}f + 1\right)R}{\frac{1}{2}fR} = \left(1 + \frac{2}{f}\right)$$

15. The energy of the first excited state of hydrogen atom is 10.20 eV. Calculate the temperature required to excite hydrogen atom to its first excited level.

Solution: Given that,

Kinetic energy $= \frac{1}{2}mv^2 = 10.2 \times 1.6 \times 10^{-12}$ erg

According to kinetic theory of gases, the average kinetic energy per molecule at temperature T

$$\frac{3}{2}kT = \frac{1}{2}mv^2 = 10.2 \times 1.6 \times 10^{-12}$$

So, $$T = \frac{10.2 \times 1.6 \times 10^{-12} \times 2}{3 \times 1.38 \times 10^{-16}}$$

$$= 78.76 \times 10^3 \text{ °K}$$

16. Find out the temperature at which the mean speed of molecules of hydrogen is four times the mean speed of oxygen at 27°C.

Solution: We know that the mean speed, $\bar{v} = 1.59 \times \sqrt{\frac{k \times T}{m}}$

so mean speed of hydrogen molecule, $\bar{v}_H = 1.59 \times \sqrt{\dfrac{k \times T_1}{m_H}}$

and mean speed of oxygen molecule, $\bar{v}_O = 1.59 \times \sqrt{\dfrac{k \times T_2}{m_O}}$

Given that,

$$\bar{v}_H = 4 \times \bar{v}_O$$

So,
$$\frac{T_1}{m_H} = 16 \times \frac{T_2}{m_O}$$

or
$$T_1 = 16 \times \frac{T_2}{m_O} \times m_H = 16 \times \frac{300}{16}$$

or
$$= 300°\text{K}$$

17. A room contains 200 m^3 of air. The temperature is 298°K and atmospheric pressure is 2 bars. If air is an ideal gas, find out the mass of air. The gas constant for atmospheric air is 287.1 J/kg°K.

Solution: We know that ideal gas equation

$$pV = mRT$$

So mass of air
$$m = \frac{pV}{RT}$$

$$= \frac{2 \times 10^5 \times 200}{287.1 \times 298}$$

$$= \frac{2 \times 10^5 \times 200}{287.1 \times 298} = 467.0 \text{ kg}$$

Exercise

Short Answer Questions

1. What do you mean by kinetic theory of gases?
2. Write the statement of equipartition theorem.
3. Give the postulates of kinetic theory of gases.
4. What do you mean by degree of freedom?
5. Discuss the Brownian motion.

6. What is the significance of pressure applied by an ideal gas?
7. Draw the Maxwell-Boltzmann distribution curve of speed.
8. What do you mean by RMS speed, average speed and most probable speed of molecules of a gas?
9. What do you mean by atomicity of the gas?
10. What is specific heat of a gas? Write the differences between specific heat at constant volume and specific heat at constant pressure.
11. What do you understand by mean free path?
12. Prove that the mean kinetic energy of translation of gas molecules is $\frac{3}{2}kT$.
13. What is the conclusion of Maxwell's law of speed distribution?
14. Give the kinetic interpretation of temperature.
15. What is low temperature behavior of specific heat of ideal gases?
16. Show that the mean kinetic energy per degree of freedom of ideal gas is $\frac{1}{2}kT$.

Long Answer Questions

1. Discuss the Brownian motion. Which important factors affect it?
2. Describe the kinetic theory of gases. Write its basic postulates and obtain the mathematical expression for pressure exerted by a perfect gas.
3. Define the term atomicity of gases and find out the atomicity of monoatomic, diatomic and triatomic gas molecules.
4. Prove that the mean speed, root mean square speed and most probable speed of the gas molecules are inversely proportional to the square root of molecular weight.
5. Explain the effect of temperature on the behavior of specific heat of ideal gases.
6. Explain the degree of freedom and establish the relation between the specific heat and degree of freedom of gas.
7. Define the Brownian motion and explain the Langevin's theory of Brownian motion.
8. Explain the law of equipartition theorem and using it show that, the ratio of two specific heats is given by $\gamma = \left(1 + \frac{2}{f}\right)$. Find out the value of γ for monoatomic, diatomic and triatomic gases and compare the values with experimental data.
9. Explain the theory of estimation of Avogadro's number.
10. Prove that the ratio between root mean square speed v_{rms}, mean speed $\overline{v}$ and most probable speed v_{mp} of the molecules of a gas is given by $v_{rms} : \overline{v} : v_{mp}$ $= 1.73 : 1.59 : 1.41$.
11. Prove that for a perfect gas, the internal energy is a temperature-dependent function.

12. Explain the principle of kinetic theory of ideal gases and prove that the pressure exerted by an ideal gas is directly proportional to the density of gas.
13. Define most probable speed of gas molecules and find an expression for it by applying Maxwell's law of distribution of speeds of molecules.
14. Show that the root mean square speed of molecules of a gas is related to most probable speed by the relation $v_{rms}^2 = \frac{3}{2} v_{mp}^2$.
15. Explain the Maxwell's distribution law of molecular speed of gas.
16. Discuss the temperature dependence of pressure of an ideal gas contained in a container.
17. Define the root mean square speed v_{rms}, mean speed $\overline{v}$ and most probable speed v_{mp} of the molecule of a gas and prove that $v_{rms} > \overline{v} > v_{mp}$.

Numerical Problems

1. Estimate the mean kinetic energy of 5 gm oxygen at N.T.P.
   [Ans. 531]
2. Find out how many grams of helium gas have kinetic energy 849.7 J at N.T.P.
   [Ans. 1 gm]
3. The root mean square speed of mercury vapor is 1.84×10^2m/s. Find out the temperature of mercury vapor. The value of gas constant is 8.31 in S.I. units.
   [Ans. 300°K]
4. Find out the temperature at which the root mean square speed of nitrogen molecules is just half of root mean square speed of hydrogen molecules at 200°K. Molecular weights of hydrogen and nitrogen molecules are given 2 and 28, respectively. **[Ans.** 700°K]
5. The kinetic energy of hydrogen molecules at 0°C is 5.64×10^{-14} ergs and molar gas constant is 8.32×10^7 ergs. Find out the value of Avogadro's number.
   [Ans. $N = 6.08 \times 10^{23}$]
6. The number of degrees of freedom of a gas is 5, specific heat at constant pressure is $C_P = 29.08$ J/mol/°K and the specific heat at constant volume is $C_v = 20.77$ J/mol/°K. Calculate the value of atomicity of the gas.
7. Calculate the temperature at which the mean speed of hydrogen molecules is equal to the mean speed of nitrogen molecules at 45°C.
8. The root mean square speed of hydrogen molecules at 20°C is 1908 m/s. Calculate the density of hydrogen gas at N.T.P. **[Ans.** 0.089 gm/ltr]
9. Find out the temperature at which root mean square speed of helium molecules is equal to the speed of earth's satellite ($v = 8$ km/sec).
   [Ans. 10.28×10^{23} °K]

10. Calculate the most probable speed of a molecule of gas at 12000°C. Gas constant $R = 8.31 \times 10^7$ ergs/gm/mol/°K and Avogadro's number $N = 6.023 \times 10^{23}$.
11. Calculate the temperature at which the root mean square speed of hydrogen molecules is 18.4×10^2m/sec. [**Ans.** 273°K]
12. Find out the temperature at which the root mean square speed of oxygen molecules is double their root mean square speed at 300°K. [**Ans.** 1200°K]
13. Find out the rotational and translational energy of 2 gm of hydrogen gas molecules at 200°K and 300°K. If the molecules of a gas are dumb-bell shaped, find out the specific heat at constant volume and constant pressure at the above two temperatures.
14. 1 litre of hydrogen gas at room temperature weighs 0.086 gm. Calculate the root mean square speed of hydrogen molecules. [**Ans.** 18.4×10^2 m/sec]
15. A particle executing Brownian motion at 300°K has root mean square speed of 10^2 m/sec. Calculate the mass of that particle in gram. [**Ans.** 1.24×10^{12} gm]

Multiple-Choice Questions

1. The equipartition theorem was given by
   (i) Newton (ii) Maxwell
   (iii) Boltzmann (iv) Einstein
2. The pressure exerted by an ideal gas molecule is due to molecules of gas is
   (i) Attracted by the walls of vessel
   (ii) Continuously decreasing their energy
   (iii) Having momentum change due to strike on the walls of vessel
   (iv) Continuously striking to the vessel
3. If the temperature of a gas is doubled to its original value, its root mean square speed becomes
   (i) $\frac{2}{\sqrt{2}}$ times (ii) half
   (iii) double (iv) $\sqrt{2}$ times
4. The ratio of specific heat at constant pressure to specific heat at constant volume is
   (i) 1.40 (ii) 1.10
   (iii) 3 (iv) 1.59
5. The root mean square speed of a gas at 800°K is
   (i) Same as at 400°K (ii) Half the value at 400°K
   (iii) Twice the value at 200°K (iv) Six times the value at 200°K

6. The relation between density ρ, pressure P and root mean square velocity u is given by

(i) $P = \frac{1}{3}\rho^2 u$ (ii) $P = \frac{1}{3}\rho u$

(iii) $P = \frac{1}{3}\rho u^2$ (iv) $P = \frac{1}{4}\rho^2 u$

7. The mean velocity of the gas molecules at equilibrium is

(i) equal to zero (ii) proportional to T

(iii) proportional to $\frac{1}{T}$ (iv) proportional to T^2

8. The perfect gas molecules have

(i) only potential energy

(ii) both kinetic energy and potential energy

(iii) only kinetic energy

(iv) no energy

9. The ratio of C_P and C_V for a diatomic molecules of a gas is

(i) 1.67 (ii) 1.00

(iii) 1.29 (iv) 1.33

10. The relation between the kinetic energy of an ideal gas and pressure P is given by

(i) $\frac{3}{2}P$ (ii) $\frac{3}{4}P$

(iii) $\frac{1}{2}P$ (iv) $\frac{1}{4}P$

11. The value of Boltzmann constant in MKS is

(i) 1.38 J/K (ii) 1.38×10^{23} J/K

(iii) 1.38×10^{26} J/K (iv) 1.28×10^{23} J/K

12. The mean translational energy of ideal gas molecules at temperature T is

(i) $\frac{3}{2}$kT (ii) $\frac{1}{2}$kT

(iii) $\frac{7}{2}$kT (iv) $\frac{1}{4}$kT

13. The relation between the degree of freedom f and ratio of two specific heats is given by

(i) $\left(1+\frac{2}{f}\right)$ (ii) $\left(1+\frac{1}{f}\right)$

(iii) $\left(\frac{1}{f}\right)$ (iv) $\left(1+\frac{f}{2}\right)$

14. Which of the following statement is correct according to the kinetic theory of gases?

(i) The pressure exerted by a gas is proportional to the square root of density.

(ii) The pressure of gas molecules is proportional to the mean square speed of gas molecules.

(iii) The average kinetic energy of molecules of a gas is proportional to the pressure of the gas.

(iv) Root mean square velocity is proportional to the energy density.

15. The temperature at which the kinetic energy of an ideal gas molecule is zero

(i) 0°C (ii) –273°K

(iii) 273°K (iv) –273°C

16. The ratio of C_P and C_V for a triatomic molecules of a gas

(i) 1.62 (ii) 1.33

(iii) 1.39 (iv) 1.41

17. The relation between root mean square speed and most probable speed of gas molecules is given by

(i) $v_{rms} = \sqrt{3}v_{mp}$ (ii) $v_{rms} = \sqrt{\frac{3}{2}}v_{mp}$

(iii) $v_{rms} = \sqrt{2}v_{mp}$ (iv) $v_{rms} = \frac{1}{\sqrt{3}}v_{mp}$

18. The mean speed of the molecules of gas is given by

(i) $\sqrt{\frac{3\,kT}{m}}$ (ii) $\sqrt{\frac{2\,kT}{m\pi}}$

(iii) $\sqrt{\frac{3\,kT}{m\pi}}$ (iv) $\sqrt{\frac{8\,kT}{m\pi}}$

19. A perfect gas has kinetic energy E at 300°K. Its kinetic energy at 600°K will be

(i) $2E$ (ii) $4E$

(iii) $\sqrt{3}E$ (iv) $\sqrt{\frac{E}{3}}$

20. If we increase the temperature of a gas by four times of its original value, the root mean square speed of molecules will be

(i) Unchanged (ii) Quadrupled

(iii) Doubled (iv) Tripled

21. The ratio of two speeds, most probable and average, is given by

(i) $\sqrt{\frac{3\pi}{4}}$ (ii) 2

(iii) $\sqrt{\frac{5}{4\pi}}$ (iv) $\sqrt{\frac{\pi}{2}}$

22. The correct relation among the v_p, v_{av}, v_{rms} for the Maxwell distribution is

(i) $\dfrac{v_p}{1} = \dfrac{v_{av}}{\sqrt{\frac{4}{\pi}}} = \dfrac{v_{rms}}{\sqrt{\frac{3}{2}}}$ (ii) $\dfrac{v_p}{4} = \dfrac{v_{av}}{\sqrt{\frac{6}{\pi}}} = \dfrac{v_{rms}}{\sqrt{\frac{5}{2}}}$

(iii) $\dfrac{v_p}{\frac{2}{\sqrt{3}}} = \dfrac{v_{av}}{\sqrt{\frac{\pi}{2}}} = \dfrac{v_{rms}}{\sqrt{\frac{3}{2}}}$ (iv) $\dfrac{v_p}{\frac{2}{\sqrt{3}}} = \dfrac{v_{av}}{\sqrt{\frac{3\pi}{2}}} = \dfrac{v_{rms}}{\sqrt{\frac{3}{2}}}$

Answers

1. (ii) 2. (iii) 3. (iv) 4. (i) 5. (iii) 6. (iii) 7. (i) 8. (iii)
9. (i) 10. (i) 11. (ii) 12. (i) 13. (i) 14. (ii) 15. (iv) 16. (ii)
17. (ii) 18. (iv) 19. (i) 20. (iii) 21. (iv) 22. (i)

2

Study of Real Gases and Properties of Pure Substances

2.1 INTRODUCTION: BEHAVIOR OF REAL GASES

A matter generally exists in three states—solid, liquid and gas. All real gases, like hydrogen, nitrogen, helium, oxygen, carbon dioxide, etc., do not follow the gas laws at low temperature and high pressure. At low pressure, real gases follow the ideal gas equation, $PV = RT$. The liquid and gas states are non-continuous. At high pressure, the gas state changes into liquid as the forces of intermolecular attraction become perceptible and molecules of gas come closer to each other. The internal energy of ideal gases is temperature-dependent but internal energy of real gases is pressure- and temperature-dependent. Heating of ice also shows such behavior.

2.2 LIQUEFACTION STATE OF CARBON DIOXIDE: ANDREW'S EXPERIMENT

Isothermals of carbon dioxide at different temperatures were studied by Andrew in his experiment. The liquefaction state of carbon dioxide was found out and the behavior of carbon dioxide was seen at high pressure and low temperature. The isothermals of carbon dioxide at various temperatures are given in Figure 2.1.

The following conclusions were obtained from isothermals of CO_2:

1. The gaseous state of CO_2 is represented by PQ at temperature 13.1°C. QR represents the transition from gasoues state to liquid state. From Q to R, enormous decrease in volume is obtained with increase in pressure. The liquid state of CO_2 is represented by RS.
2. Similar curve is obtained at temperature 21.5°C. The horizontal portion of the curve, QR decreases to UV.
3. At 31.1°C, the critical isothermal for CO_2 is observed and horizontal portion of the curve vanishes. At temperatures like 35°C and 48.1°C, the gases follow Boyle's law and horizontal portion completely vanishes. The isothermal curve is similar to isothermal curve for air.

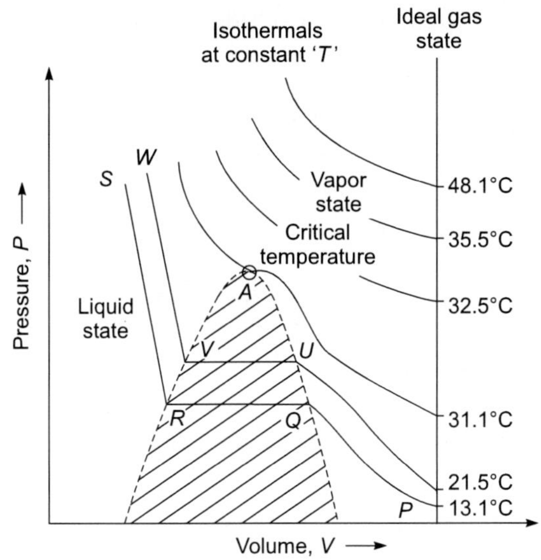

Figure 2.1 Isothermals for CO_2 and air at constant temperature

4. The critical isothermal curve of CO_2 is divided into two regions. Below the critical isothermal, the carbon dioxide gas can be liquefied by applying the pressure alone and above the critical isothermal curve, it cannot be liquefied, however the applied pressure is large enough. The liquid gas region is represented by dotted curve and highest point *A* is called the critical point of the curve and the temperature of this isothermal (31.1°C) is called the critical temperature.

2.3 REAL GASES AND VAN DER WAALS EQUATION OF STATE FOR REAL GASES

The ideal gas equation can be applied to real gases up to few atmospheres pressure. At higher densities or higher pressure, the deviation of real gas from the ideal gas becomes important. The interaction between the composite molecules is responsible for these deviations. The ideal gas equation is derived by assuming that (i) there is no force of intermolecular attraction, and (ii) the size of the molecules of gas is negligibly small. But at high pressure, the size of the molecules cannot be neglected and the force of intermolecular attraction is also significant.

There are two corrective factors in van der Waals gas equation. The ideal gas equation has been modified on the basis of these two corrections on pressure and volume. The first correction $\left(\frac{a}{V^2}\right)$ alerts the pressure in the perfect gas equation. It accounts for the forces of intermolecular attraction between gas molecules. The magnitude of a is indicative of the strength of the intermolecular attractive force. The larger the value of a, the greater the attractions between the particles. The second correction $(-b)$ accounts for the volume tied up by the gas molecules. The constant b is correction in finite molecular size and its value corresponds to the total volume occupied by gas molecules per mole, b is generally much smaller than a. The values of constants a and b generally increase with the complexity and size of the molecules. The constants a and b could be used to calculate the radii of atoms or molecules.

Thus, the van der Waals equation of state for a real gas is

$$\left(P+\frac{a}{V^2}\right)(V-b)RT \tag{2.1}$$

where, constants a and b are the van der Waals constants. At very low pressure, the particles are very far apart and the gas behaves like an ideal gas, the van der Waals equation reduces to the perfect or ideal gas equation.

2.4 COMPARISON WITH EXPERIMENTAL CURVES

The van der Waals equation of state for a real gas is given by

$$\left(P+\frac{a}{V^2}\right)(V-b)=RT$$

or the gas equation can be written as

$$P=\frac{RT}{(V-b)}-\left(\frac{a}{V^2}\right) \tag{2.2}$$

We can test the validity of van der Waals equation of real gases by comparing its PV curve with the experimental data. The theoretical pressure volume curves at various temperatures are plotted using equation (2.2) in Figure 2.2.

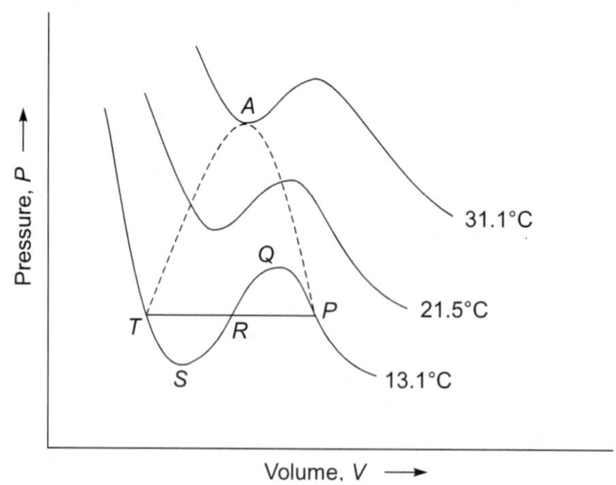

Figure 2.2 *P-V* diagram at various temperatures

It was observed that for large values of volume V, pressure P would be small and for volume $V \to \infty$, $P \to 0$. Again if volume V is very small and $V \to b$, $P \to \infty$. The graph between pressure P and volume V is not exactly similar to the experimental graph of CO_2 obtained by Andrew. In the theoretical graph between P and V, the horizontal portion is absent but in place of it the maxima and minima are obtained as shown in graph in the curve $PQRST$. In the P-V diagram, the portion PQ represents super cooling of the vapors and the portion TS represents super heating of the liquid states but the portion QRS cannot be explained due to decrement in volume with decrement in pressure, which cannot happen in practice. However, at higher temperature, both theoretical and experimental P-V diagrams are similar and thus van der Waals equation of real gases explains the experimental results.

2.5 ESTIMATION OF VAN DER WAALS CONSTANT

If we plot a graph of pressure as a function of volume, the critical isothermal PQR will be as shown in Figure 2.3. At the point Q, curve is horizontal and at this point the tangent also crosses the curve, so this is also the point of inflection for the PV diagram.

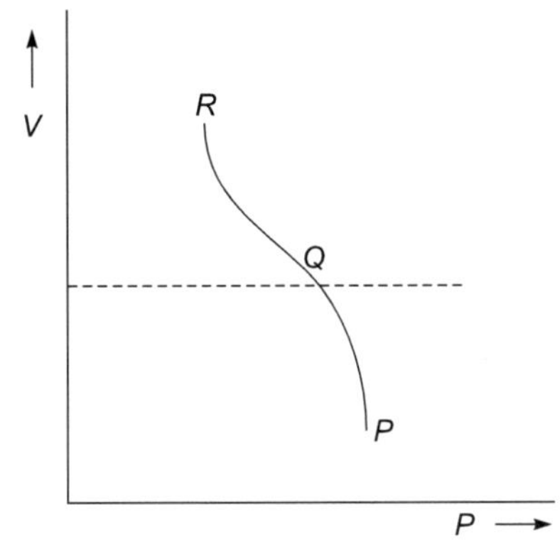

Figure 2.3 Isothermal *PQR*

Therefore, we may write at the point of inflection

$$\frac{\partial P}{\partial V} = 0$$

$$\frac{\partial^2 P}{\partial V^2} = 0$$

van der Waals equation (2.2) can also be written as

$$P = \frac{RT}{(V - b)} - \left(\frac{a}{V^2}\right) \tag{2.3}$$

So, the first order derivative of the above equation will be

$$\frac{\partial P}{\partial V} = -\frac{RT}{(V-b)} + 2\left(\frac{a}{V^3}\right) \tag{2.4}$$

and the second order derivative of the above equation will be

$$\frac{\partial^2 P}{\partial V^2} = +\frac{2RT}{(V-b)^3} - 6\left(\frac{a}{V^4}\right) \tag{2.5}$$

and (2.5) can be written as

$$P_C = \frac{RT_C}{(V_C-b)} - \left(\frac{a}{V_C^2}\right) \tag{2.6}$$

$$\frac{\partial P}{\partial V} = -\frac{RT}{(V-b)} + 2\left(\frac{a}{V^3}\right) = 0$$

or $$\frac{RT_C}{(V_C-b)} = 2\left(\frac{a}{V_C^3}\right) \tag{2.7}$$

and $$\frac{\partial^2 P}{\partial V^2} = +\frac{2RT}{(V-b)^3} - 6\left(\frac{a}{V^4}\right) = 0$$

or $$\frac{2RT_C}{(V_C-b)^3} = 6\left(\frac{a}{V_C^2}\right) \tag{2.8}$$

On dividing equation (2.7) by (2.8), we get

$$\frac{\dfrac{RT_C}{(V_C-b)}}{\dfrac{2RT_C}{(V_C-b)^3}} = \frac{2\left(\dfrac{a}{V_C^3}\right)}{6\left(\dfrac{a}{V_C^4}\right)}$$

or $$\frac{1}{2}(V_C-b) = \frac{1}{3}V_C$$

or $$3(V_C-b) = 3V_C$$

or $$[V_C = 3b] \tag{2.9}$$

On substituting the value $V_C = 3b$ in equation (2.7), we obtain

$$RT_C \frac{1}{(3b-b)^2} = 2\frac{a}{(3b)^3}$$

or
$$\frac{RT_C}{4(b)^2} = \frac{2a}{27(b)^3}$$

or
$$\left[T_C = \frac{8}{27}\left(\frac{a}{bR}\right)\right] \tag{2.10}$$

On substituting the values of V_C and T_C from equations (2.9) and (2.10) in equation (2.5), we obtain

$$P_C = \frac{8Ra}{27(3b-b)bR} - a\frac{1}{(3b)^2}$$

or
$$= \frac{4a}{27b^2} - a\frac{1}{9b^2}$$

or
$$= \frac{a}{9b^2}\left(\frac{4}{3}-1\right)$$

or
$$= \frac{a}{27b^2} \tag{2.11}$$

2.6 DETERMINATION OF VAN DER WAALS EQUATION'S CONSTANT *a* AND *b*

We know that from equation (2.11)

$$P_C = \frac{1}{27}\frac{a}{b^2}$$

or
$$\left(\frac{a}{b^2}\right) = 27P_C \tag{2.12}$$

But from equation (2.10), we get

$$\left[T_C = \frac{8}{27}\left(\frac{a}{bR}\right)\right]$$

or
$$\left(\frac{a}{b}\right) = \frac{27}{8}RT_C \tag{2.13}$$

On dividing equation (2.13) by (2.12), we get

$$\left(\frac{a}{b}\right)\times\left(\frac{b^2}{a}\right)=\frac{\frac{27}{8}RT_C}{27P_C}$$

or $$b=\left(\frac{RT_C}{8P_C}\right) \tag{2.14}$$

On substituting the value of b in equation (2.13), we obtain

$$a\left(\frac{8P_C}{RT_C}\right)=\frac{27}{8}RT_C$$

or $$a=\frac{27}{8}RT_C\times\frac{RT_C}{8P_C}$$

or $$=\frac{27}{64}\left(\frac{R^2T_C^2}{P_C}\right) \tag{2.15}$$

If we know the values of T_C and P_C for different gases, then we can calculate the values of a and b from equations (2.14) and (2.15), respectively.

2.7 CRITICAL COEFFICIENT OF REAL GASES

The critical coefficient of gases is defined by $\frac{RT_C}{P_CV_C}$ and its calculated value is same for all gases. We can calculate its value by the values of T_C, P_C and V_C.

$$\text{Critical coefficient} = R\left(\frac{T_C}{P_CV_C}\right)=R\left(\frac{\frac{8a}{27bR}}{\frac{a}{27}b^2\times 3b}\right)$$

or $$=\left(\frac{8a}{27b}\right)\times\left(\frac{27b^2}{3ab}\right)$$

$$\text{Critical coefficient} = \frac{8}{3}=2.66 \tag{2.16}$$

The experimental values of critical constant and critical coefficient for different specific gases are given in Table 2.1. Here, V_C = specific volume × molecular weight. The theoretical value of critical coefficient 2.667 is lower than the experimental value of all gases.

TABLE 2.1 Critical pressure, critical temperature and critical coefficient of selected gases

Name of Gases	Critical Pressure P_C in atm. in cm^3/g in K	Critical Temperature T_C in K	Specific Volume of Gases	Critical Coefficient $R\frac{T_C}{P_C V_C}$
Hydrogen	12.8	33.20	32.2	3.28
Helium	2.25	5.19	15.4	3.13
Nitrogen	33.5	126.2	3.21	3.42
Oxygen	49.8	154.6	2.32	3.42
Carbon dioxide	72.8	304.19	2.17	3.48
Water	217.7	647.096	3.181	4.30

2.8 DRAWBACKS OR LIMITATIONS OF VAN DER WAALS EQUATION OF REAL GASES

Limitations of van der Waals equation are as follows:

(i) In van der Waals theory, the values of constants a and b are assumed to be constant, but they are not constant and vary with pressure and temperature. At very high temperature, they approach to zero.

(ii) The value of V_C is not equal to $3b$, but actually it is equal to $3b$ in some cases and in other cases, $2b$. It is found to be dependent on the nature of gas.

(iii) The calculated value of critical coefficient is $8/3 = 2.667$ for all gases. But experimentally, it is usually more than 3 for most of the gases and this constant depends on the structure of the gas.

Thus, real gases deviate from van der Waals equation at high pressure or too low temperature.

2.9 VAN DER WAALS REDUCED EQUATION OF STATE

The reduced pressure, reduced volume and reduced temperature of gases are defined as

$$\text{Reduced pressure, } P_r = \frac{\text{Actual pressure, } P}{\text{Critical pressure, } P_C}$$

$$\text{Reduce volume, } V_r = \frac{\text{Actual volume, } V}{\text{Critical volume, } V_C}$$

$$\text{Reduced temperature, } T_r = \frac{\text{Actual pressure, } T}{\text{Critical pressure, } T_C}$$

So from the above definition we can write

$$\begin{bmatrix} P = P_C P_r = \left(\dfrac{a}{27b^2}\right) P_r \\ V = V_C V_r = (3b) V_r \\ T = T_C T_r = \left(\dfrac{8a}{27Rb}\right) T_r \end{bmatrix} \tag{2.17}$$

We know that van der Waals equation of real gases is

$$\left(P + \frac{a}{V^2}\right)(V - b) = RT$$

Substituting the values of pressure, volume and temperature from equation (2.17) in van der Waals equation, we get

$$\left[\left(\frac{a}{27b^2}\right) P_r + \left(\frac{a}{9b^2}\right)\frac{1}{V_r^2}\right] [(3b)V_r - b] = R\left(\frac{8a}{27Rb}\right) T_r$$

or

$$\left(\frac{a}{27b^2}\right)\left[P_r + \frac{3}{V_r^2}\right](3b)\left[V_r - \frac{1}{3}\right] = \left(\frac{8a}{27b}\right) T_r$$

Hence,

$$\left[P_r + \frac{3}{V_r^2}\right]\left[V_r - \frac{1}{3}\right] = \frac{8T_r}{3} \tag{2.18}$$

The above equation is said to be van der Waals reduced equation of state. The equation is independent of the constants a and b.

2.10 JOULE'S LAW OF PERFECT GAS AND JOULE'S EXPERIMENT

In the free expansion of fixed mass of a gas, no heat is allowed to enter or leave the gas, i.e., gas expands without doing any external work on it. Due to internal molecular attraction between the molecules of a gas, an internal work will be done by the gas in pulling the molecules apart, which will expand the internal energy or heat energy of the gas. If the molecules of gas repel one another, gas expands and a rise in temperature is obtained. For no intermolecular attraction, the temperature would be same.

An experiment was performed by Joule to study the intermolecular attraction. In his experimental setup, P and Q are two copper cylinders, which are connected by a tube and jointed with a stopcock or nozzle, T (Figure 2.4). Cylinder P is filled with a dry gas at high pressure and cylinder Q was completely evacuated. Cylinders P and Q are surrounded by a water bath tub and its temperature was measured as a very sensitive temperature. The stopcock was opened and the gas started to flow from P to Q.

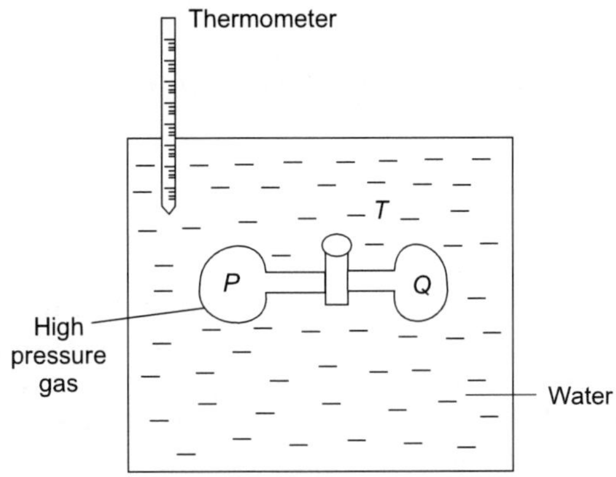

Figure 2.4 Experimental setup for Joule's experiment

If work had been done by the gas during expansion, cooling should be observed, but no cooling was observed. Joules explained that the gas suffers free expansion into vacuum, so cooling was not observed. In his experiment, the volume and pressure of the gas changed but water bath did not show change in temperature, i.e., the temperature remained constant.

In Joule's experiment, it was also observed that the internal energy is independent of pressure or volume and it is a function of temperature only. Hence, if the temperature of the gas remains constant, internal energy of the gas also remains unchanged. On the basis of these experimental results, Joule's law was explained as "the internal energy of a gas depends only on temperature and is independent of pressure and volume." Mathematically, internal energy can be defined as

$$U = f(T)$$

The first order derivative of internal energy with volume and pressure at constant temperature will be

$$\left(\frac{\partial U}{\partial P}\right)_T = 0, \left(\frac{\partial U}{\partial V}\right)_T = 0 \quad (2.19)$$

Equation (2.19) explains that the internal energy does not change with pressure or volume at constant temperature.

and also the first order derivative of internal energy at constant temperature will be

$$\left(\frac{\partial U}{\partial T}\right)_V = \left(\frac{\partial U}{\partial T}\right)_P \tag{2.20}$$

Equation (2.20) implies that the change in internal energy with respect to temperature is same, it does not depend on whether volume is taken constant or pressure is kept constant.

2.11 ESTIMATION OF JOULE'S COEFFICIENT

The free expansion of real gases in Joule's experiment is due to intermolecular forces. As in Joule's experiment, the intermolecular separation between the molecules changes and so the gas does some internal work against this force in separating the molecules. The potential energy of the system increases and kinetic energy of the gas decreases, as due to rapid expansion no heat can enter or leave the gas. The decrement in kinetic energy of the system shows the fall of temperature. Hence, for attractive intermolecular forces and free expansion, cooling effect is observed in real gases. Due to repulsive intermolecular forces, heating of the gas is observed. Experiment shows that in Joule's expansion of real gases, only cooling effect is observed and thus the intermolecular force between the gas molecules is always attractive.

The free expansion of a gas is measured in the form of a coefficient known as Joule's coefficient, which is defined as

$$\eta = \left(\frac{dT}{dV}\right) \text{ at constant internal energy} \tag{2.21}$$

We can calculate the value of Joule's coefficient using van der Waals equation for free gases. For 1 kg mole of a real gas, if dV is change in volume due to free expansion and change in temperature is dT. Then, the internal work done against the internal pressure in free expansion will be

$$dW = p'dV = \left(\frac{a}{V^2}\right)dV \tag{2.22}$$

where, $p' = \left(\frac{a}{V^2}\right)$ is internal pressure developed due to intermolecular forces, a is van der Waals constant, and V is the initial volume of the real gas. As we know that internal kinetic energy of the gas is utilized to do this internal work, so internal potential energy of the system increases. Hence,

Change in internal kinetic energy = Change in internal potential energy

$$= -dW = -\left(\frac{a}{V^2}\right)dV \tag{2.23}$$

We can also write internal kinetic energy in terms of molar specific heat at constant volume.

Change in internal kinetic energy $= C_V dT$ (2.24)

On comparing equation (2.23) with equation (2.24), we obtain

$$C_V dT = -\left(\frac{a}{V^2}\right) dV$$

or

$$\frac{dT}{dV} = -\left(\frac{a}{V^2}\right) \times \frac{1}{C_V}$$

So Joule's coefficient of free expansion of gas is

$$\left[\eta = \frac{dT}{dV} = -\left(\frac{a}{V^2}\right) \times \frac{1}{C_V}\right] \tag{2.25}$$

As Joule's coefficient of free expansion is a negative quantity, a cooling effect is observed due to free expansion in real gases, which is in agreement with the experimental results.

2.12 JOULE-THOMSON OR JOULE-KELVIN EFFECT

In the free adiabatic expansion of real gas from constant high pressure region to constant low pressure region, an inversion in temperature of the gas is observed, either in cooling form or in heating form, this effect is called Joule – Thomson effect or Joule-Kelvin effect.

This change in temperature, pertaining to the Joule-Kelvin effect, can occur when a flowing real gas passes through a regulator, which acts as a valve, throttling device, or porous plug. Here, a temperature change is not necessarily desirable.

At normal temperatures, all real gases except H_2 show cooling effect but hydrogen shows heating effect. At very low temperatures, all real gases show a cooling effect.

2.13 POROUS PLUG EXPERIMENT

William Thomson and James Prescott Joule worked in collaboration and designed an experimental device to conduct and study the Joule-Thomson effect. The apparatus used for the Joule-Thomson porous plug experiment is shown in Figure 2.5. The experimental setup consists of a cylindrical long wooden tube, which is surrounded by an insulating material, such as cotton wool or asbestos, to avoid gain or loss of heat from the surroundings. At the center of cylindrical tube, a plug of compressed silk or cotton wool is sandwiched between two brass perforated discs, $A1$ and $A2$. The experimental gas, which is compressed to a known high pressure, is allowed to pass through a spiral tube immersed in a water bath of known constant temperature. Water bath is used to maintain the constant temperature or to absorb the heat released due to compression. To measure the initial and final temperatures of the emerging gas, two sensitive platinum-resistant thermometers are used. A pressure gauge is used to read the high constant pressure.

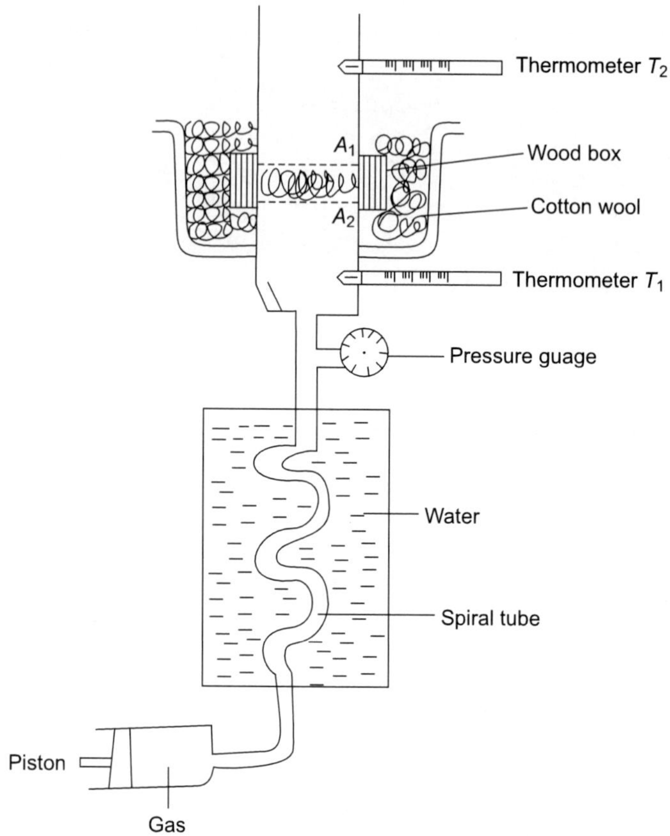

Figure 2.5 Apparatus for porous plug experiment

As the high constant pressure is supplied using a compression pump, the gas starts to pass through the porous plug. This pressure is measured by the pressure gauge. The experimental gas is throttled during its passage through the porous plug and its volume increases. In the whole process, there is no loss or gain and so the gas expands adiabatically.

Outcome of Experiments

The experiment was performed on a number of gases at different inlet temperatures. After the successful completion of the experiment, the following results are obtained:

1. It is found that at sufficiently low temperatures, in the range of 4°C to 100°C, all gases show a cooling effect whereas helium and hydrogen gases show heating effect.

2. The fall in temperature of gases is found to be directly proportional to the difference in pressure of gas on two sides.
3. At a given difference in pressure and at lower initial temperature of the gas, the fall in temperature of gas will be greater. For every experimental gas, there is a fixed transition temperature that is called its inversion temperature. If the gas is initially above its inversion temperature, heating effect will be observed, and if gas is initially lower than its inversion temperature, cooling effect will be observed, after passing through the porous plug. At the inversion temperature of gases, no change in temperature will be observed.

Theory of Experiment

A unit mass of the experimental gas is enclosed in the porous plug on the high pressure side. Let V_1 be volume of gas and P_1 be the pressure of gas on this side. Q_1 and Q_2 are two pistons at high pressure side and low pressure side of the porous plug. Let Q_1 be in contact with the porous plug. As the experimental gas in the lower part passes through the porous plug, it presses the piston Q_2 in forward direction. The piston Q_2 is pushed against a constant pressure P_2. After passing through the porous plug to the low pressure side, the volume of the experimental gas is V_2.

When a piston Q_1 is moved through the distance dy , the piston Q_2 also moves by the same distance dy. Q_1 pushes the experimental gas in the lower side of pressure completely through the porous plug and the amount of work done on the gas by piston $Q_1 = P_1V_1$. When the experimental gas enters the low pressure side of the porous plug and pushes the piston Q_2 back, an amount of work done by the gas on the piston $Q_2 = P_2V_2$.

Hence, the resultant external work done by the experimental gas,

$$W = P_2V_2 - P_1V_1$$

Let the gas do the work W' in separating the molecules of gas against the intermolecular attraction between them. The net amount of work done by the experimental gas is $W + W'$, i.e.,

$$W' + [P_2V_2 - P_1V_1]$$

No heat energy is gained or lost from outside or the surrounding since the entire experimental setup is thermally insulated. The heating or cooling of a gas after passing through the porous plug will depend on the work done against intermolecular attraction and the deviation from the Boyle's law of perfect gas. Three different possibilities are observed:

(a) At the temperature lower than the Boyle temperature,

$$P_2V_2 \rangle P_1V_1$$

so, $P_2V_2 - P_1V_1$ is positive and W' must be either zero or positive. Hence, a resultant positive work has been done by the experimental gas and cooling effect will be observed after passing through the porous plug.

(b) At the temperature higher than the Boyle temperature,

$$P_2V_2 \langle P_1V_1$$

so $P_2V_2 - P_1V_1$ is negative hence, there are two possibilities:

(i) If $[P_2V_2 - P_1V_1] \rangle W'$, heating effect will be observed.

(ii) If $[P_2V_2 - P_1V_1] \langle W'$, cooling effect will be observed.

(c) At the temperature equal to the Boyle temperature

$$P_2V_2 = P_1V_1$$

So, $$P_2V_2 - P_1V_1 = 0$$

The net amount of work done by the gas is W'. Hence, at this temperature the work done by the gas against the intermolecular attraction is responsible for the cooling effect.

2.14 TEMPERATURE OF INVERSION

Let us assume that the gas obeyed the van der Waals equation, $\left(P + \frac{a}{V^2}\right)(V - b) = RT$. Therefore, the pressure p due to the intermolecular forces of attraction between the molecules is $\frac{a}{V^2}$. Due to the expansion of gas, if the volume increases from V_1 to V_2, the work done W' by the gas against the intermolecular forces is given by

$$W' = \int_{V1}^{V2} p dV$$

or
$$= \int_{V1}^{V2} \left(\frac{a}{V^2}\right) dV$$

$$= -\left[\frac{a}{V}\right]_{V_1}^{V_2}$$

$$= \left(\frac{a}{V_1} - \frac{a}{V_2}\right)$$

This work done by the gas against the intermolecular attraction is stored as potential energy so that the increase in potential energy,

$$W' = (U_2 - U_1) = \left(\frac{a}{V_1} - \frac{a}{V_2}\right)$$

If V_1 and V_2 are the gm molecular volumes of the gas on the high pressure side and low pressure side of porous plug, the external work done by the gas is

$$(P_2V_2 - P_1V_1)$$

Therefore, the net amount of work done by the gas,

$$W = W' + (P_2V_2 - P_1V_1)$$

$$= \left(\frac{a}{V_1} - \frac{a}{V_2}\right) + (P_2V_2 - P_1V_1)$$

From the van der Waals equation of real gas, we can write

$$PV + \left(\frac{a}{V}\right) - bP - \left(\frac{ab}{V^2}\right) = RT$$

$$PV + \left(\frac{a}{V}\right) - bP - \left(\frac{ab}{V^2}\right) = RT$$

or $$PV = RT - \left(\frac{a}{V}\right) - \left(\frac{ab}{V^2}\right) + bP$$

So, total work done by the gas,

$$W = \left[RT_2 - \left(\frac{a}{V_2}\right) - \left(\frac{ab}{V_2^2}\right) + bP_2\right] - \left[RT_1 - \left(\frac{a}{V_1}\right) - \left(\frac{ab}{V_1^2}\right) + bP_1\right] + \left(\frac{a}{V_1}\right) - \left(\frac{a}{V_2}\right)$$

$$= (RT_2 - RT_1) + 2a\left(\frac{1}{V_1} - \frac{1}{V_2}\right) + (bP_2 - bP_1) + ab\left(\frac{1}{V_2^2} - \frac{1}{V_1^2}\right)$$

The terms $\frac{ab}{V_2^2}$ and $\frac{ab}{V_1^2}$ can be neglected as they are small when compared with other terms. We assume here that the temperatures T_2 and T_1 remain approximately constant at T.

So, $$W = 2a\left(\frac{1}{V_1} - \frac{1}{V_2}\right) + (bP_2 - bP_1)$$

As we know that $PV = RT$, hence,

$$V_1 = \frac{RT}{P_1}, V_2 = \frac{RT}{P_2}$$

So work done, $W = 2a\left(\frac{P_1}{RT} - \frac{P_2}{RT}\right) + (bP_2 - bP_1)$

$$= \frac{2a}{RT}(P_1 - P_2) - b(P_1 - P_2)$$

$$= \left(\frac{2a}{RT} - b\right)(P_1 - P_2) \tag{2.26}$$

Let us assume that after passing through the porous plug, change in temperature is dT, work done is given by

$$W = JQ = JC_p dT \tag{2.27}$$

where, C_P is the specific heat at constant pressure, J is Joule's constant. On comparing equations (2.6) and (2.7) of work done, we get

$$JC_p dT = \left(\frac{2a}{RT} - b\right)(P_1 - P_2)$$

$$dT = \left(\frac{2a}{RT} - b\right)\frac{(P_1 - P_2)}{JC_p}$$

$$\frac{dT}{dP} = \left(\frac{2a}{RT} - b\right)\frac{1}{JC_p} \tag{2.28}$$

In the porous plug experiment, there is a decrease in pressure so that dP is positive, the change in temperature will depend on the values of $\frac{2a}{RT}$ and b. There are three different possibilities:

(i) For $\left(\frac{2a}{RT}\right) = b,$

If the initial temperature of the experimental gas is such that $\left(\frac{2a}{RT}\right) = b$, dT will be zero, and, this temperature is known as inversion temperature, T_i.

Hence, $dT = 0,\ T_i = \left(\frac{2a}{RT}\right)$

At the inversion temperature, no cooling or heating is observed after the gas is passed through the porous plug.

(ii) For $\left(\frac{2a}{RT}\right) < b,$

If the initial temperature of the experimental gas is such that $\left(\frac{2a}{RT}\right) < b$, dT will be negative, i.e., heat is produced.

In this case $\left(\frac{2a}{RT}\right) < b$

or $$b > \left(\frac{2a}{RT}\right)$$

or $$T > \left(\frac{2a}{Rb}\right)$$

or $$T > T_i$$

If the gas is at a initial temperature larger than the inversion temperature, heating is observed when the gas is allowed to pass through the porous plug.

(iii) For $\left(\frac{2a}{RT}\right) > b,$

If the initial temperature of the experimental gas is such that $\left(\frac{2a}{RT}\right) > b$, dT will be positive, i.e., cooling is produced

In this case $\left(\frac{2a}{RT}\right) > b$

or $$b < \left(\frac{2a}{RT}\right)$$

or $$T < \left(\frac{2a}{Rb}\right)$$

or $$T < T_i$$

At the temperature below the inversion temperature, instead of heating, cooling is observed after the gas is passed through the porous plug. This effect is also known as Joule-Kelvin cooling effect or regenerative cooling. This effect has been used in the liquefaction of gases like oxygen, hydrogen, helium, and nitrogen.

2.15 COMPARISON AMONG ADIABATIC EXPANSION, JOULE-THOMSON EXPANSION AND JOULE EXPANSION

For all the three processes, heat neither enters nor it is allowed to leave the system, i.e., $dQ = 0$. In other words, all are thermally isolated from external systems and are adiabatic.

In adiabatic expansion of gas, the external work against the atmospheric pressure is done by the expanding gas, i.e., the system of gas is mechanically not isolated and also the pressure of gas is not constant. In this case, the work done is given by

$$W = \int PdV$$

In Joule-Thomson expansion of gas, the total external work done by the gas in passing through the porous plug, i.e., in this case also, the system of gas is mechanically not isolated but both pistons of high pressure side and low pressure side are kept at constant pressures P_1 and P_2. In this case, the external work done is given by

$$W = (P_2V_2 - P_1V_1)$$

But in Joule's expansion of gas, mechanical isolation from the external system is also observed. In this case, the total amount of external work done on the gas or by the gas is zero, i.e.,

$$W = 0$$

In all the three processes, the gas follows the first law of thermodynamics, i.e.,

$$dQ = (dU + dW)$$

As we know that in all the three processes $dQ = 0$, so (i) in adiabatic expansion of gas, $dU = -\int PdV$; (ii) in Joule-Thomson expansion of gas, $dU = -dW = (P_1V_1 - P_2V_2)$; and (iii) in Joule expansion of gas, $dU = 0$.

In adiabatic expansion of gas, external work on the gas is done by the charge of internal energy of the gas. Therefore in this case, fall of temperature is large, or in this expansion cooling is affected by the external work and its internal energy.

In Joule-Thomson expansion of gas, external work is done by the gas while passing through the porous plug. In this process, cooling or heating effect depends on the values of P_1V_1 and P_2V_2.

In Joule's expansion of gas, internal work is done by the gas against intermolecular attraction forces in free expansion. This work is done by the charge of kinetic energy of the gas and hence, cooling effect is always observed in all real gases.

2.16 NATURE OF GASES AT HIGH PRESSURE AND BOYLE TEMPERATURE OF THE GAS

All real gases at constant temperature and at high pressure follow the relation

$$PV = B_1 + B_2 \cdot P + B_3 \cdot P^2 + B_4 \cdot P^3 + \ldots, \tag{2.29}$$

where B_1, B_2, B_3, B_4, etc., are constants at a given temperature which depend on the nature of the given gas, these constants are known as virial coefficients. B_1, B_2, B_3 ... are known as the first virial coefficient, second virial coefficient, third virial coefficient and so on, respectively. The magnitude of these coefficients has the values in descending order, i.e., $B_1 > B_2 > B_3 > B_4 \ldots$ At very low pressure, i.e., as $P \to 0$, constant B_1 will be equal to RT and the second, third and fourth terms become very small and negligible in comparison to the first term. Hence, at low pressure, gases follow the Boyle's law.

The coefficient B_2 is very important for all gases, it has negative value at very low temperatures, increases gradually up to zero and as the temperature increases, it becomes positive. The specific temperature of gas at which the value of coefficient B_2 is zero and a real gas behaves like an ideal gas over the wide range of very low pressures is known as the Boyle temperature denoted by T_B. At this temperature, the other constants, B_3, B_4 ... are negligible.

At Boyle temperature,

$$B_2 = \frac{d}{dP}(PV) = 0 \tag{2.30}$$

and also, $(PV) = B_1 =$ constant.

$$\left(P + \frac{a}{V^2}\right)(V - b) = RT$$

or

$$PV - Pb + \left(\frac{a}{V}\right) - \left(\frac{ab}{V^2}\right) = RT$$

or

$$PV = RT + Pb - \left(\frac{a}{V}\right) - \left(\frac{ab}{V^2}\right)$$

Substituting the values of $V = \frac{RT}{P}$ in the above equation, we get

$$PV = RT + Pb - P\left(\frac{a}{RT}\right) + P^2\left(\frac{ab}{R^2T^2}\right)$$

or

$$= RT + P\left(b - \frac{a}{RT}\right) + P^2\left(\frac{ab}{R^2T^2}\right) \tag{2.31}$$

On comparing the coefficients of P and P^2 in equations (2.29) and (2.31), we get

$$B_1 = RT,\ B_2 = \left(b - \frac{a}{RT}\right) \text{ and } B_3 = \left(\frac{ab}{R^2T^2}\right)$$

As we know that at Boyle temperature, T_B, $B_2 = 0$, hence,

$$0 = \left(b - \frac{a}{RT_B}\right)$$

or, Boyle temperature, $T_B = \left(\frac{a}{Rb}\right)$ (2.32)

Equation (2.32) gives the value of Boyle temperature for all real gases. From the expression of Boyle temperature, it is clear that T_B depends on the values of constants a and b and these values of constants depend on the nature of gas. Therefore, the value of Boyle temperature is different for different gases.

The behavior of gases can be expressed by plotting the graph between pressure P and PV for different temperatures (Figure 2.6). The ordinary behavior of the graph is same for all gases but differences are obtained above Boyle temperature, at Boyle temperature and below the Boyle temperature.

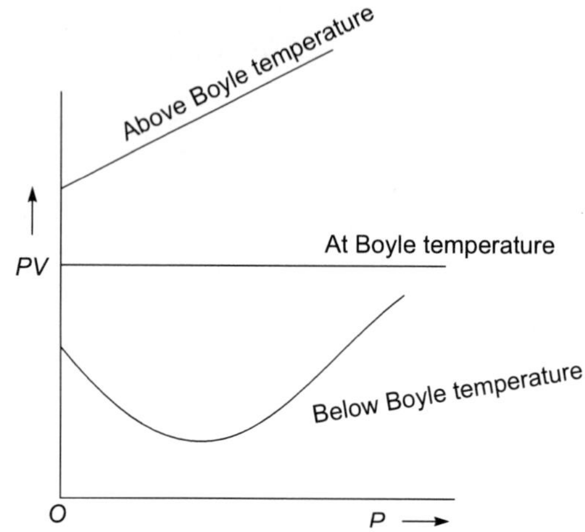

Figure 2.6 Behavior of gases below, above and at the Boyle temperature

(i) Above Boyle Temperature ($T > T_B$): The value of coefficient B_2 is positive above Boyle temperature and the intermolecular attraction is of less importance. In the curve, the value of PV continuously increases with P.

(ii) At Boyle Temperature ($T = T_B$): The value of coefficient B_2 is zero at Boyle temperature. In the curve, the product PV is approximately constant over the wide range of pressure.

(iii) Below Boyle Temperature ($T < T_B$): The value of coefficient B_2 is negative below Boyle temperature and the molecules come closer to each other. Hence, intermolecular attraction between the molecules becomes stronger below T_B and the gases are more compressible. At very high pressure, we cannot neglect the terms $B_3 P_2$, $B_4 P_3$, etc., of equation (2.29) and these terms become positive and more effective. In the curve, the value of PV starts decreasing with increase in the value of P and attains a minimum value at a specific pressure and then starts increasing.

2.17 COMPARISON AMONG TEMPERATURE OF INVERSION, BOYLE TEMPERATURE AND CRITICAL TEMPERATURES

We know that temperature of inversion, critical temperatures and Boyle temperature are defined by,

Inversion temperature, $$T_i = \left(\frac{2a}{Rb}\right),$$

Critical temperature, $$T_C = \left(\frac{8a}{27Rb}\right)$$

Boyle temperature, $$T_B = \left(\frac{a}{Rb}\right)$$

From the above expressions of inversion temperature and Boyle temperature, we obtain

$$\frac{T_i}{T_B} = 2$$

and also from the relation of temperature of inversion and critical temperature, we get

$$\frac{T_i}{T_c} = \frac{\left(\frac{2a}{Rb}\right)}{\left(\frac{8a}{27R\,b}\right)} = \left(\frac{27}{4} = 6.75\right)$$

From the above relation between temperature of inversion and critical temperature, it is clear that the ratio between T_i and T_c is independent of the nature of actual gas. The experimental value of the ratio of these two is less than 6 for actual gases.

2.18 NATURE OF VAN DER WAALS FORCES

The atoms and molecules present in the matter possess an attraction force on each other but at short distances between the molecules, there must be existence of repulsion forces also. Van der Waals assumes that the molecules of the matter are impenetrable and hard and also, they are surrounded by strong attractive field. The van der Waals forces exist between the molecules of same matter or substances. The van der Waals forces between the atoms or molecules of substance or gas may be divided in the following three types:

(i) **Dispersion forces** exist between the nonpolar molecules of gas such as Cl_2. Instantaneous electric dipole moments are responsible for these forces. The instantaneous dipole moments are generated due to rotational motion.

(ii) **Dipole-dipole forces or orientation forces** exist in polar molecules due to their electric dipole moment which is permanent. The electrons are shared unequally between the polar molecules, for example, HCl molecule is polar.

(iii) **Hydrogen bonding or polarization forces** exist between nonpolar and polar molecules due to dipole-dipole interaction. Hydrogen bonding is dipole- dipole interaction that exists between a hydrogen atom and any one of fluorine, oxygen, nitrogen, etc. For example, the bonding between HF, H_2O and NH_3....., is hydrogen bonding. The hydrogen bonding force is extremely strong as compared to other dipole forces, e.g., HCl.

2.19 COMPRESSIBILITY FACTOR OF REAL GASES

The behavior of real gases obtained from van der Waals equation gives accurate results under certain limits of temperature and pressure. Also, the constants used in equation have to be determined experimentally. All these problems are resolved by the common equation of state given by

$$PV = ZRT \tag{2.33}$$

Here, Z is known as compressibility factor.

Compressibility factor provides the measurement of gas deviation from perfect to real. For perfect or ideal gases, $Z = 1$ for all temperatures and pressures, and for real gases, either $Z > 1$ or $Z < 1$, according to the temperature and pressure.

The compressibility factor Z is a function of reduced temperature and reduced pressure, i.e.,

$$Z = f(T_r, P_r)$$

As we know that

$$P = P_C P_r$$

$$T = T_C T_r$$

and

$$V = V_C V_r$$

so from equation (2.33), we get

$$Z = \frac{P \times V}{RT}$$

or

$$= \frac{P_C P_r \times V_C V_r}{RT_C T_r}$$

or

$$= \left(\frac{P_C V_C}{RT_C}\right) \times \left(\frac{P_r V_r}{T_r}\right)$$

or

$$= Z_C \times \left(\frac{P_r V_r}{T_r}\right) \tag{2.34}$$

where, $Z_C = \left(\frac{P_C V_C}{RT_C}\right)$ is the compressibility factor at the critical points. The value of Z is one for higher values of P_C and lower values of T_C and the gas behaves like a perfect gas at ordinary pressure and temperature.

Numerical Examples

1. The volume of CO_2 gas is 44 m^3/kg at 373°K. Determine (i) pressure exerted by CO_2 gas using van der Waals equation, and (ii) pressure exerted by CO_2 gas using the ideal gas equation. The van der Waals gas constants are $a = 365.5 \times 10^3 \,Nm^4/mole^2$, and $b = 4.27 \times 10^2 \,m^3/mole$. Gas constant is $R = 8.312 \,Nm/kg\, mole°\, K$.

Solution:

i. The well-known van der Walls equation is

$$\left(P + \frac{a}{V^2}\right)(V - b) = RT$$

So,

$$P = \frac{RT}{(V-b)} - \frac{a}{V^2}$$

$$P = \frac{8.312 \times 373}{(44 - 0.0427)} - \frac{365.5}{44 \times 44}$$

$$= 70.554 - 0.18875$$

$$= 70.365 \text{ kPascal}$$

ii. Ideal gas equation is $PV = RT$

or

$$P = \frac{8.312 \times 373}{44}$$

$$P = 70.48 \text{ kPascal}$$

2. Calculate the critical pressure of oxygen gas using van der Waals equation of gas. Given that $V_C = 7.44 \times 10^{-2} \,m^3/kg\, mol$ and $T_C = -118.11°C$.

Solution: We know that the relation between critical constants is given by

$$R = \frac{8}{3}\left(\frac{V_C P_C}{T_C}\right)$$

So,

$$P_C = \frac{3}{8}\left(\frac{RT_C}{V_C}\right)$$

or

$$= \frac{3}{8}\left(\frac{8.312 \times 154.89}{7.44 \times 10^{-2}}\right)$$

or

$$P_C = 64.812 \times 10^2 \text{ kN/m}^2$$

or

$$P_C = 64.812 \text{ bars}$$

3. The nitrogen gas is contained in a vessel of 31 m^3 volume at –73°C temperature and 149.99 bar pressure. Calculate the mass of nitrogen present inside the container by the ideal gas equation and compressibility equation. The value of compressibility factor, $Z = 0.86$.

Solution: The ideal gas equation in terms of mass is given by

$$PV = \left(\frac{m}{M}\right) \times RT$$

or $$m = \frac{M(P \times V)}{RT}$$

or $$= \frac{28 \times \left(\dfrac{149.99 \times 10^5}{1000} \times 32\right)}{8.312 \times 200}$$

$$= 8109.11 \text{ kg}$$

We know that from compressibility equation

$$V = \frac{RTZ}{P}$$

$$= \frac{\dfrac{8.312}{28} \times 200 \times 0.86 \times 1000}{149.99 \times 10^5}$$

$$= 3.3899 \times 10^{-3} \text{ m}^3/\text{kg}$$

So, mass of nitrogen, $$m = \frac{32}{3.3899 \times 10^{-3}}$$

$$= 9439.80 \text{ kg}$$

4. Hydrogen gas is contained in a spherical balloon of diameter 1 m. The temperature of gas is 298°K and pressure is 1 bar. Calculate the mass of hydrogen gas inside the balloon. The critical pressure and temperature of H_2 gas are given as $P_C = 12.96$ bar and $T_C = -33.11$°K.

Solution: We know that the reduced pressure and temperature is given by

$$P_r = \frac{P}{P_C} \text{ and } T_r = \frac{T}{T_C}$$

So, $$P_r = \frac{1}{12.96} = 0.07711$$

and $$T_r = \frac{298}{33.11} = 9.00$$

Volume of container, $V = \frac{4}{3}\pi r^3$

$$= \frac{4 \times 3.14 \times 0.5 \times 0.5}{3} = 0.523 \text{ m}^3$$

As the values of P_C and T_C are high so the compressibility factor will be one, i.e., $Z = 1$.

So, $$PV = mRTZ$$

or $$m = \frac{PV}{RTZ}$$

$$= \frac{1 \times 10^5 \times 0.52}{\left(\frac{8312}{2}\right) \times 298 \times 1}$$

$$= 0.0423 \text{ kg}$$

5. Determine the critical temperature for carbon dioxide gas. The van der Waals constants are given as $a = 36.55 \times 10^{-2}$ Nm^4 / mole^2 and $b = 4.71 \times 10^{-6}$ m^3/mole.

Solution: The critical temperature of a gas is given by

$$T_C = \frac{8}{27}\left(\frac{a}{Rb}\right)$$

$$= \frac{8 \times 36.55 \times 10^{-2}}{27 \times 8.32 \times 42.71 \times 10^{-6}}$$

$$= 300.59°\text{K}$$

6. The 1.21×10^{-3} m^3/mole oxygen gas is contained in a vessel at 27°C. Calculate the pressure of gas. Van der Waals constant are given as $a = 0.133$ Nm^4/mole^2 and $b = 0.3123 \times 10^{-4}$ m^3 / mole.

Solution: We know that the van der Waals equation is given by

$$\left(P + \frac{a}{V^2}\right)(V - b) = RT$$

So, $$P = \frac{RT}{(V - b)} - \frac{a}{V^2}$$

$$P = \frac{8.32 \times 300}{(1.21 \times 10^{-3} - 0.3123 \times 10^{-4})} - \frac{0.133}{1.21 \times 10^{-3} \times 1.21 \times 10^{-3}}$$

$$= 2.0369 \times 10^6 \text{ N/m}^2$$

7. The critical constants of He gas are given as $P_C = 2.20 \times 10^5 \text{ N/m}^2$, $T_C = 5.31°\text{K}$. Calculate the values of van der Waals constans a and b.

Solution: The van der Waals constant a is defined in terms of critical constant as

$$a = \frac{27}{64} \times \left(\frac{R^2 T_C^2}{P_C}\right)$$

$$= \frac{27}{64} \times \left(\frac{8.32 \times 10^3 \times 8.32 \times 10^3 \times 5.31 \times 5.31}{2.20 \times 10^5}\right)$$

$$= 3.701 \text{ Nm}^4/\text{mol}^2$$

$$b = \frac{1}{8} \times \left(\frac{RT_C}{P_C}\right)$$

$$= \frac{1}{8} \times \left(\frac{8.32 \times 10^3 \times 5.31}{2.20 \times 10^5}\right)$$

$$= 2.51 \times 10^{-2} \text{ m}^3/\text{mole}$$

8. A certain amount of hydrogen is contained in a vessel. Determine the Boyle temperature. The van der Waals constants for H_2 are given as $a = 2.41 \times 10^{-2} \text{ Nm}^4/\text{mol}^2$ and $b = 2.66 \times 10^{-5} \text{ m}^3/\text{mole}$.

Solution: We know that the Boyle temperature is given by

$$T_B = \frac{27}{8} \times T_C$$

where,

$$T_C = \frac{8}{27}\left(\frac{a}{Rb}\right)$$

$$= \frac{8}{27}\left(\frac{2.41 \times 10^{-2}}{8.32 \times 2.66 \times 10^{-5}}\right)$$

$$= 33.139°\text{K}$$

So Boyle temperature, $T_B = \frac{27}{8} \times 33.139$

$$T_B = 111.844°K$$

9. Determine the critical temperature of 1 mole of helium gas at N.T.P. at one atmospheric pressure. The values of van der Waals constant are $a = 6.14 \times 10^{-5}$ and $b = 9.96 \times 10^{-4}$.

Solution: We know that at N.T.P., $P = 1$, $V = 1$ and $T = 273°K$

The van der Waals equation

$$\left(P + \frac{a}{V^2}\right)(V - b) = RT$$

So,
$$R = \frac{\left(P + \frac{a}{V^2}\right)(V - b)}{T}$$

$$= \frac{\left(1 + \frac{6.14 \times 10^{-5}}{1}\right)(1 - 9.96 \times 10^{-4})}{273}$$

$$= 3.64 \times 10^{-3}$$

We know that the critical temperature is given by

$$T_C = \frac{8}{27}\left(\frac{a}{Rb}\right)$$

$$= \frac{8}{27}\left(\frac{6.14 \times 10^{-5}}{3.64 \times 10^{-3} \times 9.96 \times 10^{-4}}\right)$$

$$= 4.98°K$$

10. Determine the inversion in temperature of CO_2 gas after passing through porous plug at 27°C. The pressure on both the sides of porous plug is 51 and 2 atm, respectively. The values of van der Waals constants are $a = 0.368\ \text{Nm}^4/\text{mole}^2$ and $b = 4.28 \times 10^{-5}\ \text{m}^3/\text{mole}$ and $C_P = 8.76\ \text{cal mole}^{-1}\text{K}$.

Solution: We know that the change in temperature of gas after passing through the porous plug,

$$dT = \frac{(P_1 - P_2)}{C_P}\left(\frac{2a}{RT} - b\right)$$

$$= \frac{(51-2)}{8.76 \times 4.18}\left(\frac{2 \times 0.368}{8.32 \times 300^\circ\,K} - 4.28 \times 10^{-5}\right)$$

$$= 33.68^\circ\,K$$

So, $$dT = T_1 - T_2 = 33.68^\circ K$$

As change in temperature is positive, so cooling effect is observed.

11. Helium gas suffers Joule-Thomson expansion at –174°K. The pressure difference on two sides of porous plug is 20 atm. Determine the change in temperature after passing through the porous plug. The values of constants are $a = 0.0342$ litre^2 atoms/ mole^2 and $b = 0.0236$ litre/ mole and $C_P = \frac{5}{2}R$.

Solution: Given that

$T = 99°K$

$dP = P_1 - P_2 = 20\text{ atm} = 20 \times 10^5\ \text{Nm}^{-2}$

$a = 0.0342\ \text{litre}^2\ \text{atoms/ mole}^2 = 0.0342 \times 10^{-1}\ \text{Nm}^4/\text{mol}^2$ and

$b = 0.0236\ \text{litre/ mole} = 0.0236 \times 10^{-3}\ \text{m}^3/\text{mole}$

$$C_P = \frac{5}{2}R = \frac{5}{2} \times 8.32\ \text{J/ mol}^\circ\ \text{K}$$

The change in temperature in Joule expansion

$$dT = \frac{(P_1 - P_2)}{C_P}\left(\frac{2a}{RT} - b\right)$$

$$= \frac{20 \times 10^5}{\frac{5}{2} \times 8.32}\left(\frac{2 \times 0.0342 \times 10^{-1}}{8.32 \times 99} - 0.0236 \times 10^{-3}\right)$$

$$= -1.498^\circ\ K$$

Exercise

Short Answer Questions

1. What do you mean by the critical coefficient of gas?
2. Explain why the Boyle temperature is different for different gases.
3. What are virial coefficients?
4. What is the reason for modification of ideal gas equation?
5. Write the differences between vapor and gas using Andrew's experiment on CO_2.

6. What are the limitations of van der Waals equations?
7. What are the conclusions of porous plug experiment? Define the inversion temperature.
8. Write the differences between perfect gas and real gas.
9. Write the important differences between Joule expansion of gases and Joule-Thomson expansion.
10. What is the principle of regenerative cooling.
11. Explain about the van der Waals forces.
12. Compare the adiabatic and Joule-Thomson expansion.
13. Why helium and hydrogen gases show heating effect?
14. What are the conditions for heating and cooling of real gases?
15. Write the differences between adiabatic cooling and Joule-Thomson cooling.
16. Show that the ratio of critical temperature and inversion temperature is independent of the nature of gas.
17. What are the limitations of ideal gas equations?

Long Answer Questions

1. Establish the van der Waals equation of state of real gases. Compare the results obtained by van der Waals equation with Andrew's isothermals.
2. Define Joule-Thomson effect. Establish it experimentally and interpret the effect.
3. Explain the porous plug experiment. What is inversion temperature? Discuss the conclusions obtained from porous plug experiment.
4. What are the corrections introduced in van der Waals equation? Prove that the relation among critical temperature, pressure and volume is given by

$$R\left(\frac{T_C}{P_C V_C}\right) = \frac{8}{3}$$

5. Prove that in the porous plug experiment, the enthalpy of system remains constant.
6. Explain the theory of Joule-Thomson effect. Discuss the heating effect of hydrogen and helium.
7. Establish the reduced equation of a real gas from van der Waals equation. Prove that if the reduced temperatures of two gases are same, their reduced pressure and volume will also be the same.
8. Define the critical constants of a real gas. Establish the relation between critical constant and constants of van der Waals equation.
9. What are Boyle temperature and inversion temperature? Discuss the behavior of real gases in the reference of Boyle temperature.

10. Explain Andrew's experiments on CO_2. What are the important results obtained from it? Discuss that the liquid and gases are two distinct stages of a continuous process.
11. Define viral coefficients. Write the value of the first viral coefficient. Discuss the dependence of the second viral coefficient on temperature.
12. Explain the theory of the porous plug experiment and establish the conditions for cooling and heating of real gases.
13. Describe Joule-Thomson expansion. Show that this expansion is an isenthalpic process.
14. Explain the experiment to determine the value of critical constants of a real gas.
15. Discuss the porous plug experiment. Discuss the negative Joule-Thomson effect of hydrogen gas.
16. Draw and discuss Andrew's curve for carbon dioxide gas and prove that the ratio of critical temperature and inversion temperature is given by

$$\frac{T_C}{T_i} = \frac{4}{27}.$$

Numerical Problems

1. Calculate the value of critical temperature of a real gas. The van der Waals constants are given as $a = 6.88 \times 10^{-4}\ \text{Jm}^3/\text{mole}^2$ and $b = 2.50 \times 10^{-7}\ \text{m}^3/\text{mole}$.
2. The value of van der Waals constants for CO_2 gas are $a = 1.31 \times 10^4\ \text{Nm}^4/\text{mole}^2$ and $b = 3.65 \times 10^{-5}\ \text{m}^3/\text{mole}$. Determine the values of critical temperature and pressure.
3. In the porous plug experiment, the pressure difference on both the sides of the plug is 20 atm. Calculate the inversion in temperature of helium gas when it is expanded at –100°K. The values of van der Waals constants are $a = 3.42 \times 10^{-2}$ litre^2 atoms/mole^2 and $b = 2.369 \times 10^{-2}$ litre/mole.

   [**Ans.** $dT = -1.49$°K]
4. Determine the fall in temperature due to adiabatic throttling process of 1 gm mole of O_2 when pressure is reduced by 60 atm. The initial temperature of real gas is 300 °K. Given that $a = 1.31$ litre^2 atoms/mole^2 and $b = 3.15510^{-3}$ litre/mole and $R = 8.31$ J/K mole.
5. One mole of CO_2 gas is contained in 1.30 dm^3 vessel at 47.99°C temperature. Determine the pressure exerted by CO_2. The van der Waals constants for CO_2 are given as $a = 3.58 \times 10^2$ dm^3 atoms/mole^2 and $b = 4.27 \times 10^{-2}$ dm^3/mole.

   [**Ans.** 18.49 atm]

6. The value of critical constants of a real gas are given as $P_C = 39.8$ atm and $V_C = 8.86 \times 10^{-2}\ \text{dm}^3/\text{mole}$. Determine the values of van der Waals constant of the gas.
7. Determine the fall in temperature when N_2 gas suffers expansion after passing through the porous plug at –193°C. The pressure difference between the two sides of the porous plug is 30 atm. Van der Waals constants for nitrogen gas are $a = 3.58 \times 10^2\ \text{lt}^2\ \text{atoms}/\text{mole}^2$ and $b = 2.694 \times 10^{-5}\ \text{lt}/\text{mole}$.
8. Find out the temperature at which cooling effect is observed for a certain gas. The van der Waals constants for that gas are $a = 0.1327 \times 10^{13}\ \text{dyne cm}^4/\text{mole}^2$ and $b = 0.319 \times 10^2\ \text{cm}^3/\text{mole}$.

   [**Ans.** 725.88°C]
9. Determine the critical temperature of a real gas for which van der Waals coefficient is $b = 0.060 \times 10^2\ \text{cm}^3/\text{mole}$ and $P_C = 200$ atm.
10. A real gas suffers Joule-Thomson expansion from 43.3 atm to 1.6 atm. If the initial temperature of gas is 0°C, find out the inversion in temperature. The values of van der Waals constants are $a = 1.58 \times 10^2\ \text{lt}^2\ \text{atoms}/\text{mole}^2$ and $b = 26.94 \times 10^{-6}\ \text{lt}/\text{mole}$.
11. Methane gas is confined to 4.99 dm^3 volume at a pressure of 5.01 atm. Calculate the reduced volume, pressure and temperature of methane gas. The values of critical constants are P_C = 54.59 atm, V_C = 9.87 × 10^{-12}dm^3/mole and T_C = –82.2°K

    [**Ans.** $V_r = 50.405, P_r = 0.1094, T_r = 2.087$]
12. Determine the critical pressure of a gas for which the van der Waals constant $b = 50 \times 10^{-6}\ \text{m}^3/\text{mole}$ and critical temperature is 214.2°C.

    [**Ans.** 100 atm]

Multiple-Choice Questions

1. In Joule-Thomson effect, the temperature of gas
   (i) increases
   (ii) decreases
   (iii) no change
   (iv) none of these
2. Two real gases at the same reduced pressure and reduced temperature are taken. The reduced volume of the gases
   (i) Will be same
   (ii) Higher molecular weight gas have larger reduced volume
   (iii) Higher molecular weight gas have smaller reduced volume
   (iv) None of them

3. The correction in volume $\frac{a}{V^2}$ in van der Walls equation represents
   (i) Force of attraction
   (ii) Effective area
   (iii) Mean velocity
   (iv) Volume engaged by the molecules
4. Joule-Thomson effect of real gases is based on
   (i) Sudden expansion in gases
   (ii) Sudden compression in gases
   (iii) Heating in gases
   (iv) Cooling in gases
5. The gas having critical temperature 31.10°C is
   (i) Hydrogen
   (ii) Carbon dioxide
   (iii) Nitrogen
   (iv) Oxygen
6. Which one is the wrong statement in following
   (i) In Joule's expansion, only internal work is done
   (ii) Joule-Thomson effect always shows cooling
   (iii) Joule's expansion always shows cooling effect
   (iv) In Joule-Thomson effect, external work is done by the gas and on the gas.
7. The value of van der Waals critical coefficient is
   (i) 2
   (ii) 4
   (iii) 1/3
   (iv) 8/3
8. The van der Waals gas equation is
   (i) $\left(P+\frac{a}{V^2}\right)(V-b)=RT$
   (ii) $\left(P-\frac{a}{V^2}\right)(V+b)=RT$
   (iii) $PV=RT$
   (iv) $\left(P+\frac{a}{V^2}\right)(V+b)=RT$
9. Consider the given statements
   (a) External work is not done in Joule's expansion
   (b) Internal work is not done in Joule's expansion
   (c) In Joule's expansion, gases show expansion in vacuum
   (d) Joule's expansion is a free expansion of gas

   Which of the statement given above is/are correct
   (i) only (a) is true
   (ii) (a), (b) and (c) are true
   (iii) only (b) is true
   (iv) all of the above are true

10. In the van der Walls equation of gases, the value of critical volume V_C is

(i) b (ii) $2\,b$

(iii) $4\,b$ (iv) $3\,b$

11. If the value of constants a and b are 0.00874 atoms cm^6 and 0.0023 cm^3, respectively, the critical temperature of gas will be

(i) 34.6°C (ii) 32.7°C

(iii) 39°C (iv) 41°C

12. The dimensions of the van der Waals constant b is

(i) volume (ii) pressure × volume

(iii) pressure/volume (iv) pressure

13. Compressibility factor is defined by

(i) $z = \frac{PV}{RT}$ (ii) $z = \frac{PV}{MT}$

(iii) $z = \frac{Pv}{RT}$ (iv) $z = \frac{PN}{RT}$

14. The value of compressibility factor for a perfect gas is

(i) 2 (ii) 4

(iii) 1 (iv) 9

15. According to Joule's law, the specific internal energy of molecules of a gas depends only on the

(i) Volume of the gas

(ii) Temperature of the gas

(iii) Pressure of the gas

(iv) Volume and pressure of the gas

16. The forces of attraction between the molecules of a gas become high at

(i) High temperature and low pressure

(ii) Low temperature only

(iii) Low temperature and high pressure

(iv) Low pressure only

17. In Jolue-Kelvin effect, heating will take place if

(i) The temperature of inversion is less than the temperature of the gas

(ii) The temperature of inversion is greater than the temperature of the gas

(iii) The temperature of inversion is equal to the temperature of the gas

(iv) None of these

18. The ratio of temperature of inversion and critical temperature of a gas is equal to

(i) 4 (ii) 6.23

(iii) 7 (iv) 6.65

19. The critical volume of a van der Waals gas is 58×10^{-8} m^3/mol. The value of van der Waals constant b will be

(i) 20.20×10^{-8} m^3/mol (ii) 19.33×10^{-8} m^3/mol

(iv) 12.24×10^{-8} m^3/mol (iv) 16.2×10^{-8} m^3/mol

20. The temperatrure of inversion for hydrogen gas is

(i) 74°C (ii) 80°C

(iii) 20°C (iv) 90°C

Answers

1. (i) 2. (i) 3. (i) 4. (i) 5. (ii) 6. (ii) 7. (iv) 8. (i)
9. (ii) 10. (iv) 11. (ii) 12. (i) 13. (iii) 14. (iii) 15. (ii) 16. (iii)
17. (i) 18. (iv) 19. (ii) 20. (ii)

3

Transport Phenomenon in Gases and Conduction of Heat

3.1 COLLISION OF MOLECULES

According to the kinetic theory of gases, molecules of gases are ideally hard elastic spheres, indistinguishable and are of very small but definite size. These molecules move randomly and collide with each other. Between any two successive collisions, a molecule of gas moves with a fixed speed in a linear path (Figure 3.1). Such collisions are called molecular collisions. At non-equilibrium condition, the erratic or random motion of gas molecules provides the conduction, diffusion and viscosity phenomena. In these phenomena, molecules of gases try to reach at equilibrium position by transporting or transforming heat, mass and momentum, respectively.

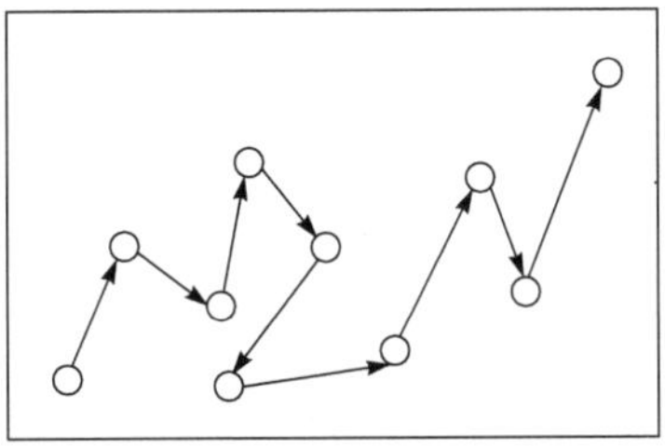

Figure 3.1 Zig-zag or random motion of a gas molecule

3.2 MEAN FREE PATH

In the kinetic theory of gases, it was assumed that the molecules continuously collide with one another. Between any two successive collisions, molecules of gases travel freely to an assured distance in a linear path. This certain distance is called free path. After every collision, the molecule changes its direction of motion and traces a zig-zag path as shown in Figure 3.1. After a number of collisions as the path does not appear constant, so distance moved by a gas molecule between any two consecutive collisions is taken as a mean distance. Hence, the average or mean distance traveled by a gas molecule between any two consecutive collisions is called a mean free path and it is represented by λ.

If the net path travelled after N successive collisions is D, the mean free path is expressed as

$$\lambda = \frac{D}{N} = \frac{\bar{v}t}{N} \tag{3.1}$$

where, $\bar{v}$ is the average velocity of the gas molecule.

The average time spent by a molecule of gas between two consecutive collisions is known as mean free time and is expressed by

$$\tau = \frac{\lambda}{\bar{v}} \tag{3.2}$$

Let us assume that the gas molecules are hard elastic identical spheres. Let d be the diameter of each molecule and the molecules of gas do not apply force on each other. We also suppose that only the gas molecule under examination is in motion and all other gas molecules are at rest. The molecule of gas can collide with those molecules whose center is within the range of distance d or less than it. Hence, it may be concluded that the collision between the molecules is possible only when the center of other molecules of gas approaches within a sphere of $2d$ diameter around the center of the molecule under examination. Such type of sphere is known as sphere of influence and it is shown in Figure 3.2.

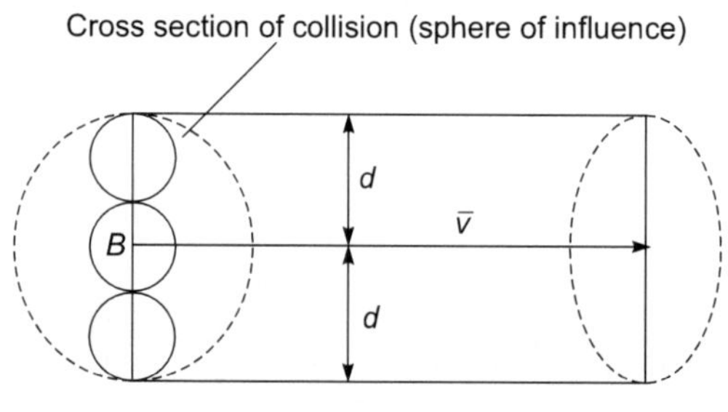

Figure 3.2 Collision of molecules and an imaginary sphere of influence

If $\bar{v}$ is the mean velocity of molecule B, then in one second this molecule will collide with all other molecules whose center approaches within a cylinder having radius d and length $\bar{v}$ and collision cross section πd^2. Hence, the volume of cylinder $= \pi d^2 \bar{v}$.

Let n be the number of molecules per cc.

Number of molecules inside the cylinder $= \pi d^2 \bar{v} n$ = number of collisions made by molecule in one second. The expression $\pi d^2 \bar{v} n$ is also known as collison frequency.

Hence, collision frequency, $\quad f = \pi d^2 \bar{v} n = \dfrac{\bar{v}}{\lambda}$

and time for one collision $= \frac{1}{(\pi d^2 \bar{v} n)}$

Hence, average time between two consecutive collisions

$$= \frac{1}{(\pi d^2 \bar{v} n)} \text{ seconds}$$

So, average distance travelled between two consecutive collisions

$$= \frac{1}{(\pi d^2 \bar{v} n)} \times \bar{v}$$

So, the mean free path

$$\lambda = \frac{1}{(\pi d^2 n)} \tag{3.3}$$

The expression in equation (3.3) is known as Claussius's expression for mean free path. In the derivation of the above expression, we assumed that only one molecule of gas under examination is moving and all other molecules are at rest. Maxwell assumes that all the molecules are moving so we have to take relative velocity of the molecules, the modified formula of mean free path is given by

$$\lambda = \frac{1}{\sqrt{2}\,(\pi d^2 n)} \tag{3.4}$$

The above expression is known as Maxwell's expression of mean free path. If the number of molecules per unit volume (n) and mean free path (λ) are known, one can calculate the molecular diameter, d.

We know that the pressure of a gas is expressed by

$$P = n \cdot k \cdot T$$

or

$$n = \frac{P}{k \cdot T}$$

Hence, the mean free path

$$\lambda = \frac{1}{\sqrt{2} \cdot \left(\pi d^2 \frac{P}{k \cdot T} \right)}$$

$$= \frac{k \cdot T}{\sqrt{2} \cdot (\pi d^2 P)} \tag{3.5}$$

From equation (3.5), it is clear that the mean free path is directly proportional to the absolute temperature (T) and inversely proportional to the pressure (P).

3.3 TRANSPORT PHENOMENON IN GASES

Kinetic theory of gases states that the gas molecules can be thermally agitated and will move with different velocities in different directions. During their motion, the molecules of gas collide with each other and create resistance in the motion. The gas gets an equilibrium or steady state by transporting mass, heat and momentum, which give rise to the diffusion, conductivity and viscosity phenomena, respectively. These phenomena are called transport phenomena in gases. These transport phenomena are possible only in the non-steady state of gases.

3.4 DIFFUSION PHENOMENON (TRANSPORT OF MASS)

Transportation of mass from higher concentration region to lower concentration region to get an equilibrium condition is known as diffusion or self-diffusion phenomenon. Let us consider different layers of a gas to define concentration gradient in the direction perpendicular to layer *AB* as shown in Figure 3.3.

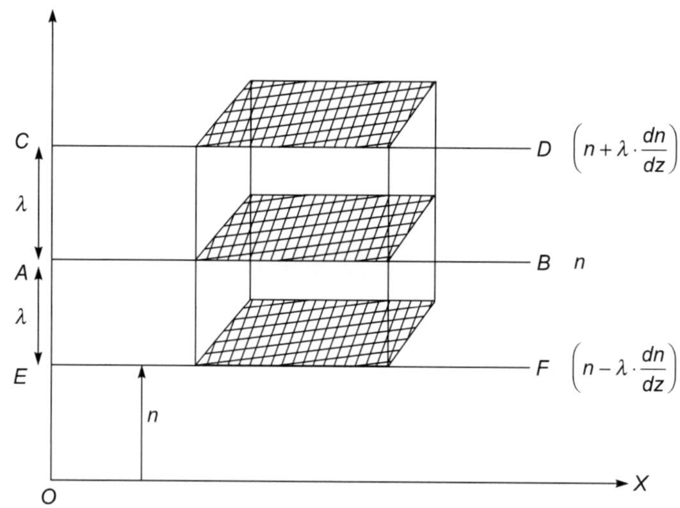

Figure 3.3 Transport of mass (diffusion)

Let us suppose that n is the number of molecules present in unit volume of gas, concentration gradient at layer *AB* will be $\frac{dn}{dz}$. Consider two layers *CD* and *EF* at a distance mean free path λ above *AB* and below *AB*, respectively. Therefore,

$$\text{concentration at upper layer } CD = n + \lambda\cdot\frac{dn}{dz},$$

$$\text{concentration at lower layer } EF = n - \lambda\cdot\frac{dn}{dz}$$

Hence, the total number of transported molecules (coming from layer *CD*) crossing layer *AB* per unit area per unit time

$$= \frac{1}{6}\bar{v}\cdot\left(n + \lambda\cdot\frac{dn}{dz}\right)$$

The total number of transported molecules (coming from layer *EF*) crossing layer *AB* per unit area per unit time

$$=\frac{1}{6}\bar{v}\cdot\left(n-\lambda\cdot\frac{dn}{dz}\right)$$

Hence, the total number of transported molecules crossing layer *AB* per unit area per second

$$=\frac{1}{6}\bar{v}\cdot\left(n+\lambda\cdot\frac{dn}{dz}\right)-\frac{1}{6}\bar{v}\cdot\left(n-\lambda\cdot\frac{dn}{dz}\right)$$

or $$=\frac{1}{6}\bar{v}\cdot\left(n+\lambda\cdot\frac{dn}{dz}-n+\lambda\cdot\frac{dn}{dz}\right)$$

or $$=\frac{1}{3}\bar{v}\cdot\lambda\cdot\frac{dn}{dz}$$

Let us suppose that *m* is the mass of a gas molecule, net mass transported by layer *AB* in per unit area per second is given by,

$$\mathcal{M}=\frac{1}{3}m\cdot\bar{v}\cdot\lambda\cdot\frac{dn}{dz}$$

or $$=\frac{1}{3}\cdot\bar{v}\cdot\lambda\cdot\frac{d(m\cdot n)}{dz}$$

$$=\frac{1}{3}\cdot\bar{v}\cdot\lambda\cdot\frac{d\rho}{dz} \tag{3.6}$$

Here, $\frac{d\rho}{dz}$ is the mass concentration gradient.

The ratio of the number of molecules transported per unit area per unit time to the concentration gradient is known as coefficient of diffusion, *D*. Hence,

$$D=\frac{\frac{1}{3}\cdot\bar{v}\cdot\lambda\cdot\frac{dn}{dz}}{\frac{dn}{dz}}$$

$$=\frac{1}{3}\cdot\bar{v}\cdot\lambda \tag{3.7}$$

From equations (3.6) and (3.7), we get

$$\mathcal{M}=D\cdot\frac{d\rho}{dz} \tag{3.8}$$

3.5 EFFECT OF TEMPERATURE AND PRESSURE ON DIFFUSION COEFFICIENT

We know that the mean free path is defined by

$$\lambda = \frac{kT}{\sqrt{2}\cdot(\pi d^2 P)}$$

and mean velocity $\bar{v} = \sqrt{\left(\frac{8kT}{m\pi}\right)}$

Hence, $$D = \frac{1}{3}\cdot\bar{v}\,.\,\lambda = \frac{1}{3}\cdot\sqrt{\left(\frac{8kT}{m\pi}\right)}\cdot\frac{kT}{\sqrt{2}\cdot(\pi d^2 P)}$$

$$= \frac{2}{3}\cdot\left(\frac{kT}{\pi}\right)^{3/2}\cdot\frac{1}{\sqrt{m}\cdot(d^2 P)} \tag{3.9}$$

or $$D \propto T^{3/2} \text{ and } D \propto \frac{1}{P}$$

Therefore, diffusion coefficient is directly proportional to $T^{3/2}$ and inversely proportional to P.

3.6 THERMAL CONDUCTIVITY PHENOMENON (TRANSPORT OF HEAT ENERGY)

The property of a substance to conduct heat is known as thermal conductivity. The value of temperatures for the different layers of gas is different. The transportation of heat energy starts from higher temperature region to lower temperature region to get equilibrium state.

Consider different layers of a gas to define concentration gradient in the direction perpendicular to layer AB as shown in Figure 3.4. Consider two layers CD and EF at a distance mean free path λ above AB and below AB, respectively. Thermal conductivity of gas is measured in terms of temperature gradient. Let us consider T as temperature along the layer AB and $\frac{dT}{dz}$ as temperature gradient. Therefore,

the temperature of layer $CD = \left(T + \lambda\cdot\frac{dT}{dz}\right)$

and the temperature of layer $EF = \left(T - \lambda\cdot\frac{dT}{dz}\right)$

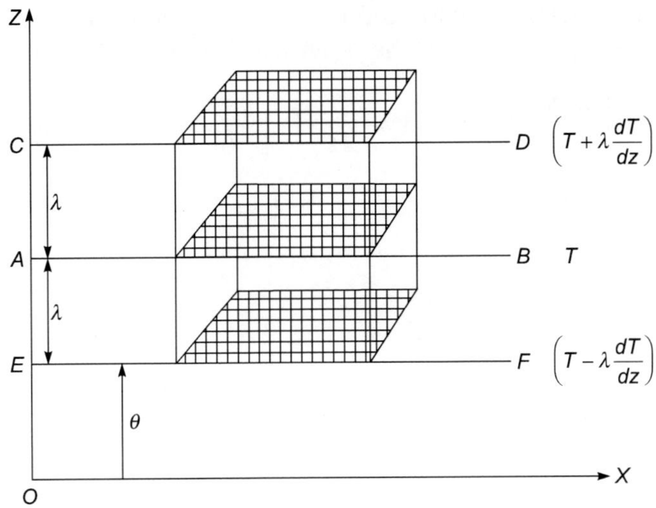

Figure 3.4 Transport of heat energy (conductivity)

Hence, the gas layer above *AB* has higher temperature and below *AB* has lower temperature. So the kinetic energy of molecules coming from *CD* to *EF* will be more in comparison to molecules moving from *EF* to *CD*. Therefore, the mean energy of molecules present in layer *EF* increases continuously to get equilibrium state.

The gas molecules travel in all directions, so the mean number of gas molecules in positive or negative direction of any one axis will be $\frac{1}{6}$ of the molecules present in the gas. If n is the number of molecules present in per unit volume of gas and $\overline{v}$ is the average velocity of molecules, the total number of molecules passing through unit area of *AB* per second, either in downward or upward direction, will be $\frac{n\overline{v}}{6}$. If mass of each molecule of gas is m, total mass of gas passing through unit area of *AB* per second will be $\frac{mn\overline{v}}{6}$. Let us suppose that C_V is specific heat at constant volume.

Heat energy transported by molecules of gas in passing through unit area of gas layer *AB* in per second in the downward direction

= mass × specific heat at constant volume × temperature of gas layer

$$= \frac{mn\overline{v}}{6} \times C_V \times \left(T + \lambda \cdot \frac{dT}{dz}\right)$$

Heat energy transported by molecules of gas in passing through unit area of gas layer *AB* in per second in the upward direction

$$= \frac{mn\overline{v}}{6} \times C_V \times \left(T - \lambda \cdot \frac{dT}{dz}\right)$$

Hence, total heat energy transported by molecules of gas in passing through unit area of gas layer AB per second in the downward direction

$$Q = \left[\frac{mn\bar{v}}{6} \times C_V \times \left(T + \lambda \cdot \frac{dT}{dz}\right)\right] - \left[\frac{mn\bar{v}}{6} \times C_V \times \left(T - \lambda \cdot \frac{dT}{dz}\right)\right]$$

$$= \frac{mn\bar{v}}{6} \times C_V \left(T + \lambda \cdot \frac{dT}{dz} - T + \lambda \cdot \frac{dT}{dz}\right)$$

$$= \frac{mn\bar{v}}{3} \times C_V \cdot \lambda \cdot \frac{dT}{dz} \tag{3.10}$$

$$= \frac{\rho\bar{v}}{3} \times C_V \cdot \lambda \cdot \frac{dT}{dz} \tag{3.11}$$

Here, $\rho = mn$ is density of gas.

or
$$Q = K \cdot \frac{dT}{dz}$$

where,
$$K = \frac{\rho\bar{v}}{3} C_V \cdot \lambda \tag{3.12}$$

K is coefficient of thermal conductivity which is defined as the amount of heat that is transported per unit area per second per unit temperature gradient.

3.7 EFFECT OF TEMPERATURE AND MASS OF MOLECULE ON COEFFICIENT OF THERMAL CONDUCTIVITY

The coefficient of thermal conductivity is defined as

$$K = \frac{\rho\bar{v}}{3} C_V \cdot \lambda = \frac{mn\bar{v}}{3} C_V \cdot \lambda \qquad (\text{as } \rho = mn)$$

$$= \frac{mn\bar{v}}{3} C_V \cdot \frac{1}{\sqrt{2}\,(\pi d^2 n)} \qquad \left(\text{as } \lambda = \frac{1}{\sqrt{2}\,(\pi d^2 n)}\right)$$

or
$$= \frac{m\bar{v}}{3\sqrt{2} \cdot \pi d^2} C_V \tag{3.13}$$

or
$$K \propto \bar{v}$$

As we know that average speed $\bar{v} = \sqrt{\dfrac{8kT}{m\pi}}$

So,
$$K \propto \sqrt{T} \text{ and } K \propto \frac{1}{\sqrt{m}}$$

Hence, thermal conductivity coefficient (K) is proportional to the square root of absolute temperature, and inversely proportional to the mass of the molecule of the gas. In other words, we can say that coefficient K decreases with decrease in temperature and increases with decrease in mass of the molecule of gas. As hydrogen molecule has least mass among diatomic gases, so coefficient K of hydrogen gas will be highest at a particular temperature.

3.8 VISCOSITY PHENOMENON (TRANSPORT OF MOMENTUM)

Let us consider the motion of molecules of a gas in the form of parallel layers which have different velocities. Hence, there will be a relative motion between the layers and the layer traveling rapidly will impart the momentum to layer traveling slowly to get an equilibrium state. This transportation of momentum between the layers of gases is known as viscosity phenomenon.

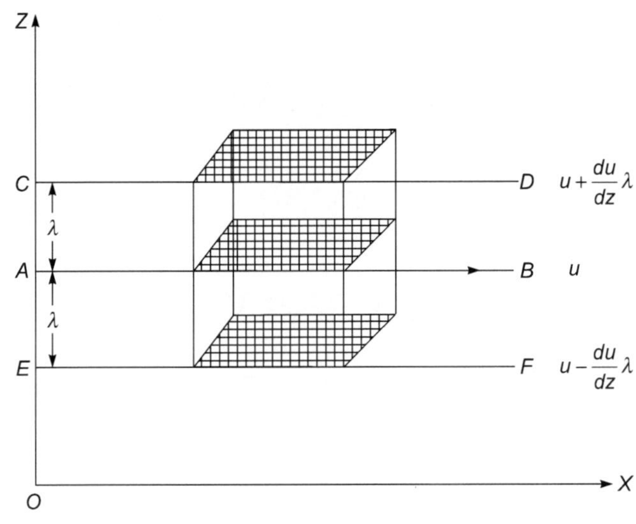

Figure 3.5 Transportation of momentum (viscosity)

Suppose that a gas is flowing in the OX direction. The initial velocity of the gas layer in connection with the OX is zero and will increase in the OZ direction with velocity gradient $\frac{du}{dz}$. Consider a gas layer AB traveling with a velocity u. All the molecules of gas present in this layer have the same velocity. Let us consider that there are two more gas layers CD and EF above and below the layer AB at a distance equal to the mean free path, λ (Figure 3.5).

$$\text{Velocity of the gas layer } CD = u + \frac{du}{dz} \cdot \lambda$$

$$\text{Velocity of the gas layer } EF = u - \frac{du}{dz} \cdot \lambda$$

Let us suppose that n is the number of molecules present in per unit volume of gas and $\bar{v}$ is the average velocity of molecules of gas, the total number of gas molecules passing through the unit area of AB per second either in downward or upward direction will be $\frac{n\bar{v}}{6}$. If mass of each gas molecule is m, total mass of gas passing through the unit area of AB per second will be $\frac{mn\bar{v}}{6}$. Hence, the momentum transformed per unit area per second from the layer CD

$$= \frac{mn\bar{v}}{6} \cdot \left(u + \frac{du}{dz} \cdot \lambda \right)$$

and the momentum transformed per unit area per second from the layer EF

$$= \frac{mn\bar{v}}{6} \cdot \left(u - \frac{du}{dz} \cdot \lambda \right)$$

Therefore, net momentum transported in downward direction per unit area per second

$$= \frac{mn\bar{v}}{6} \cdot \left(u + \frac{du}{dz} \cdot \lambda \right) - \frac{mn\bar{v}}{6} \cdot \left(u - \frac{du}{dz} \cdot \lambda \right)$$

$$= \frac{mn\bar{v}}{6} \cdot \left[u + \frac{du}{dz} \cdot \lambda - u + \frac{du}{dz} \cdot \lambda \right]$$

$$= \frac{mn\bar{v}}{3} \cdot \frac{du}{dz} \cdot \lambda \tag{3.14}$$

This momentum is equivalent to force applied on the lower surface of layer by the top moving surface of layer.

$$F = \frac{mn\bar{v}}{3} \cdot \frac{du}{dz} \cdot \lambda$$

This dragging force is working in unit area and equivalent to tangential force $\eta \cdot \frac{du}{dz}$ or viscous.

$$\eta \cdot \frac{du}{dz} = \frac{mn\bar{v}}{3} \cdot \frac{du}{dz} \cdot \lambda$$

where η is coefficient of viscosity. From the above expression, it is clear that coefficient of viscosity is defined as the ratio of tangential stress on the gas layer to the velocity gradient.

So,
$$\eta = \frac{mn\bar{v}}{3} \cdot \lambda \tag{3.15}$$

or
$$= \frac{\rho\bar{v}}{3} \cdot \lambda \tag{3.16}$$

where $mn = \rho$.

We know that mean free path is defined as

$$\lambda = \frac{1}{\sqrt{2}\,(\pi d^2 n)}$$

So from equation (3.15), we get

$$\eta = \frac{mn\overline{v}}{3} \cdot \frac{1}{\sqrt{2}\,(\pi d^2 n)}$$

$$= \frac{m\overline{v}}{3\sqrt{2}\,(\pi d^2)} \tag{3.17}$$

Equations (3.15), (3.16) and (3.17) give the expression for coefficient of viscosity.

3.9 EFFECT OF TEMPERATURE AND PRESSURE ON COEFFICIENT OF VISCOSITY

From equation (3.16), the coefficient of viscosity is given by

$$\eta = \frac{\rho\overline{v}}{3} \cdot \lambda$$

The density of the gas decreases with increase in gas temperature but mean free path of gas molecule increases in equal amount, the value of $\rho\lambda$ remains constant. The mean speed $\overline{v}$ is directly proportional to the square root of absolute temperature of gas, the coefficient of viscosity is also proportional to $\sqrt{T}$.

Hence, $$\eta \propto \sqrt{T}$$

Therefore, the coefficient of viscosity of gases increases with increase in temperature.

At a constant temperature or at constant velocity $\overline{v}$, the value of $\rho\lambda$ remains constant and the mean speed $\overline{v}$ is also independent of pressure. Hence, the coefficient of viscosity η is independent of pressure.

3.10 RELATION BETWEEN COEFFICIENT OF VISCOSITY (η) AND THERMAL CONDUCTIVITY (K)

We know that the coefficient of viscosity is defined by

$$\eta = \frac{\rho\overline{v}}{3} \cdot \lambda$$

and the coefficient of thermal conductivity of gas is defined by

$$K = \frac{\rho\overline{v}}{3} C_V \cdot \lambda$$

Comparing these two equations of viscosity and thermal conductivity, we get

$$K = \eta \cdot C_V \tag{3.18}$$

The relation is valid only when the supplied heat is changed into translational kinetic energy of gas molecules.

In all other cases, the corrected value is

$$K = \varepsilon\eta \cdot C_V$$

Here, $\varepsilon = \frac{(9\gamma - 5)}{4}$. Hence, the coefficient of thermal conductivity can also be defined as

$$K = \frac{\rho\bar{v}}{3}\varepsilon C_V \cdot \lambda \tag{3.19}$$

3.11 RELATION BETWEEN COEFFICIENT OF VISCOSITY (η) AND COEFFICIENT OF DIFFUSION (D)

We know that the coefficient of viscosity is defined as

$$\eta = \frac{\rho\bar{v}}{3} \cdot \lambda$$

and the coefficient of diffusion is given by

$$D = \frac{1}{3} \cdot \bar{v} \cdot \lambda$$

Comparing these two equations, we get

$$D = \frac{1}{\rho} \cdot \eta \tag{3.20}$$

3.12 RELATION BETWEEN COEFFICIENT OF THERMAL CONDUCTIVITY (K) AND COEFFICIENT OF DIFFUSION (D)

We know that the relation between the coefficient of viscosity and coefficient of conductivity is given by

$$K = \eta \cdot C_V \tag{3.21}$$

Also, the relation between the coefficient of diffusion and coefficient of viscosity is given by

$$D = \frac{1}{\rho} \cdot \eta \tag{3.22}$$

Substituting the value of $\eta = \frac{K}{C_V}$ from equation (3.21) to equation (3.22), we get

$$D = \frac{1}{\rho} \cdot \frac{K}{C_V}$$

or

$$= \frac{K}{\rho \cdot C_V} \qquad (3.23)$$

Numerical Examples

1. The diameter of the molecules of nitrogen is 3.2×10^{-8} cm. The number of molecules at 1 atmospheric pressure and 273°K temperature is $2.68 \times 10^{19}/\text{cm}^3$. Find out the mean free path of nitrogen molecules.

Solution: We know that the mean free path is given by

$$\lambda = \frac{1}{\sqrt{2}\,(\pi d^2 n)}$$

$$= \frac{1}{\sqrt{2} \times (3.14 \times 3.2 \times 10^{-8} \times 3.2 \times 10^{-8} \times 2.69 \times 10^{19})}$$

$$= \frac{1}{122.32 \times 10^3}$$

$$= 0.008175 \times 10^{-3} \text{ cm}$$

$$= 8.175 \times 10^{-6} \text{ cm}$$

2. The mean free path of a gas molecule is 2.12×10^{-5} cm and the molecular diameter is 2.01×10^{-8} cm. Calculate the number of molecules per cm^3.

Solution: The mean free path of a gas molecule is given by

$$\lambda = \frac{1}{\sqrt{2}\,(\pi d^2 n)}$$

So,

$$n = \frac{1}{\sqrt{2}\,(\pi d^2 \lambda)}$$

$$= \frac{1}{1.44 \times (3.14 \times 2.18 \times 10^{-8} \times 2.18 \times 10^{-8} \times 2.12 \times 10^{-5})}$$

$$= 3 \times 10^{19} \text{ /cm}^3$$

3. The diameter of oxygen molecules is 3×10^{-8} cm. If the wavelength of sound wave is equal to the mean free path of oxygen at N.T.P., calculate the frequency of sound wave.

Solution: We know that the mean free path is given by

$$\lambda = \frac{1}{\sqrt{2}\,(\pi d^2 n)}$$

At N.T.P., the value of number of molecules, $n = \frac{6.023 \times 10^{23}}{22400}$

and diameter, $d = 3 \times 10^{-8}$ cm,

So mean free path, $\lambda = \frac{1}{1.44 \times \left(3.14 \times 3 \times 10^{-16} \times 3 \times 10^{-16} \times \frac{6.023 \times 10^{23}}{22400}\right)}$

$$= \frac{22400}{1.44 \times (3.14 \times 9 \times 10^{-32} \times 6.023 \times 10^{23})}$$

$$= 9.369 \times 10^{-6} \text{ cm}.$$

4. A gas is contained in a vessel of 10^{-5} mm of Hg pressure at 0°C. 1 gram gas molecule occupies 22.4 ltr at N.T.P. If the molecular diameter is 1.99 Å, determine the mean free path of molecules of gas.

Solution: We know that the number of molecules in 22.4 litres of gas at 760 mm Hg and 0°C or 273°K temperature

$$= 6.023 \times 10^{23}$$

Hence, the molecules per cm^3 in the vessel at 273°K temperature and 10^{-6} mm of pressure

$$n = \frac{6.023 \times 10^{23} \times 10^{-5}}{224000 \times 760}$$

$$= 3.5399 \times 10^{10} \text{ molecules cm}^{-3}$$

and $\quad d = 1.99 \times 10^{-8}$ cm

We know that mean free path is given by

$$\lambda = \frac{1}{\sqrt{2}\,(\pi d^2 n)}$$

So, $\quad \lambda = \frac{1}{1.44 \times (3.14 \times 1.99 \times 10^{-8} \times 3.5399 \times 10^{10})}$

$$= 19.68 \times 10^{3} \text{ cm}$$

5. Determine the coefficient of viscosity for O_2 at N.T.P. Given that: $\overline{v} = 4.545 \times 10^{2}$ m/sec, $\lambda = 9.45 \times 10^{8}$ m and $\rho = 1.24$ kg/m^3.

Solution: We know that the mean free path in terms of viscosity is given by

$$\lambda = 3\left(\frac{\eta}{\rho\bar{v}}\right)$$

So, coefficient of viscosity, $\eta = \frac{1}{3}(\lambda\rho\bar{v})$

$$\eta = \frac{1}{3} \times 9.45 \times 10^{8} \times 1.24 \times 4.545 \times 10^{2}$$

Viscosity per unit velocity gradient

$$\eta = 1.7867 \times 10^{-5}\,\text{N/m}^2$$

6. The root mean square velocity of gas molecules is $4.49 \times 10^2\,\text{m/sec}$ and the density of gas is 1.249 gm/litre. The viscosity of gas per unit gradient at N.T.P is 1.659×10^{-4} dynes/cm^2. Determine (a) mean free path, (b) frequency of collisions of molecules, and (c) diameter of gas molecules.

Solution:

(a) We know that mean free path is given by

$$\lambda = 3\left(\frac{\eta}{\rho\bar{v}}\right)$$

$$= \left(\frac{3 \times 1.659 \times 10^{-4}}{1.249 \times 10^{-3} \times 4.49 \times 10^{4}}\right)$$

$$= 8.962 \times 10^{-8}\,\text{m}$$

(b) The number of collisions per second is defined as the frequency of collision and we know that

$$\lambda = \frac{v_{rms}}{N}$$

$$N = \frac{v_{rms}}{\lambda}$$

$$= \frac{4.49 \times 10^{2}}{8.962 \times 10^{-8}}$$

$$= 4.99 \times 10^{9}$$

(c) We know that Maxwell's relation of mean free path

$$\lambda = \frac{1}{\sqrt{2}\,(\pi d^2 n)}$$

$$d^2 = \frac{1}{\sqrt{2}\,(\pi n \lambda)}$$

$$d = \left[\frac{1}{\sqrt{2}\,(\pi n \lambda)}\right]$$

$$= \frac{1}{\left[1.44 \times 3.14 \times \left(\frac{6.023 \times 10^{23}}{22400}\right) \times 8.962 \times 10^{-6}\right]^{1/2}}$$

$$= 2.94 \times 10^{-10} \text{ m}$$

7. The $3.01 \times 10^{25}/\text{m}^3$ molecules of oxygen are present in a container. The diameter of oxygen molecule is 3.61×10^{-8} cm. Determine the frequency of collision of molecules and mean free path.

Solution: We know that the mean speed of the molecules of a gas is given by

$$\bar{v} = \sqrt{\frac{8kT}{m\pi}}$$

$$= \sqrt{\frac{8 \times 1.38 \times 10^{-23} \times 273}{\frac{32}{6.023 \times 10^{26}} \times 3.14}}$$

$$= 4.259 \times 10^{2} \text{ m/sec}$$

We know that collision frequency of molecules of gas

$$= \pi d^2 \bar{v} n$$

$$= 3.14 \times 3.61 \times 10^{-10} \times 3.61 \times 10^{-10} \times 4.259 \times 10^{2} \times 3.01 \times 10^{25}$$

$$= 5.199 \times 10^{9}$$

We know that the mean free path is given by

$$\lambda = \frac{1}{\sqrt{2}\,(\pi d^2 n)}$$

$$= \frac{1}{3.14 \times 3.61 \times 10^{-10} \times 3.61 \times 10^{-10} \times 3.01 \times 10^{25}}$$

$$= 5.821 \times 10^{-6} \text{ cm}$$

8. Determine the frequency of sound wave at which its wavelength becomes half of the mean free path of oxygen molecules at 273°K and one atmospheric pressure. The diameter of oxygen molecules is 3.01×10^{-6} m.

Solution: We know that the mean free path is given by

$$\lambda = \frac{1}{\sqrt{2}\,(\pi d^2 n)}$$

$$= \frac{1}{1.41 \times 3.14 \times 3.01 \times 10^{-6} \times 3.01 \times 10^{-6} \times \frac{6.023 \times 10^{23}}{22400}}$$

$$= 9.369 \times 10^{-8}\ \text{m}$$

Let us suppose that the wavelength of sound wave is λ',

$$\lambda' = \frac{\lambda}{2}$$

The frequency of sound wave is given by

$$\nu = \frac{u}{\lambda'} = \frac{u}{\frac{\lambda}{2}} = \frac{2u}{\lambda}$$

$$= \frac{2 \times 332}{9.369 \times 10^{-8}}$$

$$= 3.5389 \times 10^{9}\ \text{Hz}$$

9. The radius of oxygen molecule is 3.035×10^{13} cm and mass is 5.30×10^{-23} gm. Determine the coefficient of thermal conductivity at 0°C. Given that specific heat at constant pressure is 20.90×10^{3} J/kilo-mole°K.

Solution: We know that the coefficient of thermal conductivity is given by

$$K = \frac{mvC_V}{3\sqrt{2}\,(\pi d^2)}$$

Where v is root mean square velocity given by

$$v = v_{rms} = \sqrt{\frac{3kT}{m}}$$

So we get

$$K = \frac{5.31 \times 10^{-26} \times (\sqrt{3 \times 1.38 \times 10^{23} \times 273}) \times 20.90 \times 10^{3}}{3 \times 1.41 \times 3.14 \times 3.035 \times 10^{13} \times 3.035 \times 10^{13}}$$

$$= 24.02 \times 10^{-3}\ \text{J/m sec}\,^{\circ}\text{K}$$

10. Diameter of a molecule is 1.99×10^{-10} cm at a pressure 1 atm and 292°K temperature. Determine (i) mean free path, and (ii) frequency of collision. Given that the velocity of molecules is 510×10^{2} cm/sec.

Solution:

(i) We know that the ideal gas equation

$$PV = RT = NkT$$

$$n = \frac{N}{V} = \frac{P}{k \times T}$$

$$= \frac{1 \times 10^5}{1.38 \times 10^{-23} \times 292}$$

$$= 2.512 \times 10^{25} \text{ molecules/ m}^3$$

The mean free path is given by

$$\lambda = \frac{1}{\sqrt{2}\,(\pi d^2 n)}$$

$$= \frac{1}{1.41 \times 3.14 \times 1.99 \times 10^{-10} \times 1.99 \times 10^{-10} \times 2.512 \times 10^{25}}$$

$$= 2.249 \times 10^{-5} \text{ cm}$$

(ii) We know that the collision frequency is defined as

$$\nu = \frac{\overline{v}}{\lambda}$$

$$= \frac{510}{2.249 \times 10^{-7}}$$

$$= 2.269 \times 10^9 \text{ / sec}$$

11. The mean free path of a gas molecule is 2.12×10^{-7} m and the number of molecules are 3×10^{19}/cc. Determine the diameter of molecules.

Solution: The mean free path of a gas molecule is given by

$$\lambda = \frac{1}{\sqrt{2}\,(\pi d^2 n)}$$

$$d = \left[\frac{1}{\sqrt{2}\,(\pi \lambda n)}\right]^{1/2}$$

$$= \left[\frac{1}{1.41 \times 3.14 \times 2.12 \times 10^{-5} \times 3 \times 10^{19}}\right]^{1/2}$$

$$= 2.01 \times 10^{-10} \text{ m}$$

12. Determine the mean free path of nitrogen molecule. The diameter of nitrogen molecules is 3.22×10^{-10} m and the number of molecules present in per m^3 is 2.7×10^{25}.

Solution: We know that the mean free path is given by

$$\lambda = \frac{1}{\sqrt{2}\,(\pi d^2 n)}$$

$$= \frac{1}{1.41 \times 3.14 \times 3.22 \times 10^{-10} \times 3.22 \times 10^{-10} \times 2.7 \times 10^{25}}$$

$$= 8.01 \times 10^{-10} \text{ cm}$$

13. Determine the mean free path of molecules of nitrogen at 1 atmospheric pressure and 26°C temperature. The diameter of nitrogen molecules is 3.49×10^{-10} m.

Solution: We know that mean free path in terms of pressure is given by

$$\lambda = \frac{k \times T}{\sqrt{2}\,(\pi d^2 P)}$$

$$= \frac{1.38 \times 10^{23} \times 299}{1.41 \times 3.14 \times (76 \times 10^{-3} \times 13.6 \times 10^3 \times 9.8) \times 3.49 \times 10^{-10} \times 3.49 \times 10^{-10}}$$

$$= 7.499 \times 10^{-10} \text{ cm}$$

14. The 2.789×10^{19} molecules of benzene are present in per cc. The mean free path of benzene molecules is 2.19×10^{-8} m. determine the diameter of benzene molecules.

Solution: We know that the mean free path is given by

$$\lambda = \frac{1}{\sqrt{2}\,(\pi d^2 n)}$$

$$d^2 = \frac{1}{\sqrt{2}\,(\pi \lambda n)}$$

$$= \frac{1}{1.41 \times 3.14 \times 2.19 \times 10^{-10} \times 2.789 \times 10^{19}}$$

$$= 36.2501 \times 10^{-20} \text{ m}$$

$$d = 6.04 \times 10^{-10} \text{ m}$$

15. The mean free path of molecules of a gas is 8.85×10^{-10} cm. The coefficient of viscosity of gas molecules is 1.67×10^{-5} N sec/m^2 and density of gas is 1.24 k gm^{-3}. Determine the average speed of gas molecules.

Solution: We know that the mean free path in terms of viscosity is given by

$$\lambda = \frac{3\eta}{\overline{v}\rho}$$

$$\overline{v} = \frac{3\eta}{\lambda\rho}$$

$$= \frac{3 \times 1.67 \times 10^{-5}}{8.85 \times 10^{-10} \times 1.24}$$

$$= 4.49 \text{ cm/sec}$$

16. The number of oxygen molecules present in per cc at N.T.P. is 3.01×10^{19}. The mean free path of oxygen molecules is 8.360×10^{-8} m. Determine the diameter of oxygen molecules and time between collisions of molecules of oxygen gas.

Solution: The mean free path is given by

$$\lambda = \frac{1}{\sqrt{2}\,(\pi d^2 n)}$$

$$d^2 = \frac{1}{\sqrt{2}\,(\pi \lambda n)}$$

$$= \frac{1}{1.41 \times 3.14 \times 8.360 \times 10^{-6} \times 3.01 \times 10^{19}}$$

$$d = \frac{1}{[1.41 \times 3.14 \times 8.360 \times 10^{-6} \times 3.01 \times 10^{19}]^{1/2}}$$

$$= 2.99 \times 10^{-8} \text{ cm}$$

The mean time between two successive collisions is given by

$$t = \frac{\lambda}{\overline{v}}$$

where,

$$\overline{v} = \sqrt{\frac{8kT}{m\pi}}$$

$$= \sqrt{\frac{8 \times 1.38 \times 10^{23} \times 273}{\left(\frac{32}{6.023 \times 10^{23}}\right) \times 3.14}}$$

$$= 4.249 \times 10^{2} \text{ m/sec}$$

Hence, the average time between collisions

$$t = \frac{\lambda}{\bar{v}}$$

$$= \frac{8.360 \times 10^{-6}}{4.249 \times 10^{4}}$$

$$= 19.67 \times 10^{-11} \text{ sec}$$

17. Determine the coefficient of viscosity of oxygen molecules at 0°C. The radius of oxygen molecules is 1.5×10^{-10} m and mass of oxygen molecules is 5.32×10^{-23} gram.

Solution: We know that the coefficient of viscosity is given by

$$\eta = \frac{1}{3\sqrt{2}}\left(\frac{m\bar{v}}{\pi d^2}\right)$$

The mean velocity is given by

$$\bar{v} = \sqrt{\frac{8kT}{m\pi}}$$

$$= \sqrt{\frac{8 \times 1.38 \times 10^{23} \times 273}{\left(\frac{32}{6.023 \times 10^{23}}\right) \times 3.14}}$$

$$= 4.249 \times 10^{2} \text{ m/sec}$$

So, coefficient of viscosity

$$\eta = \frac{1}{3 \times 1.41}\left(\frac{5.32 \times 10^{-26} \times 4.249 \times 10^{2}}{3.14 \times 1.5 \times 10^{-10} \times 1.5 \times 10^{-10}}\right)$$

$$= 1.921 \times 10^{-5} \text{ N/m}^2$$

18. Oxygen and hydrogen gases are kept at similar temperature and pressure. Diameters of oxygen molecules and hydrogen molecules are 3.01 Å and 2.49 Å, respectively. Determine the ratio of coefficient of viscosity of oxygen and hydrogen.

Solution: The coefficient of viscosity is given by

$$\eta = \frac{1}{3}(mn\bar{v}\lambda)$$

where,

$$\lambda = \frac{1}{\sqrt{2}\,(\pi d^2 n)}$$

$$\bar{v} = \sqrt{\frac{8kT}{m\pi}}$$

So, the ratio of coefficient of viscosity of oxygen and hydrogen is given by

$$\frac{\eta_O}{\eta_H} = \sqrt{\frac{m_O}{m_H}} \times \left(\frac{d_H^2}{d_O^2}\right)$$

$$= \sqrt{\frac{32}{2}} \times \left(\frac{2.49}{3.01}\right)^2$$

$$= 4 \times (0.827)^2$$

$$= 4 \times 0.6841 = 2.736$$

$$= 2.736$$

19. The mean free path of gas molecules is 4×10^{-8} m at temperature 400°K and pressure P. Determine the mean free path at (a) P, 800°K, (b) $P/2$, 800°K, and (c) $2P$, 400°K.

Solution: The mean free path is given by

$$\lambda = \frac{1}{\sqrt{2}\,(\pi d^2 n)}$$

$$= \frac{1}{\sqrt{2}\left[\pi d^2 \times \left(\frac{P}{kT}\right)\right]}$$

$$= T \times \frac{k}{\sqrt{2}\,[\pi d^2 \times P]}$$

At 400°K and pressure P mean free path,

$$\lambda = 400 \times \frac{k}{\sqrt{2}\,[\pi d^2 \times P]}$$

(a) Mean free path at pressure P and temperature 800°K

$$\lambda = 800 \times \frac{k}{\sqrt{2}\,[\pi d^2 \times P]} = 2\lambda = 2 \times 4 \times 10^{-8}\text{m} = 8\ \text{Å}$$

(b) Mean free path at pressure $P/2$ and temperature 800°K,

$$\lambda = 800 \times \frac{k}{\sqrt{2}[\pi d^2 \times P/2]}$$

$$= 1600 \times \frac{k}{\sqrt{2}\,[\pi d^2 \times P]} = 4\lambda$$

$$= 4 \times 4 \times 10^{-8}\, m = 16 \text{ Å}$$

(c) The mean free path at pressure $2P$ and temperature 400°K,

$$\lambda = 400 \times \frac{k}{\sqrt{2}[\pi d^2 \times 2P]}$$

$$= 200 \times \frac{k}{\sqrt{2}\,[\pi d^2 \times P]} = \lambda/2$$

$$= = \frac{1}{2} \times 4 \times 10^{-8}\, m = 2 \text{ Å}$$

20. A perfect gas is contained in a vessel at N.T.P. The diameter of the molecules of gas is 3 Å. Determine the mean free path of gas molecules.

Solution: We know that the gas equation for perfect gas is given by

$$PV = RT = NkT$$

$$\frac{N}{V} = \frac{P}{kT}$$

Let us suppose that n is the number of gas molecules present in per m^3 volume,

$$n = \frac{N}{V} = \frac{P}{kT}$$

We know that the mean free path is given by

$$\lambda = \frac{1}{\sqrt{2}\,(\pi d^2 n)}$$

$$= \frac{1}{\sqrt{2}\left[\pi d^2 \times \left(\frac{P}{kT}\right)\right]}$$

$$= T \times \frac{k}{\sqrt{2}\,[\pi d^2 \times P]}$$

$$= \frac{273 \times 1.38 \times 10^{-23}}{[1.41 \times 3.14 \times 3 \times 10^{-10} \times 3 \times 10^{-10} \times (0.76 \times 13.6 \times 10^3 \times 9.8)]}$$

$$= 0.933 \times 10^{-7} \text{ m}$$

Exercise

Short Answer Questions

1. Define the term mean free path of gas molecules.
2. What do you mean by the collision cross section and the sphere of influence?
3. Show the variation of mean free path with temperature and pressure.
4. What is transport phenomenon in gases?
5. What do you mean by frequency of collision?
6. What do you mean by degree of freedom?
7. Obtain the relation between coefficient of viscosity and coefficient of diffusion.
8. Discuss the effect of temperature and pressure on the coefficient of thermal conductivity.
9. Obtain the Clausius expression of mean free path.
10. Derive the Maxwell relation of mean free path.
11. What do you mean by mean free time of gas molecules.
12. The mean free path of gas molecules at temperature T and pressure P is λ. Calculate the mean free path if
    i. The temperature remains constant and the pressure is doubled to its original value.
    ii. The pressure is kept constant and temperature is doubled to its initial value.
13. Discuss the effect of temperature and pressure on the coefficient of viscosity.
14. Obtain the relation between the coefficient of viscosity and the coefficient of thermal conductivity.
15. Discuss the effect of temperature and pressure on the coefficient of diffusion.
16. Obtain the relation between the coefficient of diffusion and the coefficient of thermal conductivity.

17. Discuss the effect of temperature and pressure on the mean free path of gases.
18. Prove that on the basis of kinetic theory of gases, Maxwell established the relation for mean free path which is given by $\lambda = \dfrac{1}{\sqrt{2}\,(\pi d^2 n)}$, where n is the number of molecules per unit volume, and d is the molecular diameter.
19. What do you mean by the transport phenomenon?
20. Define the mean free path and the mean free time.
21. What do you mean by the mean free path of gas molecules?
22. Find out the expression of mean free path on the basis of Clausius.
23. What is the effect of pressure and temperature on the mean free path?
24. Discuss the effect of temperature and pressure on the coefficient of thermal conductivity.
25. What is the effect of temperature and pressure on the coefficient of viscosity?
26. Establish the relation between the coefficient of diffusion and coefficient of viscosity.
27. What is the effect of temperature and pressure on the coefficient of diffusion?
28. Derive the relation between the coefficient of viscosity and coefficient of thermal conductivity.
29. What do you mean by the frequency of collision?
30. Establish the relation between the coefficient of diffusion and coefficient of thermal conductivity.

Long Answer Questions

1. Deduce an expression for the coefficient of viscosity of gas in terms of the mean free path of molecules. Prove that it depends on the temperature (T) of the gas but independent of pressure, P.
2. What do you mean by the transport phenomenon in gases? Explain briefly the self-diffusion, conductivity and viscosity on the basis of kinetic theory of gases.
3. Explain the theory of the mean free path and derive the Clausius expression for the mean free path on the basis of kinetic theory of gases.
4. Obtain an expression for thermal conductivity of a gas on the basis of kinetic theory model of gases. Show that the hydrogen has the largest coefficient of thermal conductivity among all diatomic molecules.
5. Discuss the mean free path of gas molecules. Establish expression for λ and show that it is directly proportional to the absolute temperature of the gas and inversely proportional to the pressure.
6. Discuss the diffusion phenomenon in gases. Derive the expression for the coefficient of self-diffusion and prove that it is inversely proportional to pressure and directly proportional to $T^{3/2}$.

7. Discuss the collision of molecules in gases. Compare the three transport phenomena of gases on the basis of pressure and temperature.
8. Define the mean free path of gas molecules and derive the expression for it. Prove that mean free path is directly proportional to the absolute temperature and inversely proportional to the pressure of the gas.
9. What do you mean by viscosity? Obtain the expression for coefficient of diffusion and prove that it is directly proportional to $T^{3/2}$.
10. What is mean free path? Show that mean free path of a gas molecule is given by the expression

$$\lambda = \frac{1}{\sqrt{2}\,(\pi d^2 n)}$$

where, d is the diameter of the gas molecule and n is the number of molecules present in per unit volume of gas.
11. Define the thermal conductivity. Obtain the expression for the coefficient of thermal conductivity on the basis of kinetic theory.
12. Discuss mean free path and derive the Clausius' expression of mean free path of molecule of gases.
13. What do you mean by the phenomenon of diffusion? Obtain the expression for coefficient of self-diffusion on the basis of kinetic theory of gases.
14. Establish the expression for the coefficient of thermal conductivity and prove that hydrogen molecule has the largest coefficient of thermal conductivity.
15. Obtain the expression for coefficient of self-diffusion and show that it is inversely proportional to the pressure of the gas.
16. Define the term mean free path, mean free time of molecule of gas and collision frequency. Find out the variation of mean free path with temperature and time.

Numerical Problems

1. The number of molecules present in per cm^3 of a gas are 3×19^{19} and the mean free path of molecules is 2×10^{-8} cm, find out the value of the diameter of gas molecules. **[Ans.** 7.28×10^{-7} cm]
2. The mean free path of molecules of a gas at temperature 200°K and pressure P is 2×10^{-5} m. Calculate the value of mean free path (i) $2P$, 200°K (ii) $P/2$, 400°K (iii) P, 400°K **[Ans.** (i) 1×10^{-5} cm (ii) 8×10^{-5} cm (iii) 4×10^{-5} cm]
3. The coefficient of viscosity for nitrogen is 1.66×10^{-5} in MKS units and its density at N.T.P. is 1.25 kg /m^3. Calculate the mean free path of a molecule of the gas. **[Ans.** 8.08×10^{-5} m]

4. The diameter of a gas molecules is 3×10^{-10} m. Determine the mean free path λ at N.T.P. Given that Boltzmann constant $k=1.38\times10^{-23}$ J/K.

[**Ans.** 1×10^{-7} cm]

5. Determine the mean free path of argon molecules at 298°K and 1 atmospheric pressure. Given that the diameter of argon molecule is 2.56×10^{-10} m and Boltzmann constant $k=1.38\times10^{-23}$ J/K. [**Ans.** 1.39×10^{-7} m]

6. The radius of oxygen molecules is 1.66×10^{-8} cm. Calculate the temperature at which their mean free path at 1 atm. pressure is 9.95×10^{-6} cm. [**Ans.** 373°K]

7. The mean free path for benzene is 2.2×10^{-8} m. Calculate the diameter of benzene molecules. Given that the number of molecules per cc is 2.78×10^{19}.

[**Ans** 6.06×10^{-10} m]

8. The diameter of the molecule of oxygen gas is 3.95×10^{-8} cm. If the viscosity of gas is 169 micropoise, find out the temperature of gas. Avogadro's number $=6.023\times10^{23}$ and Boltzmann constant $k=1.38\times10^{-23}$ J/K. [**Ans.** 289°C]

9. The mean free path of a gas molecule is $=8.78\times10^{-6}$ cm and the number of molecules of the gas is $=2.5\times10^{19}$ per cm^3. Calculate the diameter of gas molecules. [**Ans.** 3.2 Å].

10. The mean free path of oxygen molecules at 100°C and one atmospheric pressure is 9.960×10^{-6} cm. Determine the radius of oxygen molecules.

[**Ans.** 1.694×10^{-8} cm]

11. The diameter of the gas molecules is 3.969×10^{-8} cm. Calculate the coefficient of viscosity of molecules of gas at 289°K. [**Ans.** 169 micropoise]

12. The mean free path of a molecule of gas is 8.790×10^{-6} cm. If the diameter of gas molecules is 3.19 Å. Calculate the value of number of molecules in per unit volume. [**Ans.** 2.49×10^{25}]

13. Determine the mean free path of nitrogen present in hydrogen at N.T.P. The cross-sectional area for slow neutron in H_2 is 19.90×10^{-28} m^2.

[**Ans.** 13.39×10^{2} cm]

14. Density of oxygen molecules at N.T.P. is 1.45 kg/m^3. Determine the root mean square velocity of oxygen molecules at 30°C.

15. The diameter of molecules of a gas is 2.969×10^{-8} cm. Find out the coefficient of viscosity of molecules of gas at temperature 300°K.

16. The number of molecules of a gas present in per cm^3 is 2.99×10^{19} and the diameter of molecules of gas is 7.279×10^{-9} m. Determine the mean free path of gas molecules. [**Ans.** 1.99×10^{-10} m]

17. The mean free path of a gas molecule is 3×10^{-5} cm at temperature 300°K and pressure P. Determine the mean free path of molecule of gas (i) $P/2$, 600°K (ii) P, 600°K (iii) $2P$, 300°K.

18. The diameter of oxygen molecules is 3.09 Å and mass of oxygen molecules is 5.31×10^{-23} gm. Calculate the coefficient of viscosity at 0°C.

19. Oxygen and nitrogen gases are kept at similar temperature and pressure. Diameters of oxygen molecules and nitrogen molecules are 3.01 Å and 2.89 Å, respectively. Determine the ratio of coefficient of viscosity of oxygen and nitrogen.

20. The number of oxygen molecules present in per cc at N.T.P. is 3.34×10^{19}. The mean free path of oxygen molecules is 7.360×10^{-8} m. Determine the diameter of oxygen molecules and the collision frequency of molecules of oxygen gas.

21. The mean free path of a molecule of a gas is 8.89×10^{-8} m. The coefficient of viscosity of gas molecules is 1.69×10^{-5} N sec/m^2 and density of gas is 1.36 kgm^{-3}. Determine the average speed of gas molecules.

22. Determine the mean free path of molecules of nitrogen at 1 atmospheric pressure and 40°C temperature. The diameter of nitrogen molecule is 3.49×10^{-8} cm.

23. The diameter of molecule of a gas is 2×10^{-8} m at a pressure of 1 atm and 3000°K temperature. Determine (i) mean free path, and (ii) frequency of collision. Given that the velocity of molecules is 512 m/sec.

24. The radius of oxygen molecules is 3.136×10^{11} m and mass is 5.30×10^{-26} kg. Determine the coefficient of thermal conductivity at 0°C. Given that specific heat at constant pressure is 18.90×10^3 J/kilo-mole°K.

Multiple-Choice Questions

1. The mean free path of molecules of gas is
   (i) Inversely proportional to the square root of the diameter of the molecule
   (ii) Proportional to the diameter of molecule
   (iii) Inversely proportional to the square of the molecular diameter
   (iv) Proportional to the third power of molecular diameter

2. In the process of diffusion, which of the following is transported?
   (i) Mass (ii) Momentum
   (iii) Pressure (iv) Energy

3. The coefficient of viscosity is directly proportional to
   (i) T^2 (ii) Density of gas
   (iii) Temperature (iv) $T^{1/2}$

4. If the temperature is very low, the coefficient of viscosity

(i) Increases with decrease of pressure

(ii) Increases with increase of pressure

(iii) Is equal to pressure

(iv) Is independent of pressure

5. The mean free path of a molecule of a gas is

(i) $\dfrac{PT}{\sqrt{2}kTd^2}$ (ii) $\dfrac{kT}{\sqrt{2}\pi Pd^2}$

(iii) $\dfrac{\sqrt{2}kT}{\pi Pd^2}$ (iv) $\dfrac{\pi kT}{\sqrt{2}Pd^2}$

6. The phenomenon of viscosity is due to the transport of

(i) Mass (ii) Energy

(iii) Momentum (iv) None

7. The mean free path of molecule of gas is

(i) $\dfrac{1}{\sqrt{2}nd^2}$ (ii) $\dfrac{\sqrt{2}}{nd^2}$

(iii) $\dfrac{1}{\sqrt{2}\pi nd^2}$ (iv) $\dfrac{1}{\sqrt{2}n^3\pi d^2}$

8. The coefficient of self-diffusion is inversely proportional to

(i) Square of pressure (ii) Pressure

(iii) $P^{3/2}$ (iv) P^2

9. The coefficient of conductivity is directly proportional to

(i) Density of gas (ii) Square of the density

(iii) Fourth power of density (iv) Independent of density

10. The coefficient of diffusion is proportional to

(i) T^2 (ii) $T^{1/3}$

(iii) $T^{3/2}$ (iv) T

11. The coefficient of thermal conductivity is proportional to

(i) T (ii) $\sqrt{T}$

(iii) $\sqrt{2T}$ (iv) $\sqrt{T/2}$

12. The thermal conductivity is largest for

(i) Nitrogen (ii) Oxygen

(iii) Hydrogen (iv) Helium

13. The mean free path of hydrogen molecule at N.T.P. is

(i) 19.65×10^{-10} m
(ii) 18.3×10^{-8} m
(iii) 17.65×10^{-10} m
(iv) 16.3×10^{-8} m

14. For a given pressure and temperature, the mean free path of a gas molecule provides information about

(i) The nature of molecule
(ii) Intermolecular attraction
(iii) Molecular size
(iv) Molecular shape

15. Hydrogen and nitrogen gases are kept under same temperature and pressure conditions. If the ratio of their coefficient of viscosity is 4.9, the ratio of their diameters is

(i) 6/7
(ii) 5/7
(iii) 4/9
(iv) 1/3

16. The transport phenomenon in gases is due to

(i) Surface tension
(ii) Conduction, diffusion, viscosity
(iii) Convection
(iv) Radiation

17. The coefficient of viscosity of a gas is

(i) $\frac{2}{3}\rho c\lambda$
(ii) $\frac{1}{3}\rho c\lambda$
(iii) $\frac{1}{\sqrt{3}}\rho c\lambda$
(iv) $\frac{1}{3\rho}c\lambda$

18. The coefficient of diffusion of a gas is

(i) $\frac{2}{3}c\lambda$
(ii) $\frac{1}{3}c\lambda$
(iii) $\frac{\rho}{\sqrt{3}}c\lambda$
(iv) $\frac{2}{3\rho}c\lambda$

19. The coefficient of thermal conductivity of a gas is proportional to

(i) $\frac{1}{\sqrt{m}}$
(ii) $\frac{2}{\sqrt{m}}$
(iii) $\frac{1}{\sqrt{2m}}$
(iv) $\frac{3}{\sqrt{m}}$

20. Mean free path of hydrogen gas molecules at room temperature is

(i) 38.3×10^{-8} m (ii) 18.3×10^{-8} m

(iii) 63.3×10^{-8} m (iv) 58.3×10^{-8} m

Answers

1. (iii) 2. (i) 3. (ii) 4. (iv) 5. (ii) 6. (iii) 7. (iii) 8. (ii)
9. (i) 10. (iii) 11. (ii) 12. (iii) 13. (ii) 14. (iii) 15. (ii) 16. (ii)
17. (ii) 18. (ii) 19. (i) 20. (ii)

4

Basic Laws of Radiation

4.1 INTRODUCTION

We know that heat can be transferred from one place to another by three different modes, namely, convection, conduction and radiation. A material medium is required to transfer heat in convection and conduction processes but not in radiation process. The heat energy radiated by a hot body is defined as radiant energy. The propagation of heat from a hot body to a cold body through vacuum or material medium without the active role of the medium is defined as radiation. When the heat passes through a medium, inverting medium may be heated. The absorption degree of heat radiation is different for different materials. Some materials like wood, glass, metals, etc., get heated by absorbing radiation while the materials like rock salt, fluorspar, quartz, etc., do not get heated by the passage of heat radiation. Since heat is also a form of energy so it propagates in the form of waves, such as light or electromagnetic waves travel in vacuum. The heat or thermal radiation follows the following properties:

(i) Similar to light wave, thermal radiation travels in a straight path.
(ii) Heat radiation follows the inverse square law.
(iii) Like light wave, thermal radiation also exhibits interference, reflection, diffraction, polarization and refraction phenomena.
(iv) Heat radiation can travel in vacuum.

Hence, heat radiations have the same properties as electromagnetic waves. The wavelength of heat radiation is larger than red colour wavelength, i.e., it approaches to infra red region of electromagnetic spectrum and also known as infrared radiation. In electromagnetic spectrum, wavelength of heat radiation ranges from 8000 Å to 4000 Å. To detect heat or thermal radiation, bolometer or thermopile is used.

4.2 HEAT EXCHANGE THEORY BY PREVOST

Prevost gives the correct theory of transformation of heat from one body to another. He pointed out that thermal radiation is an exchange process and increase or decrease of temperature sensed by a body is due to the effect of exchange of energy with its surroundings. He stated that "every body continuously emits radiations and absorbing radiations from its surroundings at the rate depending on the nature of its surface and the temperature."

The amount of heat radiated by a body depends on its temperature and it decreases with decrease in temperature. At absolute zero temperature, the heat radiated by the body becomes zero, i.e., body will not radiate thermal energy. Hence, all bodies have a dynamical equilibrium with their surroundings.

We sense cold in the presence of a cooler body as our body loses more radiation than it receives from the cold body. Similarly, in the presence of a hot body, we sense warmth due to the absorption of more heat energy from our body in comparison to radiated heat energy by the body.

4.3 BLACK BODY

There are two types of black bodies that exist in nature. One is ideal type, which are not practically possible and another one is real.

(i) **Perfectly Black Body:** A body which absorbs the entire thermal radiations (radiation of all wavelengths) incident on it, is called a perfect black body. Since this body does not reflect light, it appears black. No existing body is said to be a perfectly black body. However, a body having diffused layer of a black ligament can be considered to be nearly a black body.

(ii) **Black Body in real Practice:** Let us now understand the distribution of energy radiation emitted from a hot body. A non-reflecting solid object having a hollow cavity is considered. The inside of the object is blackened with soot so that it acts as good as an absorber and as bad as a reflector. With the small hole in it, the radiation can enter or escape. These radiations strike the inner wall where it can be partly reflected or absorbed. The reflected part of the radiations strikes some other part of the inner wall many times and is partly absorbed at every reflection until none remains. Since the area of the hole is very small as compared to the total area of the cavity wall, the part of incoming radiation which can go out through the hole can be neglected. Hence, the entire radiations incident on the hole is absorbed. Such a body is thus known as a black body as it absorbs all the radiations entering it. At low temperatures, the hole appears black. A black body is both, an ideal absorber and an ideal radiator.

When a body gets heated, it emits radiation. The nature of the radiation is dependent on the temperature. At very low temperatures, the body emits radiation of long wavelengths in the invisible infrared region. On the other hand, at higher temperatures, shorter wavelength radiations are emitted. Furthermore, it can be said that the amount of emitted radiation differs for different wavelengths. It is very interesting to observe the distribution of energy among different wavelengths of light at different temperatures. For example, if a platinum wire gets heated to about 500°C, it appears dull red. The colour changes to cherry red at about 900°C and at about 1100°C, becomes orange red. Further, it on heating at higher temperature, say 1300°C, it appears yellow and finally, white at 1600 C°. This shows that with the increase in temperature, the radiations of shorter wavelengths are produced.

A practical black body was designed by Fery and is known as Fery's black body. In this black body, a hollow sphere was coated with lampblack on its inner surface. A fine hole is made for narrow opening in front of the pointed projection. The incident radiation enters the sphere through the narrow opening and falls on point projection. After suffering multiple reflections by the pointed projection, the radiation is totally absorbed (Figure 4.1). This body worked as an absorber.

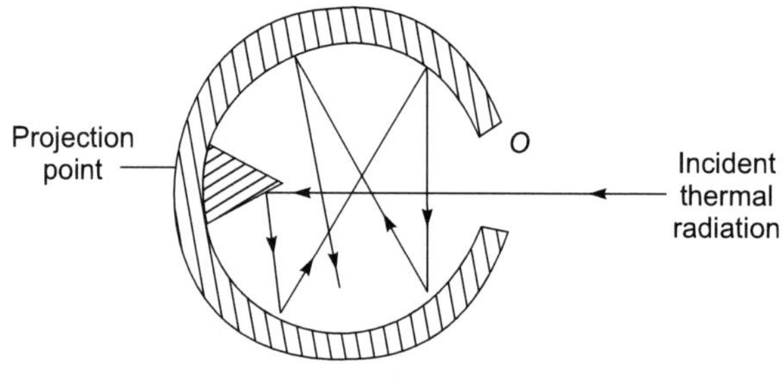

Figure 4.1 Fery's black body (absorber)

When this spherical body is placed in a bath maintained at a fixed temperature, all the heat radiations come outside of the hole and the hole works as a radiator (Figure 4.2).

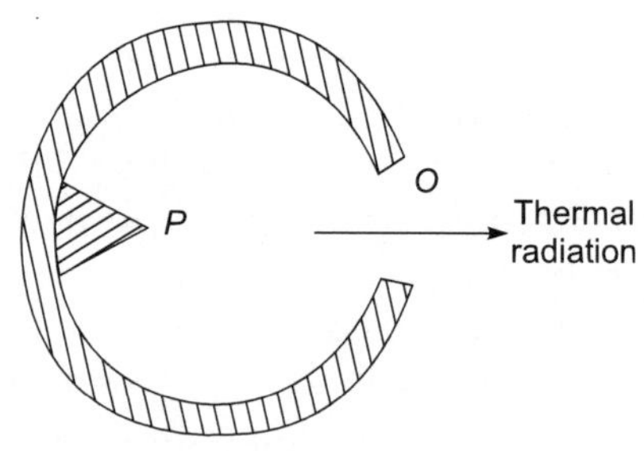

Figure 4.2 Fery's black body (reflector)

In Fery's black body, only the hole acts as the black body radiator. Lampblack used in its inner surface can be replaced by platinum black. Lampblack can absorb 96% of visible radiation while platinum black can absorb 98% of incident visible radiation.

Another practical black body was constructed by Wien in the form of cylinder. This black body consisted of a hollow cylindrical chamber of platinum or brass (Figure 4.3).

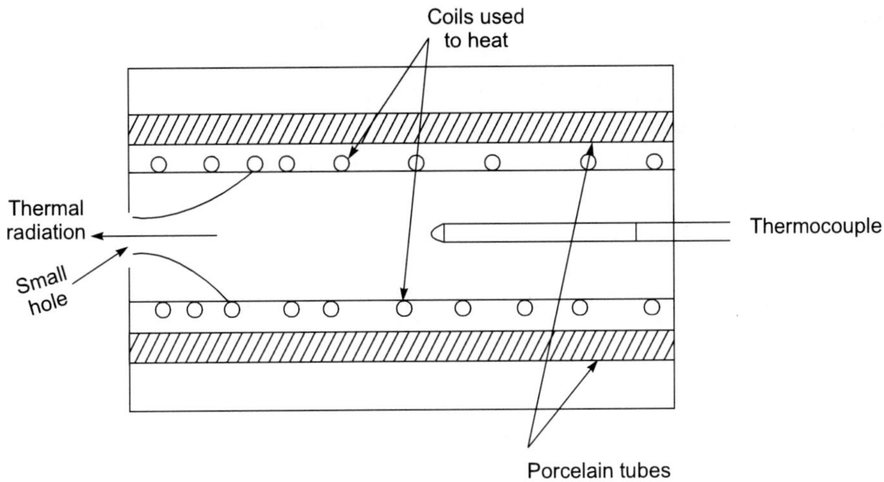

Figure 4.3 Wien's black body

4.4 ENERGY DISTRIBUTION SPECTRUM OF BLACK BODY

Two scientists Lummer and Pringsheim calculated the intensity of emitted energy at different temperatures with wavelengths radiated from a black body. The amount of radiation emitted at various wavelengths is shown in the form of energy distribution graph at each temperature in Figure 4.4.

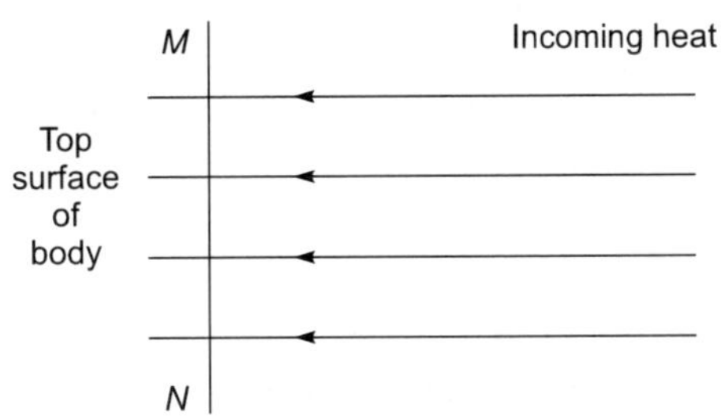

Figure 4.4 Normal incidence of radiation on the surface

These curves in the graph reveal the following features of black body radiation.

(i) At a particular temperature, the energy is not distributed uniformly in the radiation spectrum of the body.

(ii) For a given temperature, T, the emitted energy has maximum value for a particular given wavelength, λ_m.

(iii) The product $\lambda_m \times T$ remains constant.

$$\lambda_m \times T = c \; ....$$

where, c is a constant whose value is about 2.9×10^{-3} m K.

(iv) The above equation shows that with the increase in temperature T, the wavelength λ shifts to a shorter wavelength.

(v) At all wavelengths, with the increase in temperature there is an increase in energy emission. The intensity of radiation increases with increase in wavelengths. At a particular wavelength λ_m, the radiation intensity has a maximum value. Further, with increase in wavelengths, the intensity of radiation decreases.

(vi) The area under each curve depicts the total energy E radiated over all wavelengths for a particular temperature. It is found that the area of each curve is directly proportional to the fourth power of kelvin of temperature T. Therefore,

$$E \propto T^4$$

or

$$E = \sigma T^4$$

where, σ is a constant known as Stephen-Boltzmann constant and its value is $\sigma = 5.67 \times 10^{-8}$ in MKS system. Expression given in the above equation is known as Stephen-Boltzmann law.

4.5 APPLICATIONS OF THERMAL AND BLACK BODY RADIATIONS

There are a number of applications of black body radiations in our daily life. Some applications are as follows:

(i) Hot water pipes outside the rooms are painted with white color so that they do not radiate heat to their surroundings, while the same pipes inside the room are painted with black color so that they can radiate maximum heat to the room.

(ii) We prefer to wear white or light color clothes during summer and dark color clothes in winter as dark color absorbs highest heat and light colour absorbs minimum amount of heat from outside.

(iii) Electric heaters are designed to reflect maximum heat in the room by polishing them with a good reflector.

(iv) Cooking appliances are polished on the top surface while blackened at the bottom. As black colour is a good absorber, so it will absorb more heat from heating system and it top surface will not allow the heat to enter inside it.

The black bodies can be used for heating, lighting, thermal imaging, security, and also, as testing and measurement applications.

Since the intensity of the energy at different temperatures and different wavelengths is determined using the Planck's radiation law, a black body radiation source with a given temperature, or, unknown temperature is usually used for calibration and testing of the radiation thermometers.

4.6 ENERGY DENSITY

The total radiant energy per unit volume is defined as the total energy density. The total energy density (E) exists around the point where all wavelengths are taken together. If energy density is defined in a range of frequency and wavelength, then it is known as the spectral energy density and is represented by $E(\upsilon)$ or $E(\lambda)$.

The total energy density can be written as

$$E = \int_0^\infty E(\lambda)\, d\lambda$$

4.7 EMISSIVE AND ABSORPTIVE POWERS

The radiant energy per sec per unit surface area at a particular wavelength is defined as emissive power. This is represented by $e(\lambda)$. The total emissive power is represented by e and can be written as

$$e = \int_0^\infty e(\lambda)\, d\lambda$$

The ratio of the energy absorbed per unit area per unit time to the total incident energy in the same unit area at same time is defined as absorptive power. Absorptive power is represented by $a(\lambda)$.

4.8 KIRCHHOFF'S LAW AND ITS APPLICATION

The Kirchhoff's law states that "the ratio of the emissive and absorptive power for a given wavelength and temperature is equal for all bodies at the same temperature and it is also equal to the emissive power of a perfectly black body at the same temperature."

Let us consider dQ is the thermal energy in the wavelength range λ and $\lambda + d\lambda$. This heat energy is incident on unit area of a body in one second. If $a(\lambda)$ is absorptive power, the amount of heat energy absorbed by the surface $= a(\lambda)dQ$.

If $e(\lambda)$ is emissive power, then the heat energy emitted per unit area per sec in the wavelength range λ and $\lambda + d\lambda$ $= e(\lambda)d\lambda$

As we know that, heat energy absorbed by the surface = heat energy emitted by the surface.

Hence,

$$a(\lambda)dQ = e(\lambda)d\lambda \tag{4.1}$$

For a perfect black body $a(\lambda) = 1$ and $e(\lambda) = E(\lambda)$, then

$$dQ = E(\lambda)d\lambda$$

So from equation (4.1), we get

$$a(\lambda)E(\lambda)d\lambda = e(\lambda)d\lambda$$

or

$$E(\lambda) = \frac{e(\lambda)}{a(\lambda)} = \text{constant} \tag{4.2}$$

This equation (4.2) is known as Kirchhoff's law.

The great achievement of Kirchhoff's law is that it reveals that all good absorbers are also good emitters. A body which strongly absorbs a radiation of particular wavelength will also strongly emit that particular wavelength. For example, sodium vapor emits two wavelengths of yellow color, i.e., 5890 Å and 5896 Å and also absorbs both the wavelengths. The presence of Fraunhofer dark lines in sun spectrum is also successfully explained by Kirchhoff's law.

4.9 PRESSURE PRODUCED DUE TO RADIATION

When a radiation is incident on a body, it exerts some pressure on it. There are two types of pressure due to radiation on the surface of a body.

(i) Pressure due to Normal Incidence of Radiation: Let us consider a thermal radiation of energy density $E(\upsilon)$ is incident normal on the surface of a body (Figure 4.5). The radiation exerts a pressure on the surface which is equal to the energy density. Thermal radiation is similar to light wave. Let us consider a photon of energy $h\upsilon$, which moves with velocity of light c.

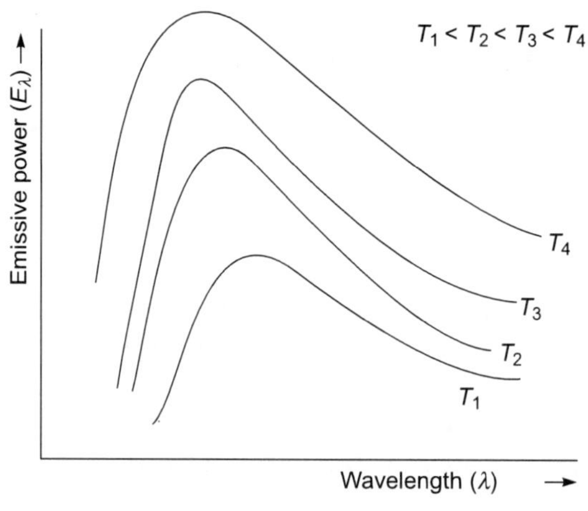

Figure 4.5 Energy distribution curve of a black body at different temperatures

The total energy passing through the area S per second $= E(\nu)cS$

So the radiation energy per unit area in per second $= \dfrac{E(\nu)cS}{S} = E(\nu)c$

According to the theory of relativity, we know that

$$mc^2 = h\nu$$

or $$m = \frac{h\upsilon}{c^2}$$

So the momentum of photon,

$$p = \frac{E(\nu)c}{c} = E(\upsilon)$$

The rate of change of momentum per unit area is defined as the pressure of radiation. If all the momentum per unit area in per second is absorbed in the process, we get: radiation pressure,

$$P = p = E(\nu) = \text{ energy density}$$

(ii) Pressure due to Diffused Radiation: Let us consider a thermal radiation incident at an angle θ with the normal to the surface of the body (Figure 4.6). Suppose that this radiation is diffused, i.e., the photons move in all directions.

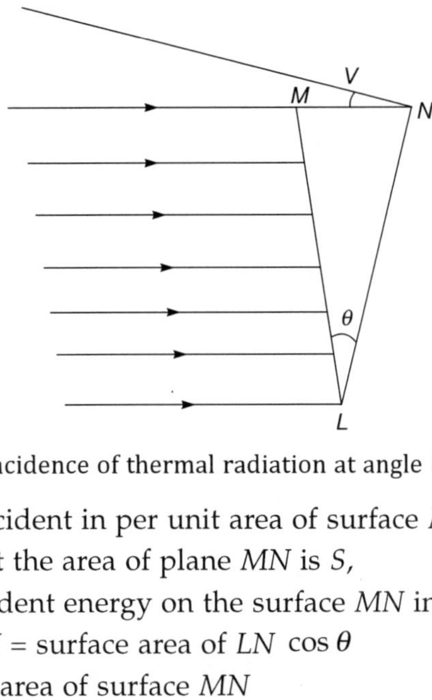

Figure 4.6 Incidence of thermal radiation at angle θ on the surface

The heat energy incident in per unit area of surface MN per second $= E(\nu)c$.
Let us suppose that the area of plane MN is S,
The amount of incident energy on the surface MN in one second $= E(\nu)cS$
Surface area of MN = surface area of LN $\cos\theta$
If S_1 is the surface area of surface MN

So, $$S = S_1 \cos\theta$$

Hence, heat energy incident on the surface of ML in one second

$$= E(\nu)cS_1 \cos\theta$$

This energy will be equal to energy incident on the plane LN in one second

$$= E(\nu)cS_1 \cos\theta$$

So, the energy incident on per unit area of LN in one second

$$= \frac{E(\nu)cS_1 \cos\theta}{S_1} = E(\nu)c\cos\theta$$

Hence, the momentum incident per unit area of LN in unit second

$$=\frac{E(\nu)c\cos\theta}{c}=E(\nu)\cos\theta$$

So, the normal component of momentum $=E(\nu)\cos\theta\times\cos\theta$

$$=E(\nu)\cos^2\theta$$

Here, radiation is incident equally from all directions. Let us suppose that N beams of radiation are incident on the surface.

Total pressure exerted by radiation on the surface

$$=\sum E(\nu)\cos^2\theta \tag{4.3}$$

To find out the value of $\sum E(\nu)\cos^2\theta$, consider a hemisphere of radius r around the center O of the plane LN (figure 4.7). In this hemisphere, consider a ring of radius $r\sin\theta$ and width $rd\theta$ between angle θ and $\theta+d\theta$.

So, the area of this ring $=(2\pi r\sin\theta)\times rd\theta$

$$=(2\pi r^2\sin\theta)\,d\theta$$

As the total number of N beams are incident on the surface of hemisphere of area $=4\pi r^2/2=2\pi r^2$, so the number of incident beams crossing the area of the ring under consideration is given by

$$dN=\frac{(2\pi r^2\sin\theta)d\theta}{2\pi r^2}\times N$$

$$=(N\sin\theta)\,d\theta \tag{4.4}$$

Hence, the fraction of incident beams between the angle θ and $(\theta+d\theta)$

$$\frac{dN}{N}=(\sin\theta)\,d\theta$$

So, the net pressure can be written as

$$P=\frac{dN}{N}\times E(\nu)\cos^2\theta$$

$$=(\sin\theta)d\theta E(\nu)\cos^2\theta$$

$$=E(\nu)\sin\theta\cos^2\theta d\theta \tag{4.5}$$

For the hemisphere, variation of θ can be taken from 0 to $\frac{\pi}{2}$, we get

$$P=\int_0^{\frac{\pi}{2}}E(\nu)\sin\theta\cos^2\theta d\theta$$

Let us assume that $\cos\theta=y$, $-\sin\theta d\theta=dy$.

So, putting these values in equation (4.5), we get

$$P = E(\nu)\int_1^0 -y^2 dy$$

$$= E(\nu)\int_0^1 y^2 dy = E(\nu)\left[\frac{y^3}{3}\right]$$

$$= \frac{E(\nu)}{3} \tag{4.6}$$

Hence, the pressure applied by diffused radiation to the surface is one third of energy density, $E(\nu)$.

4.10 PLANCK'S LAW FOR BLACK BODY RADIATION

In 1900, Max Planck found an empirical formula to explain the distribution of energy among various wavelengths in the black body radiation spectrum. Few assumptions were made to derive the Planck's radiation formula, which were known as Planck's hypothesis of quantum theory of radiation.

The radiation chamber of a black body is filled up with simple harmonic oscillators or resonators along with radiations. These resonators, having molecular dimensions, were called Planck's oscillators or resonators. These resonators can vibrate with all possible frequencies and they entail only one degree of freedom. These oscillators or resonators can neither radiate nor absorb energy continuously but the oscillators with frequency ν can absorb or radiate energy given by $h\nu$. Here, h is known as Planck's constant with value being equal to 6.63×10^{-34} J-s. This assumption explains that the exchange of energy between matter and radiation cannot occur continuously but is limited to certain discrete set of values, such as $0, h\nu, 2h\nu, 3h\nu \ldots, nh\nu$. These units are called quantum.

To derive Planck's law of radiation, let us derive the number of resonators per unit volume in the frequency range ν and $\nu + d\nu$.

Consider a hollow cubic enclosure with perfectly insulating walls. The radiations are enclosed in the enclosure. The radiation is supposed to consist of a number of waves, according to the electromagnetic theory. These waves can travel in all possible directions inside the enclosure and undergo large number of reflections from all the walls of the enclosure. Stationary vibrations are produced inside the enclosure, considering the walls of the enclosure as the nodal points, which are produced due to the interference of reflected waves with that of the incident wave. We can relate the formation of the stationary wave inside the enclosure with vibration of a stretched string, whose end points are stretched. Taking 'l' as the length of the string, the allowed wavelengths are,

$$\lambda = \frac{2l}{n}; \quad n = 1,2,3... \tag{4.7}$$

Hence, the allowed frequencies are,

$$\nu = \frac{c}{\lambda} = \frac{cn}{2l} n = 1, 2, 3... \text{ where, } c \text{ is the velocity of waves.} \tag{4.8}$$

These allowed frequencies are called modes of vibration.

Now, imagine a cube (three dimensional) with its walls as nodal planes. Let length of each side of the cube be s. If the intersecting edges of the cube are X, Y and Z axes of the Cartesian coordinate system, then the direction cosines of the propagation of the wave are $\cos\alpha$, $\cos\beta$ and $\cos\gamma$. Then, $s\cos\alpha$, $s\cos\beta$ and $s\cos\gamma$ will be the projection of edges of the cube on the direction of propagation of waves.

Similar to the case of string, we can replace l in equation (4.7) by $s\cos\alpha$, $s\cos\beta$ and $s\cos\gamma$. Hence, the allowed wavelengths in equation (4.7) will be,

$$\lambda = \frac{2s\cos\alpha}{n_1}; \tag{4.9-a}$$

$$= \frac{2s\cos\beta}{n_2} \tag{4.9-b}$$

and

$$= \frac{2s\cos\alpha}{n_3} \tag{4.9-c}$$

Here, n_1, n_2 and n_3 are positive integers.

The trigonometric condition of direction cosines is

$$\cos^2\alpha + \cos^2\beta + \cos^2\gamma = 1 \tag{4.10}$$

Putting the values of $\cos\alpha$, $\cos\beta$ and $\cos\gamma$ from equation (4.9) into equation (4.10), we get,

$$\frac{n_1^2\lambda^2}{4s^2} + \frac{n_2^2\lambda^2}{4s^2} + \frac{n_3^2\lambda^2}{4s^2} = 1$$

or

$$n_1^2 + n_2^2 + n_3^2 = \frac{4s^2}{\lambda^2}.$$

Putting, $\lambda = \frac{c}{\nu}$, the above equation reduces to,

$$n_1^2 + n_2^2 + n_3^2 = \frac{4s^2\nu^2}{c^2} = \left(\frac{2s\nu}{c}\right)^2 \tag{4.11}$$

Equation (4.11) gives the allowed modes of vibrations inside the cube in terms of frequency. According to the above equation, each value of positive integral values n_1, n_2 and n_3 corresponds to a frequency or a mode of vibration. This equation can be used to calculate the number of modes of vibration within the frequency range ν and $\nu + d\nu$.

To calculate this, let us consider a two-dimensional analogue of equation (4.11).

$$n_1^2 + n_2^2 = \left(\frac{2sv}{c}\right)^2 \tag{4.12}$$

Plotting a graph with n_1 along X-axis and n_2 along Y-axis for a particular value of v, we obtain points on a circle of radius $\left(\frac{2sv}{c}\right)$ as represented from equation (4.12).

The graph is shown in Figure 4.7 for $\left(\frac{2sv}{c}\right) = 12$. In the graph, the corresponding values are drawn on X-axis and Y-axis for all possible integral values of n_1 and n_2. Every point of intersection on abscissa and ordinate represents mode of vibration for an allowed set of values of n_1 and n_2. The points on the circle represent the value of frequency ν, and those lying inside the circle represent the frequencies less than v. Since only positive values of n_1 and n_2 are allowed, we have to consider the intersections in positive quadrants only. The lines which are drawn in the circle divide the quadrant in a number of unit squares.

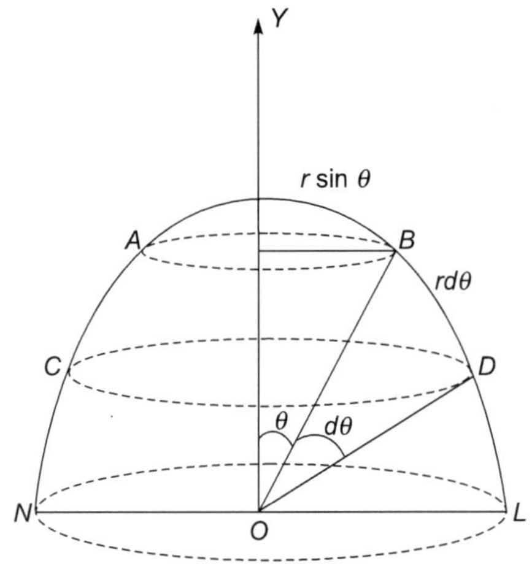

Figure 4.7

Every point of intersection can be associated with the unit square. Since the area of each square is unity, the total number of squares is equal to the area inside the quadrant of the circle. Hence, the total number of modes of vibration with number of frequencies less than v is equal to the area of quadrant of the circle represented by equation (4.12). The area of quadrant is given by,

$$\frac{1}{4}\pi r^2 = \frac{1}{4}\pi\left(\frac{2\pi\nu}{s}\right)^2$$

$$= \frac{\pi a^2 \nu^2}{s^2}$$

Hence, the number of modes of vibration within the frequency range ν and $\nu + d\nu$ is equal to the area in the positive quadrant lying between two circles of radii $\left(\frac{2s\nu}{c}\right)$ and $\left(\frac{2s\,(\nu + d\nu)}{c}\right)$. The area is calculated to be,

$$\frac{1}{4}\pi\left[\left\{\frac{2a(\nu + d\nu)}{s}\right\}^2 - \left(\frac{2a\nu}{s}\right)^2\right]$$

$$= \frac{1}{4}\pi\left(\frac{8a^2\nu d\nu}{s^2}\right) = \frac{2\pi a^2 \nu d\nu}{s^2} \tag{4.13}$$

Let us consider a three-dimensional case, where a circle would be replaced by a sphere and the mesh of squares would be replaced by mesh of cubes. Now, the number of modes of vibration within the frequency range ν and $\nu + d\nu$ is equal to the volume of the octant, i.e., $\frac{1}{8}$th of a spherical shell with the radii $\left(\frac{2s\nu}{c}\right)$ and $\left(\frac{2s\,(\nu + d\nu)}{c}\right)$.

The volume is given by,

$$\frac{1}{8}\left\{\frac{4}{3}\pi\,(r + dr)^3 - \frac{4}{3}\pi r^3\right\} = \frac{1}{8}\cdot 4\pi r^2$$

$$= \frac{1}{8}\cdot 4\pi\left(\frac{2s\nu}{c}\right)^2\left(\frac{2sd\nu}{c}\right)$$

$$= \frac{4\pi s^3 \nu^2}{c^3}d\nu$$

Putting the volume of the cube as $V = s^3$. The number of modes of vibration inside the cubical enclosure within the frequency range ν and $\nu + d\nu$ is equal to $\frac{4\pi V \nu^2}{c^3}d\nu$.

Therefore, the total number of modes of vibration per unit volume within the frequency range ν and $\nu + d\nu$ is equal to $\frac{4\pi\nu^2}{c^3}d\nu$.

These black body radiations are transverse in nature and travel with the velocity of light, c.

As there are two possible orientations for each transverse wave, the modes of vibration are twice as for longitudinal waves. Hence, for electromagnetic radiations or black body radiations, the number of modes of vibration per unit volume within the frequency range ν and $\nu + d\nu$ will be twice for transverse waves and will be given by

$$= \frac{8\pi v^2}{c^3} dv. \tag{4.14}$$

Let N be the total number of Planck's resonators and E be the total energy of the resonators, then the average energy of the Planck's oscillator will be,

$$\bar{\epsilon} = \frac{E}{N} \tag{4.15}$$

According to Maxwell's law of molecular motion, if ϵ denotes certain amount of energy, then the probability of the system having energies 0, ϵ, 2ϵ ... $m\epsilon$ will be in the ratio

$$1 : e^{-\epsilon/kT} : N_0 e^{-2\epsilon/kT} : \ldots e^{-m\epsilon/kT}$$

Let N_0 be the number of resonators having zero energy, then the number of resonators N_1 with energy ϵ will be $N_0 e^{-\epsilon/kT}$, the number of resonators N_2 with energy 2ϵ will be $N_0 e^{-2\epsilon/kT}$. Hence, in general, we can say that the number of resonators N_m with energy $m\epsilon$ will be $N_0 e^{-m\epsilon/kT}$.

Hence, we can find the total number of resonators as

$$N = N_0 + N_1 + N_2 + \ldots + N_m$$

$$= N_0 + N_0 e^{-\epsilon/kT} + N_0 e^{-2\epsilon/kT} + \ldots + N_0 e^{-m\epsilon/kT}$$

$$= N_0 \left(e^{-\epsilon/kT} + e^{-2\epsilon/kT} + \ldots + e^{-m\epsilon/kT}\right)$$

Substituting $e^{-\epsilon/kT} = z$, we get,

$$N = N_0 (1 + z + z^2 + \ldots + z^m)$$

$$N = \frac{N_0}{1 - z} \tag{4.16}$$

Hence, the total energy of the Planck's oscillator will be given by,

$$E = 0 \times N_0 + \epsilon \times N_1 + 2\,\epsilon \times N_2 + \ldots m\epsilon \times N_r$$

$$= 0_0 + \epsilon \times N_0 e^{-\epsilon/kT} + 2\epsilon \times N_0 e^{-2\epsilon/kT} + \ldots m\epsilon \times N_0 e^{-m\epsilon/kT}$$

$$= \epsilon\, N_0 \left[e^{-\epsilon/kT} + 2e^{-2\epsilon/kT} + \ldots m e^{-m\epsilon/kT}\right]$$

Again substituting $e^{-\epsilon/kT} = z$, the above equation becomes,

$$E = \epsilon N_0 \frac{z}{(1 - z)^2} \tag{4.17}$$

Hence, the average energy of the oscillator will be,

$$\bar{\epsilon} = \frac{E}{N} = \frac{\epsilon N_0 \dfrac{z}{(1-z)^2}}{\dfrac{N_0}{1-z}}$$

or
$$= \frac{\epsilon z}{1-z}$$

$$= \frac{\epsilon e^{-\epsilon/kT}}{1-e^{-\epsilon/kT}}$$

or
$$= \frac{\epsilon}{e^{\epsilon/kT}-1} \qquad (4.18)$$

Taking into account Planck's hypothesis of quantum theory, energy can be represented in terms of frequency by $\epsilon = h\nu$, hence the average energy of Planck's oscillator is given by

$$\bar{\epsilon} = \frac{h\nu}{e^{h\nu/kT}-1} \qquad (4.19)$$

The energy density $E(\nu)$ is defined as total energy per unit volume and is determined by multiplying the average energy of Planck's resonator by the number of oscillators per unit volume in the frequency range ν and $(\nu + d\nu)$. Hence, energy density is written as,

$$E(\nu)d\nu = \left(\frac{8\pi\nu^2}{c^3}d\nu\right)\left(\frac{h\nu}{e^{h\nu/kT}-1}\right)$$

or
$$E(\nu)d\nu = \frac{8\pi h\nu^3}{c^3}\left(\frac{1}{e^{h\nu/kT}-1}\right)d\nu \qquad (4.20)$$

Equation (4.20) represents ***Planck's radiation law.***

The energy density can be expressed in terms of wavelength using the relation $\nu = \frac{c}{\lambda}$. Therefore we can write,

$$d\nu = -\frac{c}{\lambda^2}d\lambda \qquad (4.21)$$

From equation (4.21), we conclude that an increase in frequency corresponds to decrease in wavelength, therefore,

$$E(\lambda)d\lambda = -E(\nu)d\nu \qquad (4.22)$$

Hence, the energy density in terms of wavelength will be,

$$E(\lambda)d\lambda = -\frac{8\pi h}{c^3}\left(\frac{c^3}{\lambda^3}\right)\left(\frac{1}{e^{hc/\lambda kT}-1}\right)\left(-\frac{c}{\lambda^2}\right)d\lambda \tag{4.23}$$

or

$$= \frac{8\pi hc}{\lambda^5}\left(\frac{1}{e^{hc/\lambda kT}-1}\right)d\lambda \tag{4.24}$$

Equation (4.24) represents Planck's radiation law in terms of wavelength.

4.11 RAYLEIGH-JEANS LAW FOR SPECTRAL DISTRIBUTION OF ENERGY

Rayleigh-Jeans formula for spectral distribution of energy in the normal spectrum is based on the principle of equipartition of energy for all possible modes of vibration (free vibrations) which are assigned to the radiation. Hence, Rayleigh–Jeans considered the average energy of an oscillator as,

$$\bar{\epsilon} = kT$$

Since, the number of modes of vibration per unit volume in the frequency range ν and $\nu + d\nu$ is

$$N(\nu)d\nu = \frac{8\pi\nu^2}{c^3}d\nu \tag{4.25}$$

Hence, we can write the energy density in the frequency range ν and $\nu + d\nu$ as,

$$E(\nu)d\nu = \frac{8\pi\nu^2}{c^3}d\nu \times kT = \frac{8\pi\nu^2 kT}{c^3}d\nu \tag{4.26}$$

Equation (4.26) represents Rayleigh-Jeans law in terms of frequency.

Rayleigh-Jeans law can be expressed in terms of wavelength as,

$$E(\lambda)d\lambda = \frac{8\pi}{c^3}\left(\frac{c^2}{\lambda^2}\right)\left(-\frac{c}{\lambda^2}\right)d\lambda \times kT$$

or

$$= \frac{8\pi kT}{\lambda^4}d\lambda \tag{4.27}$$

This law explains the experimental determination of energy distribution at longer wavelength (low frequencies) and at higher temperatures.

Experimentally, it was observed that this law agrees well at lower frequencies but at higher frequencies, this law is not valid. According to this law, the energy density $E(\lambda)$ increases continuously with the decrease in the wavelength λ and approaches to infinity when λ tends to zero.

Hence, the total energy density for all wavelengths ranging from 0 to ∞ is,

$$E = \int_0^\infty E(\lambda)d\lambda = \int_0^\infty \frac{8\pi kT}{\lambda^4}d\lambda$$

$$= 8\pi kT\left[-\frac{1}{3\lambda^3}\right]_0^\infty = \infty \tag{4.28}$$

Hence, according to this law, almost all energy should be restricted in vibrations of shorter wavelength. This is known as **ultraviolet catastrophe**.

4.12 WIEN'S DISPLACEMENT LAW

According to Wien's displacement law, "The product of the maximum incident wavelength (λ_m) and the absolute temperature (T) is constant". Hence,

$$\lambda_m \cdot T = \text{constant } = b \tag{4.29}$$

where, b is Wien's displacement constant and its value is 0.2892 cm°K. Hence, as the temperature increases, the maximum intensity of emitted radiation is shifted towards the shorter wavelength. This is the conclusion of Wien's displacement law.

Wien also deduced that the maximum energy (E_m) is proportional to the fifth power of absolute temperature, i.e.,

$$E_m \propto T^5$$

or
$$E_m = T^5 \times \text{constant}$$

or
$$\frac{E_m}{T^5} = \text{constant}$$

or
$$E_m \times T^5 = \text{constant} \tag{4.30}$$

On combining the equations (4.29) and (4.30), the combined effect will be in the form

$$E(\lambda) = C \times [\lambda^{-5} f(\lambda T)] \tag{4.31}$$

Here, C is a constant and $f(\lambda T)$ is the function of the product λT. $E(\lambda)$ represents the emissive power of black body radiation and it gets maximum value at absolute temperature.

To find out the value of $f(\lambda T)$, Wien assumes that the function $f(\lambda T)$ has the form $B\,e^{-\frac{a}{\lambda T}}$. Hence, equation (4.31) can also be written as

$$E(\lambda) = C \times \left[\lambda^{-5} B\,e^{-\frac{a}{\lambda T}}\right]$$

$$E(\lambda) = A \times \left[\lambda^{-5} e^{-\frac{a}{\lambda T}}\right]$$

where, A is another constant and its value is 4.94×10^{15} in C.G.S. system and the value of constant a is 1.43 cm-degree.

Multiplying both sides of the above equation by $d\lambda$, we get

$$E(\lambda)d\lambda = A \times \left[\lambda^{-5} e^{-\frac{a}{\lambda T}}\right] d\lambda \tag{4.32}$$

Here, $E(\lambda)d\lambda$ expresses the amount of energy dE of the spectral region in the range of wavelength λ and $\lambda + d\lambda$. Hence,

$$dE = A \times \left[\lambda^{-5} e^{-\frac{a}{\lambda T}}\right] d\lambda \tag{4.33}$$

The above expression is called Wien's distribution law of energy.

4.13 DERIVATION OF WIEN'S DISPLACEMENT LAW FROM PLANCK'S RADIATION FORMULA

Planck's radiation law is given by the formula,

$$E(\lambda)d\lambda = \frac{8\pi hc}{\lambda^5}\left(\frac{1}{e^{hc/\lambda kT} - 1}\right) d\lambda$$

Differentiating the above equation with respect to λ, we get,

$$\frac{\partial E(\lambda)}{\partial \lambda} = \left(\frac{1}{e^{hc/\lambda kT} - 1}\right)\left[\frac{-5\,(8\pi hc)}{\lambda^6}\right] + \left(\frac{8\pi hc}{\lambda^5}\right)\left(\frac{1}{e^{hc/\lambda kT} - 1}\right)^2 \left(\frac{hc}{\lambda^2 kT}\right) e^{hc/\lambda kT}$$

$$= \left(\frac{1}{e^{hc/\lambda kT} - 1}\right)\left[\frac{-40\pi hc}{\lambda^6}\right] + \left(\frac{8\pi hc}{\lambda^5}\right)\left(\frac{1}{e^{hc/\lambda kT} - 1}\right)^2 \left(\frac{hc}{\lambda^2 kT}\right) e^{hc/\lambda kT}$$

$$= \left(\frac{1}{e^{hc/\lambda kT} - 1}\right)\left[\frac{8\pi hc}{\lambda^6}\right]\left[\left(\frac{1}{e^{hc/\lambda kT} - 1}\right)\left(\frac{hc}{\lambda kT}\right) e^{hc/\lambda kT} - 5\right]$$

For maximum value of $E(\lambda)$, i.e., for maximum emission of energy, $\frac{\partial E(\lambda)}{\partial \lambda}$ must be equal to zero,

or,
$$\left(\frac{1}{e^{hc/\lambda kT} - 1}\right)\left[\frac{8\pi hc}{\lambda^6}\right]\left[\left(\frac{1}{e^{hc/\lambda kT} - 1}\right)\left(\frac{hc}{\lambda kT}\right) e^{hc/\lambda kT} - 5\right] = 0$$

or
$$\left[\left(\frac{1}{e^{hc/\lambda kT} - 1}\right)\left(\frac{hc}{\lambda kT}\right) e^{hc/\lambda kT} - 5\right] = 0 \tag{4.34}$$

Substituting $\frac{hc}{\lambda kT} = x$, equation (4.34) reduces to,

$$\left(\frac{xe^x}{e^x - 1}\right) - 5 = 0$$

or
$$\left(\frac{xe^x}{e^x - 1}\right) = 5$$

or
$$\frac{x}{1 - e^{-x}} = 5$$

or
$$\frac{x}{5} = 1 - e^{-x}$$

or
$$\frac{x}{5} + e^{-x} = 1 \tag{4.35}$$

It is clear from equation (4.35) that there must be a root in neighborhood of 5. This equation has a single root at $x = 4.965$.

or
$$\frac{hc}{\lambda kT} = 4.965$$

The wavelength λ_m at which the spectral radiancy (energy per unit range of wavelength) is maximum is given by,

$$\lambda_m \cdot T = \frac{hc}{4.965\,\text{k}} = b\,(\text{constant}) \tag{4.36}$$

This is Wein's displacement law.

4.14 STEFAN-BOLTZMANN LAW

Stefan's law states that "the rate of emission of thermal energy per unit area of a perfectly black body is proportional to the fourth power of the absolute temperature". Hence,

$$E = \sigma \cdot T^4 \tag{4.37}$$

Here, σ is known as Stefan's constant and its value is $5.67 \times {}^{-8}\text{W/m}^2/\text{K}^4$.

Stefan's law provides only emission of radiated heat by the black body itself but not to the loss of heat by the body during exchange process with the surroundings.

If a black body at absolute temperature T is covered with another black body at absolute temperature T_0, the total loss rate of thermal energy per unit area is given by

$$E = \sigma \cdot (T^4 - T_0^4) \tag{4.38}$$

The above expression is known as Stefan-Boltzmann law.

4.15 STEFAN-BOLTZMANN LAW FROM PLANCK'S LAW

According to Planck's radiation formula, the total radiant energy per unit volume is given by,

$$E = \int_0^\infty E(\lambda)d\lambda = \int_0^\infty \frac{8\pi hc}{\lambda^5}\left(\frac{1}{e^{hc/\lambda kT} - 1}\right)d\lambda \tag{4.39}$$

Since, $$\frac{hc}{\lambda kT} = x$$

Hence, $$\lambda = \frac{hc}{xkT} \tag{4.40}$$

and $$d\lambda = -\frac{hc}{x^2 kT}dx \tag{4.41}$$

Substituting the values of λ and $d\lambda$ from equations (4.40) and (4.41) to equation (4.39) we get,

$$E = -8\pi hc\int_0^\infty \left(\frac{xkT}{hc}\right)^5 \left(\frac{1}{e^x - 1}\right)\frac{hc}{x^2 kT}$$

$$= -\frac{8\pi k^4 T^4}{h^3 c^3}\int_0^\infty x^3\left(\frac{1}{e^x - 1}\right)dx$$

Using the value $\int_0^\infty x^3\left(\frac{1}{e^x - 1}\right)dx = \frac{\pi^4}{15}$, the above equation gives,

$$E = AT^4 \tag{4.42}$$

where, $A = \frac{8\pi^5 k^4}{15h^3 c^3}$ is a constant and T is the absolute temperature of the enclosure.

Putting $\sigma = \frac{1}{4}Ac$, the rate of emission of radiant energy per unit area is given by

$$E = \sigma T^4 \tag{4.43}$$

where, $\sigma = \frac{2\pi^5 k^4}{15h^3 c^2}$ is a constant known as Stefan's constant.

Equation (4.43) is well known Stefan's law.

Numerical Examples

1. Determine the radiated energy per minute from the filament of a lamp at 4000°K. Given that relative emittance is 0.86 and surface area is 0.51 cm^2.

Solution: We know that radiated energy is given by

$$E = (Ae\sigma T^4) \times t$$

Here,

$$e = 0.86$$

$$A = 0.51 \text{ cm}^2 = 5.1 \times 10^{-5} \text{ m}^2$$

$$t = 1 \text{ minute} = 60 \text{ seconds}$$

$$T = 4000°\text{K}$$

and

$$\sigma = 5.67 \times 10^{-8} \text{ Wm}^{-2} \text{ K}^{-4}$$

So radiated energy,

$$E = (Ae\sigma T^4) \times t$$

$$= 5.1 \times 10^{-5} \times 0.86 \times 5.67 \times 10^{-8} \times (4000)^4 \times 60$$

$$= 37.040 \times 10^3 \text{ k joules}$$

2. The wavelength of maximum emitted intensity from the sun is $\lambda_m = 4752 \times 10^{-8}$ cm and from moon is $\lambda_m = 13.9 \times 10^{-4}$ cm. Determine the surface temperature of the sun and moon.

Solution: We know that Wien's displacement law can be expressed as

$$\lambda_m \times T = \text{constant} = 0.2899$$

(i) The wavelength of radiation emitted by the sun

$$\lambda_m = 4752 \times 10^{-8} \text{ cm}$$

So the temperature of the sun for emitted radiation

$$T_{\text{sun}} = \frac{0.2899}{4752 \times 10^{-8}} = 6099° \text{K}$$

(ii) The wavelength of radiation emitted by the moon

$$\lambda_m = 13.9 \times 10^{-6} \text{ m}$$

So the temperature of the moon for emitted radiation

$$T_{\text{moon}} = \frac{0.2899}{13.9 \times 10^{-4}} = 206° \text{K}$$

3. Determine the loss of thermal energy radiated by a black body at temperature 647°K.

Solution: We know that radiated energy by a black body is given by

$$E = \sigma T^4$$

where, $\sigma = 5.67 \times 10^{-8}\ \text{Wm}^{-2}\text{K}^{-4}$

So,
$$E = 5.67 \times 10^{-8} \times (647)^4$$
$$E = 1.01\ \text{W/cm}^2$$

4. The value of Stefan-Boltzmann constant is $\sigma = 5.67 \times 10^{-8}\ \text{Wm}^{-2}\text{K}^{-4}$. Calculate the ratio of radiant emittance of a given black body at temperature 40°K and 4000°K.

Solution: We know that radiated energy by black body is given by

$$E = \sigma T^4$$

So,
$$\frac{E_1}{E_2} = \left(\frac{T_1^4}{T_2^4}\right)$$
$$= \left(\frac{40}{4000}\right)^4$$
$$= 10^{-8}$$

5. A cubical shape cavity of a black body emits radiation. Calculate the modes of vibration per cc in the wavelength range 5010 Å to 5050 Å.

Solution: Given that, $\lambda = 5030\ \text{Å} = 5030 \times 10^{-8}$ cm

and
$$d\lambda = (5050 - 5010)\ \text{Å} = 40\ \text{Å} = 40 \times 10^{-8}\ \text{cm}$$

We know that the modes of vibration per unit volume in the range of wavelength λ and $\lambda + d\lambda$ is given by the expression

$$N = \left(\frac{8\pi}{\lambda^4}\right) \cdot d\lambda$$

$$= \frac{8 \times 3.14 \times 40 \times 10^{-8}}{(5030 \times 10^{-8})^4}$$

$$= \frac{8 \times 3.14 \times 40 \times 10^{-8}}{(5030 \times 10^{-8})^4}$$

$$= 1.57 \times 10^{12}$$

6. A black body emitted wavelength 4699 Å with maximum energy at temperature 6175°K. Determine the temperature at which same black body emits 1.39×10^{-3} cm with same amount of maximum energy.

Solution: Let us suppose that the T_1 and T_2 are temperatures at two different maximum wavelengths.

We know that the Wien's displacement law is given by

$$\lambda_m \cdot T = \text{constant}$$

$$\frac{(\lambda_m)_1}{(\lambda_m)_2} = \frac{T_2}{T_1}$$

$$T_2 = \frac{(\lambda_m)_1 \times T_1}{(\lambda_m)_2}$$

$$= \frac{4699 \times 10^{-8} \times 6175}{1.39 \times 10^{-3}}$$

$$= 206.972° \text{ K}$$

7. Two black body radiators of spherical shape (concentric) are kept at two different temperatures, 100°K and 400°K. If vacuum is created between two spheres, determine the total rate of transfer of energy between the spheres.

Solution: We know that the rate of energy transfer is given by

$$E = \sigma \times (T_1^4 - T_2^4)$$

$$= 5.672 \times 10^{-8} \times [(400)^4 - (100)^4]$$

$$= 5.672 \times 10^{-8} \times 10^8 \times [256 - 1]$$

$$= 1446.36 \text{ watts/m}^2$$

8. A black body emits wavelength 2000 × 10^{-9} m with maximum energy at 1499°K temperature. Determine the wavelength emitted by the sun at temperature 5454°K.

Solution: We know that Wien's displacement law is given by

$$\lambda_m \cdot T = \text{constant}$$

Hence, $$1499 \times 2000 \times 10^{-9} = 5454 \times \lambda_m$$

So, $$\lambda_m = \frac{1499 \times 2000 \times 10^{-9}}{5454}$$

$$= 549.9 \times 10^{-9} \text{ m}$$

9. Determine the emittance of radiance of a perfect black body at temperature (i) 200°K (ii) 2000°K.

Solution: We know that Stefan's law

$$E = e \cdot \sigma \cdot T^4$$

For a perfectly black body e = 1 and σ = 5.67 × 10^{-8} W m^{-2} K^{-4}.

(i) For T = 200°K

$$E = 5.67 \times 10^{-8} \times (200)^4$$

$$E = 90.75 \text{ W/m}^2$$

For T = 2000°K

$$E = 5.67 \times 10^{-8} \times (2000)^4$$

$$E = 90.75 \times 10^4 \text{ W/m}^2$$

10. Determine the modes of vibration in the frequency range 3×10^{14} / sec and 3.01×10^{14} / sec. The volume of container is 100 cc.

Solution: Here, $\nu = 3 \times 10^{14}$ / sec and $\nu + d\nu = 0.01 \times 10^{14}$ / sec

We know that the modes of vibration in range ν and $\nu + d\nu$

$$= \left(\frac{8\pi\nu^2}{c^3}\right) d\nu$$

$$= \left(\frac{8 \times 3.14 \times 3 \times 10^{14} \times 3 \times 10^{14}}{(3 \times 10^{10})^3}\right)(0.01 \times 10^{14})$$

$$= 8.4375 \times 10^{10} \text{ cc.}$$

Hence, the modes of vibration present in the volume 100 cc.

$$= 8.4375 \times 10^{10} \times 100$$
$$= 8.4375 \times 10^{12}$$

11. A spherical black body of diameter 7.99 cm is heated to 227°C temperature. At the time of heating, the surrounding temperature is 27°C. If the energy is radiated at the rate of 65.27 watts, then calculate the diameter of body. The value of Stefan's constant is $\sigma = 5.99 \times 10^{-8}$ $\text{Wm}^{-2}\text{K}^{-4}$.

Solution: We know that the Stefan's law is expressed as

$$R = \sigma A e(T^4 - T_0^4)$$

$$A = \frac{R}{\sigma e(T^4 - T_0^4)}$$

$$= \frac{65.27}{5.99 \times 10^{-8} \times A \times 1 \times [(500)^4 - (300)^4]}$$

$$= 10^{-2} \text{ m}^2$$

12. The frequency of an oscillator is 5.9×10^{13} / sec at 1526°C. Determine the average energy of the oscillator if (i) oscillator is classical, and (ii) oscillator is Planck's oscillator.

Solution:

(i) We know that the average energy of an oscillator (classical) is given by

$$\bar{E} = kT$$

$$= 1.382 \times 10^{-23} \times 1799$$

$$= 24.83 \times 10^{-19} \text{ joule}$$

(ii) The mean energy of Planck's oscillator is given by

$$\bar{E} = \frac{h\nu}{e^{(h\nu/kT)} - 1}$$

Multiplying and dividing by kT in above expression we get

$$\bar{E} = \frac{(h\nu/kT)}{e^{(h\nu/kT)} - 1} \times kT$$

We can solve the above expression by obtaining the value of $(h\nu/kT)$,

$$\frac{h\nu}{kT} = \frac{6.63 \times 10^{-34} \times 5.9 \times 10^{13}}{1.382 \times 10^{-23} \times 1799}$$

$$= 1.589$$

So the mean energy of Planck's oscillator is

$$\bar{E} = \frac{(h\nu/kT)}{e^{(h\nu/kT)} - 1} \times kT$$

$$= \frac{1.589 \times 24.83 \times 10^{-19}}{e^{1.589} - 1}$$

$$= 10.5 \times 10^{-21} \text{ joule}$$

13. A unclothed man is in a hall at temperature 293°K. The surface area of the man is 2.99 m^2 and body temperature is 310°K. Determine the amount of energy lost by his body in 20 minutes. The emissivity is 0.89 and $\sigma = 5.672 \times 10^{-8}$ Wm^{-2}K^{-4}.

Solution: We know that the emitted energy is given by

$$R = \sigma A e(T^4 - T_0^4)$$

$$= 5.672 \times 10^{-8} \times 2.99 \times 0.89 \times [(310)^4 - (293)^4]$$
$$= 285.87 \text{ J/sec}$$

So, total loss of heat by the skin of man in 20 minutes

$$= 285.87 \times 20 \times 60$$
$$= 34.80 \times 10^4 \text{ joules}$$

14. At the temperature 5454°K, radiation emitted by the sun has wavelength 5500×10^{-8} cm with maximum energy. Determine the Wien's constant.

Solution: We know that Wien's displacement law

$$\lambda_m \cdot T = \text{ constant } = b$$

Given that, $\lambda_m = 5500 \times 10^{-8}$ cm and $T = 5454°$K

Hence, $b = \lambda_m \cdot T = 5500 \times 10^{-8} \times 5454$

$$= 0.2958 \text{ cm° K}$$

15. A sphere of diameter 4.98 cm is placed inside an encloser at a temperature 273°C. Determine the quantity of heat lost per second by emission of radiation at temperature 327°C. The value of Stefan's constant is given $\sigma = 5.72 \times 10^{-8}$ W/ m^{20}K^4.

Solution: Let us suppose that T_0 is the temperature of the surrounding and T is temperature of sphere. The total energy lost per unit area per unit second from the sphere is given by Stefan's law

$$E = \sigma \cdot (T^4 - T_0^4)$$

We know that the total surface area of sphere = $4\pi a^2$
So the total heat loss from the surface of sphere

$$= E \times 4\pi a^2$$

$$= 4\pi a^2 \times \sigma \cdot (T^4 - T_0^4)$$

$$= \frac{4 \times 3.14 \times (2.49)^2 \times 5.72 \times 10^{-8} \times [(600)^4 - (300)^4]}{4.2}$$

$$= 13.01 \text{ cal/sec}$$

16. A furnace of cross section 1 cm^2 radiates 1.529 × 10^5 calories per hour. The temperature of furnace is 2227°K. Determine the relative emittance of the surface. Where Stefan's constant, $\sigma = 1.36 \times 10^{-8}$ cal/ m^{2}sec°K^4.

Solution: We know the net energy lost per unit area per unit second from the sphere is given by

$$E = (Ae\sigma T^4) \cdot t$$

Given that: t = 1 hour = 60 × 60 = 3600 sec, A = 1 cm^2 = 10^{-4} m^2, and T = 2227°C = 2500°K

So, $$e = \frac{E}{(AT^4\sigma) \cdot t}$$

$$= \frac{1.529 \times 10^5}{10^{-4} \times (2500)^4 \times 1.36 \times 10^{-8} \times 3600}$$

$$= \frac{1.529 \times 10^5}{10^{-4} \times 390625 \times 10^8 \times 1.36 \times 10^{-8} \times 3600}$$

$$= 0.798$$

17. A foil is placed between two concentric perfectly black spheres. The temperatures of the spheres are 27°C and –72°C, respectively. Determine the rate of transportation of heat energy from one sphere and the foil. Given relative emittance is 0.2 and $\sigma = 5.72 \times 10^{-8}$ W/m^{20}K^4.

Solution: Let us suppose that T_1 and T_1 are the temperatures of two spheres and y is the temperature of the foil after the steady state. The total energy lost per unit area per unit second from sphere is given by Stefan's law

$$E = e\sigma \cdot (T^4 - T_0^4)$$

At steady state,

$$e\sigma \cdot (T_1^4 - y^4) = e\sigma \cdot (y^4 - T_2^4)$$

$$2y^4 = (T_1^4 - T_2^4)$$

$$y^4 = \frac{1}{2}[(300)^4 - (201)^4]$$

$$= 48.49 \times 10^8$$

$$y = 262.97 \text{ K}$$

So, the rate of transportation of heat

$$E = e\sigma \cdot (T_1^4 - y^4)$$

$$= 0.2 \times 5.72 \times 10^{-8} \times [81 \times 10^8 - 48.49 \times 10^8]$$

$$= 37.01 \text{ W/m}^2$$

Exercise

Short Answer Questions

1. What do you mean by a perfect black body?
2. Give the reason for the use of polished reflectors in electric heaters.
3. Derive the expression for pressure due to diffused radiation.
4. What do you mean by black body radiation?
5. Define the term absorptive power and emissive power.
6. Derive a relation between energy and pressure of diffused radiation.
7. What do you mean by black body? Write the differences between perfect black body and black body in practice.
8. Explain the meaning of terms absorptive power and emissive power in black body radiation?
9. Draw the curve for the distribution of energy in black body spectrum at three different temperatures.
10. State Kirchhoff's law of radiation and write its applications.
11. Find out the ratio of the emissive power to absorptive power of a substance.
12. Give the statement of Stefan-Boltzmann radiation law.
13. State the Rayleigh-Jeans law of radiation.
14. Write the statement of Wien's law and show its validity.
15. State the Planck's black body radiation law.
16. Write the important characteristics of black body radiation.
17. Show that Planck's radiation law reduces to Wien's law for smaller wavelength and to Rayleigh-Jeans law for larger wavelength.
18. Show that the diffused radiation exerted a pressure equal to (energy density/3).

Long Answer Questions

1. State and explain the Kirchhoff's law of thermal radiation with application.
2. Discuss the absorptive power and emissive power. Show that the ratio of these two powers of a substance is constant and is also equal to the power of emission of a perfect black body.
3. Explain the black body energy distribution at different temperatures by plotting a graph and also give the important facts obtained from the graph.
4. Explain the different characteristics of thermal radiation emitted by a perfect black body. Prove that the pressure exerted by diffused radiation is equal to 1/3 of its energy density.

5. Prove that the Rayleigh-Jeans law and Planck's black body radiation law become identical at very high temperature.
6. Explain the Planck's black body radiation theory and also obtain expression for Rayleigh-Jeans law and Wien's law from Planck's law.
7. Write the statement of Stefan's law and show that the energy density of emitted radiation is proportional to the fourth power of temperature.
8. Obtain Wien's distribution law from black body distribution of energy and explain the validity of law.
9. Derive and discuss the Planck's black body radiation law and explain its experimental importance.
10. Explain the limitations of Wien's and Rayleigh-Jeans law. In what way these limitations are overcome by Planck's law.
11. Obtain Wien's law of shorter wavelength from Planck's law of radiation.
12. Explain the Planck's law of radiation and prove that mean kinetic energy of Planck's oscillator is given by the expression

$$\overline{E} = \frac{h\nu}{e^{h\nu/kT} - 1}$$

where, ν is the frequency of emitted radiation, h is Planck's constant and T is its temperature.
13. State and explain the Wien's displacement law of radiation and give its experimental validity.
14. Discuss the Stefan-Boltzmann law of distribution of energy and show that

$$E = \sigma(T^4 - T_0^4)$$

where, σ is Stefan's constant and T and T_0 are temperatures of two bodies.
15. Explain the theory of Planck's black body radiation law and deduce this law in terms of wavelength.

Numerical Problems

1. Two perfect black bodies A_1 and A_2 are at temperatures 500°K and 600°K, respectively and are kept inside an enclosure at 300°K. Find out the ratio of their loss of energy. **[Ans.** A_1: A_2 = 148:405]
2. Two perfect black concentric spheres are kept at temperatures 127°C and 327°C, respectively. The net transportation rate of heat between the two spheres is 5.89 × 10^3 W/m^2. Determine the Stefan's constant σ.

[Ans. $\sigma = 5.72 \times 10^{-8}$ W/m^{20}K^4]
3. Determine the radiated energy per second from the filament of lamp at 4000°K. The surface area of filament is 10^{-2} m^2 and relative emittance is 0.439.

4. Find out the modes of vibration in a 200 cc container in the range of wavelength 4000 Å to 4004 Å.
5. If the mean energy emitted per unit area of the sun surface is 6.892×10^9 W. Determine the temperature of the surface of sun. Stefan's constant is $\sigma = 5.72 \times 10^{-8}\ \text{W}/\text{m}^{20}\text{K}^4$.
6. The four stars A_1, A_2, A_3, A_4 show maximum brightness in their spectra for the wavelengths 5500 Å, 4600 Å, 5200 Å, 6280 Å, respectively. Compare the temperatures of four stars A_1, A_2, A_3, A_4.
7. A foil is placed between two concentric perfectly black spheres. The temperatures of the spheres are 29°C and –78°C, respectively. Determine the rate of transportation of heat energy from one sphere and the foil. Given relative emittance is 0.4 and $\sigma = 5.72 \times 10^{-8}\ \text{W}/\text{m}^{20}\text{K}^4$.
8. A sphere of diameter 4.98 cm is placed inside an encloser at a temperature 273°C. Determine the quantity of heat lost per second by emission of radiation at temperature 327°C. The value of Stefan's constant is $\sigma = 5.72 \times 10^{-8}\ \text{W}/\text{m}^{20}\text{K}^4$.
9. An unclothed man is in a hall at temperature 283°K. The surface area of the man is 3.99 m^2 and body temperature is 320°K. Determine the amount of energy lost by his body in 30 minutes. The emissivity is 0.99 and $\sigma = 5.672 \times 10^{-8}\ \text{Wm}^{-2}\text{K}^{-4}$.
10. A spherical black body of diameter 8.99 cm is heated to 247°C temperature. At the time of heating, the surrounding temperature is 37°C. If the energy is radiated at the rate of 85.27 watts, then calculate the diameter of body. The value of Stefan's constant is $\sigma = 5.99 \times 10^{-8}\ \text{Wm}^{-2}\text{K}^{-4}$.
11. A black body emits wavelength 4000×10^{-9} m with maximum energy at 1599°K temperature. Determine the wavelength emitted by the sun at temperature 6454°K.
12. Determine the emittance of radiance of a perfectly black body at temperature (i) 500°K (ii) 5000°K.
13. Determine the modes of vibration in the frequency range $4 \times 10^{14}/\text{sec}$ and $5.01 \times 10^{14}/\text{sec}$. The volume of container is 200 cc.
14. Two black body radiators of spherical shape (concentric) are kept at two different temperatures 2000°K and 600°K. If vacuum is created between two spheres, determine the total rate of transfer of energy between the spheres.
15. The frequency of an oscillator is $4.95 \times 10^{13}/\text{sec}$ at temperature of 1426°C. Determine the average energy of oscillator if (i) oscillator is classical, and (ii) oscillator is Planck's oscillator.
16. A cubical shape cavity of a black body emitted radiation. Calculate the modes of vibration per cc in the wavelength range 5040 Å to 5090 Å.

17. A black body emitted wavelength 4899 Å with maximum energy at temperature 6275°K. Determine the temperature at which same black body emitted 1.49 × 10^{-4} cm the same amount of maximum energy.

18. The wavelength of maximum emitted intensity from sun $\lambda_m = 4692 \times 10^{-8}$ cm and from moon $\lambda_m = 12.89 \times 10^{-4}$ cm. Determine the surface temperature of the sun and moon.

19. A perfect black body consists of 1.407×10^{12} modes of vibration per unit volume. Determine the range of frequency. [**Ans.** 5.49 × 10^{12} Hz to 5.54 × 10^{12}]

20. The energy radiated per minute from the filament of a lamp is 11712 joules. If the area of surface is 10^{-4} m^2 and relative emittance of filament is 0.424, determine the temperature of the lamp. [**Ans.** 3000°K]

21. Two perfect black bodies B_1 and B_2 are at temperatures 400°K and 800°K, respectively and are kept inside an enclosure at 200°K. Find out the ratio of their loss of energy.

22. A cubical cavity of a black body emitted radiation. Calculate the modes of vibration per cc in the wavelength range of 4040 Å to 6090 Å.

Multiple-Choice Questions

1. If the absolute temperature of a perfect black body is doubled, the rate of emission of energy per unit area will be
   (i) 4 times (ii) 16 times
   (iii) 2 times (iv) 36 times
2. Wien's displacement law is mathematically expressed as
   (i) $\lambda_m T^2$ = constant (ii) $\lambda_m T$ = constant
   (iii) $\lambda_m \sqrt{T}$ = constant (iv) $\lambda_m T^4$ = constant
3. A perfect black body is one which
   (i) Can absorb all the incident radiation
   (ii) Is entirely black in colour
   (iii) Can radiate all of its energy
   (iv) Can convert a body in black colour
4. Rayleigh-Jeans law of radiation applies to
   (i) All wavelengths (ii) Smaller wavelengths
   (iii) Longer wavelengths (iv) Independent of wavelength
5. The statement "good absorbers are good emitters" is known as
   (i) Wien's law (ii) Planck's law
   (iii) Kirchhoff's law (iv) Rayleigh-Jeans law

6. In Planck's hypothesis, average energy of Planck's oscillator is given by

(i) $E = h\nu$ (ii) $E = h\nu/2$

(iii) $E = mc^2$ (iv) $E = \dfrac{h\nu}{e^{\left(\frac{h\nu}{kT}\right)} - 1}$

7. The absorptive power of a perfectly black body is

(i) ∞ (ii) 0

(iii) 1 (iv) 2

8. The mathematical expression of Stefan-Boltzmann law is

(i) $E = \sigma T_0^4$ (ii) $E = \sigma T^3$

(iii) $E = \sigma(T^4 - T_0^4)$ (iv) $E = \sigma/(T^4 - T_0^4)$

9. The emission of electromagnetic radiation is possible by

(i) All bodies at 0°C

(ii) Only certain bodies at all temperatures

(iii) All bodies at all temperatures

(iv) All bodies at 0°K

10. The value of Stefan's constant in CGS is

(i) 5.672×10^{-9} (ii) 5.672×10^{-5}

(iii) 5.432×10^{-9} (iv) 6.672×10^{-9}

11. Pressure of diffused radiation is

(i) 1/2 of total energy density (ii) 1/4 of total energy density

(iii) 1/3 of total energy density (iv) equal to total energy density

12. Wien's law of energy distribution states that

(i) $dE \propto \lambda^{-5}$ (ii) $dE \propto \lambda^{-3}$

(iii) $dE \propto \lambda^{-4}$ (iv) $dE \propto \lambda^{-1/2}$

13. Rayleigh-Jeans law states that

(i) $E_\lambda d\lambda = \dfrac{8\pi kT}{\lambda^3}$ (ii) $E_\lambda d\lambda = \dfrac{8\pi kT}{\lambda^4}$

(iii) $E_\lambda d\lambda = \dfrac{8\pi kT}{\lambda^5}$ (iv) $E_\lambda d\lambda = \dfrac{8\pi kT}{\lambda^2}$

14. The radiant emittance of a perfect black body at a temperature 400°K is

(i) 1329 watts/m^2 (ii) 1452 watts/m^2

(iii) 1211 watts/m^2 (iv) 1471 watts/m^2

15. Two metallic spheres have radii R_1 and R_2 and are placed in the same enclosure. If the spheres are heated at the same temperature, their rate of loss of heat is given by

(i) $\frac{R_1^2}{R_2^2}$ (ii) $\frac{R_1}{R_2^2}$

(iii) $\frac{R_1^4}{R_2^2}$ (iv) $\frac{R_1^2}{R_2^3}$

16. Two stars P and Q radiate maximum energy at 4600×10^{-10} m and 3600×10^{-10} m, respectively. The ratio of absolute temperatures of P and Q is

(i) 2/3 (ii) 4/3

(iii) 1/3 (iv) 1/6

17. Emissivity of a perfect black body is

(i) 0 (ii) 1

(iii) 4 (iv) ∞

18. The fastest process is

(i) Convection (ii) Radiation

(iii) Diffusion (iv) Conduction

Answers

1. (ii) 2. (ii) 3. (i) 4. (iii) 5. (iii) 6. (iv) 7. (iii) 8. (iii)
9. (iii) 10. (ii) 11. (iii) 12. (i) 13. (ii) 14. (ii) 15. (i) 16. (ii)
17. (ii) 18. (ii)

5

Concept of Heat, Zeroth and First Law of Thermodynamics

5.1 THERMODYNAMIC SYSTEMS: CLOSED AND OPEN SYSTEMS

A thermodynamic system may be defined as prescribed quantity of matter enclosed in a closed surface. Anything external to the system which can exchange energy with it is called the surrounding. A system is separated from its surroundings by a system boundary. The system boundary may be fixed or movable, through which heat energy may transfer. The combined system and its surrounding together is called the universe.

A gas is contained in a cylinder with a moveable piston. This gas is treated as the system. If gas is heated by supplying heat from external sources, it will expand and the piston moves up. As the piston moves up, the boundary will extend and the mechanical work is done by the system. If, in place of heating, gas is cooled, then the reverse will be obtained. If in a system, mass is fixed, i.e., transfer of mass across the boundary is not allowed but transfer of energy is allowed, such a system is known as a closed system. If in a system, mass and energy both are transferrable, then the system is known as an open system. A completely uniform system such as pure solid, liquid, gas, combination of gases or a solution is referred as a homogeneous system. A system having two or more than two phases, which are detached from one another by fixed boundary surfaces, is called heterogeneous system. A system which transfers neither matter nor energy with its surroundings is known as an isolated system. The universe is an isolated system.

5.2 CONCEPT OF TEMPERATURE AND ZEROTH LAW OF THERMODYNAMICS

Though temperature is a well-known property, defining it correctly is difficult. The concept of temperature, however, occurs from the sense of hotness or coldness, when we come in contact of a body. However, the physiological sensation is very unreliable. Because of this difficulty, we define the equality of temperature. To define the concept of equality of temperature, let us consider two blocks of copper, one block is hot and the other is cold. Both the blocks are in contact with a mercury-in-glass thermometer. If each block is brought in thermal contact, we observe that the electrical resistance of the cold block increases with time while that of the hot block decreases with time. After some time, however, no further change takes place. Similarly, the mercury column of the thermometer in the hot block drops at first, while in the cold block rises. After sufficient time, no change in either height is observed. We can, therefore say, that two bodies have equality of temperature when no change in any observable property occurs, while they are in thermal communication.

The Zeroth law of thermodynamics states that, if two systems *A* and *B* are separately in thermal equilibrium with a third system *C*, both the systems *A* and *B* are also in thermal equilibrium with each other.

Let us consider three systems *A*, *B* and *C* (Figure 5.1). *A* and *B* are insulated from each other but in thermal contact with a third system *C*. After some time, system *A* and *B* will get thermal equilibrium with *C*. If *A* and *B* are brought in thermal contact by removing insulating partition between them, no further change is observed. This implies that both the systems *A* and *B* are also in thermal equilibrium with each other.

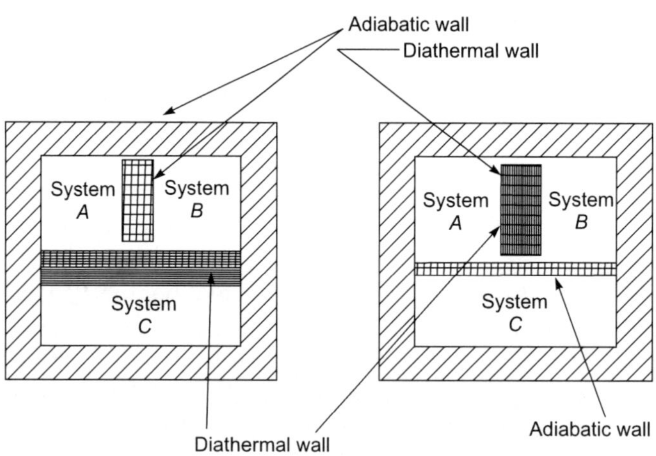

Figure 5.1 Zeroth law of thermodynamics

The Zeroth law of thermodynamics is not always true in other type of interactions between systems. This law is the basis of measurement of temperature.

5.3 CONCEPT OF HEAT AND ITS EQUIVALENCE WITH WORK

Heat is defined as the transition of energy. It can flow from one body to another. If a body is at a fixed temperature, body has both mechanical and thermal energies which cannot be separate. Hence, we cannot say about heat energy. It can only be used for the transfer of energy between the systems.

Let us consider two systems surrounded by adiabatic walls and in thermal contact. If Q_1 is the transferred heat energy by system 1 and Q_2 by system 2,

$$Q_1 = -Q_2$$

Hence, in a composite system, the loss of heat by one system is equal to the gain of heat by another system.

Work is the form of energy exchanged between a system and its surroundings due to modification in volume of the system. Any mechanical device which changes heat energy to mechanical work is called a heat engine.

A relation between work done and heat originated was given by Joule. Joule says that in the transformation of work into heat and heat into work, the work is mechanically corresponding to the heat. Hence,

$$W \propto Q$$

or
$$W = JQ \tag{5.1}$$

where, W is work done, Q is produced heat and J is Joule's constant, which are also known as mechanical equivalent of heat. Equation (5.1) can also be written as

$$J = \frac{W}{Q} = 4.18 \text{ joule/calories}$$

If $Q = 1, W = J$, the Joule's constant is defined as the net amount of work done in generating a unit quantity of heat. If W and Q both are measured in SI unit joule, then $W = Q$.

5.4 INDICATOR DIAGRAM AND WORK AS PATH-DEPENDENT FUNCTION

A graph representing the different thermodynamic state and changes in the state of a system in terms of coordinates (P, V), is known as indicator diagram or P-V curve. The work done in a thermodynamic process is equal to the area enclosed by the P-V or indicator diagram. Work done is positive, if the P-V diagram is drawn in clockwise direction and negative, if drawn in anti-clockwise direction (Figure 5.2). Let us consider a thermodynamic system of a gas. Initial state of the gas is represented by A and the final by B. A system can go from initial sate A to final state B in a number of ways as represented in Figure 5.3.

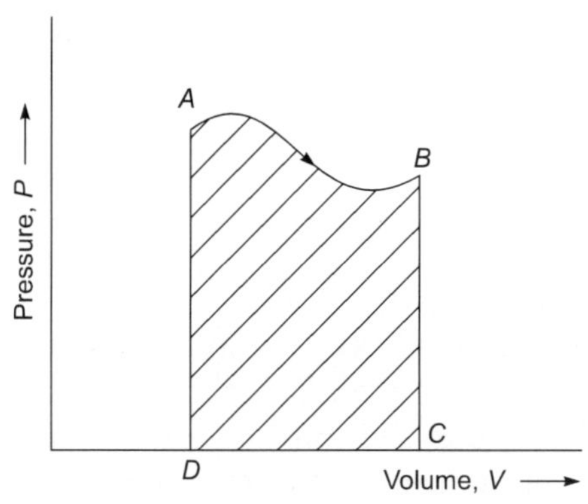

Figure 5.2

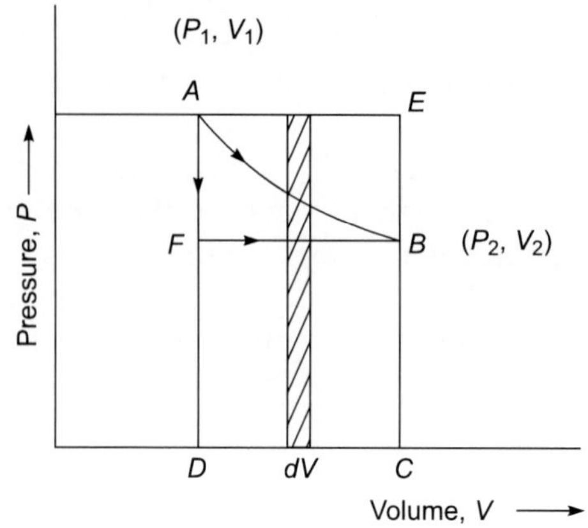

Figure 5.3

Let P_1 and V_1 be pressure and volume of position A and P_2 and V_2 pressure and volume of position B, the coordinates of A are (P_1, V_1) and B are (P_2 V_2).

(i) When pressure decreases from P_1 to P_2 along the curve AB, volume increases from V_1 to V_2. In this case, the work is done by the system and is given by

$$W_1 = \int_{V_1}^{V_2} P \cdot dV = \text{area } ABCD$$

(ii) Volume V_1 is constant in moving from A to F and the pressure decreases from P_1 to P_2. Pressure P_2 is constant from F to B. The work done in the process (along the path AFB) is given by

$$W_2 = P_2 \cdot (V_2 - V_1) = \text{area } FBCD$$

(iii) The pressure is constant in moving from A to E and the volume V_2 is also constant from E to B. The work done in the process (along the path AEB) is given by

$$W_3 = P_1 \cdot (V_2 - V_1) = \text{area } AECD$$

From the above expression of work done, it is clear that $W_1 \neq W_2 \neq W_3$. Hence, the work done depends on the path taken by the process, when the system goes from one state to another.

5.5 WORK DONE IN AN ISOTHERMAL PROCESS

If a gas expands isothermally, work is done by the gas itself. In an isothermal process, temperature remains steady throughout the process. Let us consider a gas is contained in a piston-cylinder arrangement in contact with steady temperature bath. During the expansion of gas, the constant temperature bath controls the system temperature steady by bringing it to its initial value.

The isothermal process is represented as a hyperbola curve in an indicator *P-V* diagram (Figure 5.4).

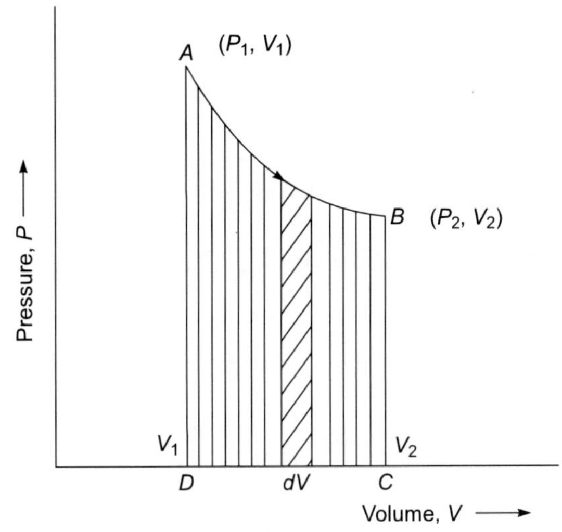

Figure 5.4

Consider, P_1, *V*1 as initial pressure and volume, and P_2, V_2 final pressure and volume, respectively.

At state *A*, $$P_1V_1 = RT_1$$

At state *B*, $$P_2V_2 = RT_2$$

As temperature is kept constant during the process, so from the above expressions we get,

$$P_1V_1 = P_2V_2 = RT = K$$

The shaded area of *P-V* diagram represents the work done for a slight change in volume *dV* and is given by

Work done $$W = \int_{V_1}^{V_2} PdV = \text{area } ABCD$$

As we know that for isothermal expansion,

$$PV = RT$$

or $$P = \left(\frac{RT}{V}\right)$$

So, $$W = \int_{V_1}^{V_2} \left(\frac{RT}{V}\right) dV$$

$$= RT \int_{V_1}^{V_2} \left(\frac{1}{V}\right) dV$$

$$= RT \log_e \left(\frac{V_2}{V_1}\right)$$

or
$$W = 2.3026 \times RT \log_{10} \left(\frac{V_2}{V_1}\right) \quad (5.2\text{-a})$$

We also know that in isothermal process,

$$P_1V_1 = P_2V_2 = RT = K$$

So equation (5.2) can also be written as

$$W = 2.3026 \times P_1V_1 \log_{10} \left(\frac{P_1}{P_2}\right) = 2.3026 \times P_2V_2 \log_{10} \left(\frac{P_1}{P_2}\right) \quad (5.2\text{-b})$$

Equation (5.2-a) and (5.2-b) represent the work done in isothermal process.

5.6 WORK DONE IN AN ADIABATIC PROCESS

In an adiabatic process, the system is thermally insulated from the surroundings, i.e., exchange of heat is not allowed. Thermally insulation is achieved either by the use of insulating material or doing the process so fast that gas does not get sufficient time to exchange heat.

Let us consider an adiabatic process represented in *P-V* diagram (Figure 5.5). Pressure, temperature and volume are variables in this process. The gas expands adiabatically from volume V_1 to V_2. P_1 and P_2 are initial and final pressures, respectively.

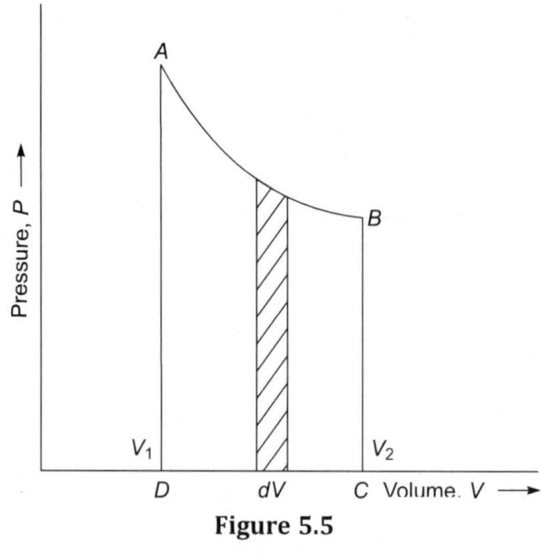

Figure 5.5

The shaded area of *P-V* diagram represents the work done by the gas for a small increase in volume *dV* and is given by

Work done, $$W = \int_{V_1}^{V_2} PdV = \text{area } ABCD$$

In an adiabatic process, $$PV^{\gamma} = K = \text{ constant}$$

or $$P = K\left(\frac{1}{V^{\gamma}}\right)$$

or $$W = K\int_{V_1}^{V_2}\left(\frac{1}{V^{\gamma}}\right)dV$$

$$= \frac{K}{(1-\gamma)}\left[\frac{1}{V_2^{(\gamma-1)}} - \frac{1}{V_1^{(\gamma-1)}}\right]$$

As *A* and *B* lie on the same adiabatic,

$$P_1V_1^{\gamma} = P_2V_2^{\gamma} = K$$

So, $$W = \frac{1}{(1-\gamma)}\left[\frac{P_2V_2^{\gamma}}{V_2^{(\gamma-1)}} - \frac{P_1V_1^{\gamma}}{V_1^{(\gamma-1)}}\right]$$

$$= \frac{1}{(1-\gamma)}[P_2V_2 - P_1V_1] \quad (5.3)$$

If T_1 and T_2 are the temperatures at initial and final points, respectively

$$P_1V_1 = RT_1, P_2V_2 = RT_2$$

On substituting these values in equation (5.3), we get,

$$W = \frac{1}{(1-\gamma)}[RT_2 - RT_1]$$

$$= \frac{R}{(1-\gamma)}[T_2 - T_1] \quad (5.4)$$

Equations (5.3) and (5.4) represent the work done in the adiabatic process, which depends only on initial and final temperatures.

5.7 WORK DONE IN AN ISOBARIC PROCESS

If the expansion of gas is at constant pressure, the process is called isobaric. Let us consider a gas contained in a cylinder. The pressure of gas can be kept constant, *yt*, putting weight on the piston. This constant pressure process is described by Boyle's law. Friction between the cylinder and piston is negligible. The temperature and volume are variables in such a process.

In a *P*-*V* diagram, the isobaric process is represented by a straight line parallel to volume axis (Figure 5.6).

At state *A*, $$P_1V_1 = RT_1$$

At state *B*, $$P_2V_2 = RT_2$$

As pressure is constant, so from the above expression, $\dfrac{V_1}{V_2} = \dfrac{T_1}{T_2}$

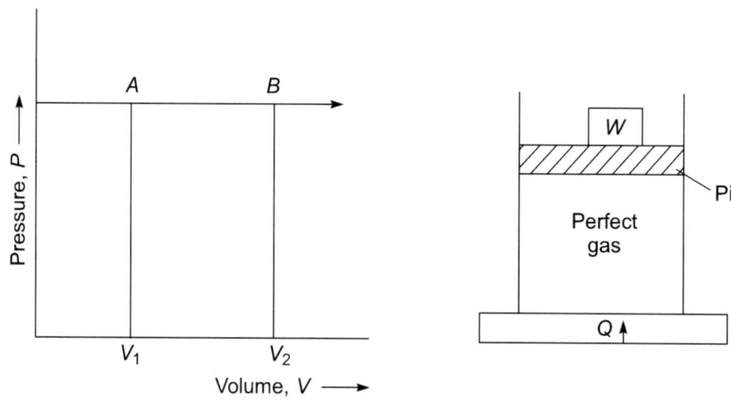

Figure 5.6

The work done dW in isobaric process for a small change in volume dV is given by

$$dW = P \cdot dV$$

Let us assume that the V_1 and V_2 are initial and final volumes of a perfect gas, enclosed in a container at constant pressure.

Work done in the expansion of gas from volume V_1 to V_2 is given by

$$W = \int_{V_1}^{V_2} P \cdot dV$$

$$= P \cdot (V_2 - V_1) \tag{5.5}$$

Equation (5.5) represents work done in isobaric process at constant pressure.

5.8 WORK DONE IN AN ISOCHORIC PROCESS

If a gas contained in a rigid container goes through a change at constant volume, the process is known as isochoric process. In the isochoric process, the temperature and pressure are variables and the volume remains constant. This process is described by Charle's law

At state A, $$P_1V_1 = RT_1$$

At state B, $$P_2V_2 = RT_2$$

As volume is constant, so from the above expression, $\dfrac{P_1}{P_2} = \dfrac{T_1}{T_2}$

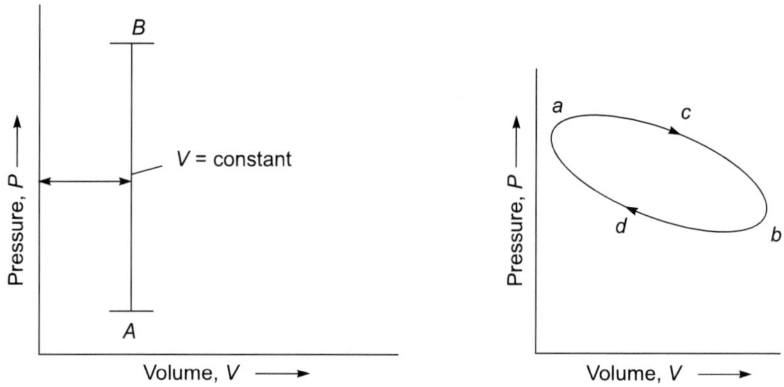

Figure 5.7

In a P-V diagram, the isochoric process is represented by a straight line (Figure 5.7). As there is no change in volume, i.e., $dV = 0$, so the work done in isochoric process is

$$W = \int_A^B P \cdot dV = \int_A^B P \cdot 0 = 0$$

or $$= 0$$

Hence, for a constant pressure process, work done is zero.

5.9 INTERNAL ENERGY OF A SYSTEM: THE THERMODYNAMIC PROPERTY

The energy stored in a system is known as internal energy and it is represented by U. The energy associated with internal energy of the system is as follows:

(i) Temperature dependent energy, i.e., translational kinetic energy, rotational kinetic energy and vibrational kinetic energy.

(ii) Phase-dependent energy, i.e., potential energy as a result of intermolecular attraction.

(iii) The electrical energy and the magnetic energy.

For ideal gases, internal energy is temperature dependent only but for all other gases, it is the function of both temperature and pressure or volume.

Since internal energy of a system is associated with the thermodynamic state of a system, hence it is a point function. As internal energy also depends on the mass of the system, therefore, it is extensive property. We cannot calculate absolute value of internal energy of a system but we can calculate changes in internal energy of a system.

In a cyclic and isothermal process, the change in internal energy is zero while in adiabatic process, we can get net amount of internal energy.

5.10 FIRST LAW OF THERMODYANMICS

The first law of thermodynamics is the principle of conservation of energy. It affirms that the net amount of flow of energy beyond the edge of a system is equivalent to modification in energy of the system. From thermodynamics point of view, only two types of energy progress across the edges. Out of these two energies, one is work done on or by the system and the second is flow or movement of heat, either by radiation or by conduction.

Let us consider dQ as the amount of heat given to a system. This amount of heat is utilized in three different processes:

(i) Part of dQ is utilized in raising the temperature of the system, i.e., increasing the internal kinetic energy.

(ii) Partial amount of dQ is utilized in doing internal work in opposition to inter-molecular attraction, i.e., in raising the internal potential energy of the system.

(iii) The remaining amount of heat is utilized in the expansion of system in opposition to external pressure, i.e., in action of external work.

Let us assume that dU_K and dU_P are the changes in internal potential energy and internal kinetic energy, respectively and dW is the external work done on the system. As we know that energy can neither be created nor destroyed but can only be converted from one form to another form, we get

$$dQ = (dU_K + dU_P + dW) \tag{5.6}$$

Here, $dU = dU_K + dU_P$ is the net increase in total internal energy of the system,

So,
$$dQ = (dU + dW) \tag{5.7-a}$$

Equation (5.7) represents the differential form of the first law of thermodynamics. The general form of the first law of thermodynamics can be written as

$$Q = (U_f - U_i) + W = \Delta U + W \tag{5.7-b}$$

On the basis of the above expression, first law of thermodynamics may be stated as:

The heat energy supplied to a system must be equivalent to addition of incresase in internal energy and external work done on the system.

In expression (5.7), dQ will be positive, if heat is given to the system, and negative, if heat is taken from the system. Similarly, the work done dW will be positive, if the external work is done by the system and negative, if the work is done on the system.

Importance of the First Law of Thermodynamics

The significance of the first law of thermodynamics is as follows:

(i) It originates the theory of internal energy.
(ii) It gives process to find out the change in the internal energy of the system.
(iii) This law is relevant to any process by which a system goes through a chemical or physical change.

Limitations of the First Law of Thermodynamics

The first law of thermodynamics is another form of law of conservation of energy. This law is applicable to each process between the equilibrium states but there are some limitations of it which are as follows:

(i) This law deals only with conservation of energy and does not tell us about direction of transformation or absorption of heat, i.e., in which direction heat flows.
(ii) It does not state about the resultant temperature of two given bodies when they are in contact.
(iii) This law does not provide the spontaneity or feasibility of system.
(iv) This law does not clear the terms under which a system can convert its heat energy to mechanical work.
(v) This law also does not tell how much amount of heat energy can be transformed into work.

5.11 IMPORTANT APPLICATIONS OF THE FIRST LAW OF THERMODYNAMICS

1. Isobaric Process at Constant Pressure

Expansion in a system under constant pressure is known as isobaric process. If Q is the heat supplied to the system at constant pressure P and due to the expansion, volume increases from V_1 to V_2, then from the first law of thermodynamics, we get

$$\begin{aligned} dQ &= (dU + dW) \\ &= (dU + PdV) \\ &= (U_2 - U_1) + P(V_2 - V_1) \\ &= (U_2 + PV_2) - (U_1 + PV_1) \\ &= (H_2 - H_1) \qquad (5.8) \end{aligned}$$

where, $H = (U + PV)$

Equation (5.8) shows that the amount of heat absorbed at constant pressure is spent to increase in quantity H, which is known as enthalpy. The enthalpy H is also known as heat function at constant pressure.

2. Isochoric Process at Constant Volume

Any thermodynamic change in a system under constant volume is known as isochoric process.

We know that from the first law of thermodynamics

$$dQ = (dU + dW)$$

As volume of the system is constant, so $dV = 0$, the work done is given by

$$dW = P \cdot dV = 0$$

So,
$$dQ = dU \tag{5.9}$$

Hence, the total supplied heat is used to increase internal energy of the system only and we can say that in isochoric process, the supplied heat should be used only to raise its internal energy only.

3. Isothermal Process at Constant Temperature

A thermodynamic process at constant temperature is known as isothermal process. In this process, system should be perfectly conducting to the atmosphere or surroundings. If the gas is compressed quickly, some heat is generated but when the compression is slow, the generated heat moves outside at once, so that the temperature remains stable, this process is said to be isothermal. If gas is expanded quickly, work is done by the working substance or the gas itself and some heat is consumed. A process of continuous supply of heat to maintain consistent temperature is known as isothermal process.

As we know that internal energy is the function of temperature and there is no change in temperature in isothermal process, so change in internal energy will be

$$dU = 0$$

We know that from the first law of thermodynamics,

$$dQ = (dU + dW)$$

$$= dW \tag{5.10}$$

Hence, in an isothermal process, heat absorbed or rejected is equal to the external work done on the gas or by the gas.

4. Adiabatic Process

A thermodynamic process under no transformation of heat is known as adiabatic process. Hence, in an adiabatic process, heat is not allowed to enter or leave the system. If a gas is expanded, it does external work by using heat energy of the gas. Hence, heat energy of the gas decreases. If gas is compressed, work is done on the gas and the energy release in this process is presented as heat energy. Therefore, heat energy of the gas rises.

As transformation of heat is not possible in adiabatic process, so

$$dQ = 0$$

We know that from the first law of thermodynamics,

$$dQ = (dU + dW)$$

$$0 = (dU + dW)$$

$$dU = -dW \tag{5.11}$$

Hence, in an adiabatic process, increase or decrease in internal energy is equal to the work done on the gas or by the gas.

Adiabatic Equation of a Ideal Gas

Let us consider one gm molecule of ideal gas stored in a entirely non-conducting vessel equipped with a non-conducting piston, as shown in Figure 5.8. Let us assume that the V, P and T are volume, pressure and temperature, respectively. Gas is compressed adibatically so that the piston shifts onwards by a distance dy,

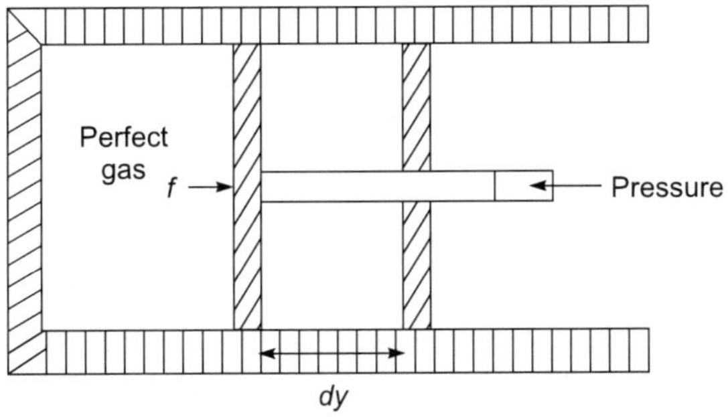

Figure 5.8

Work done by the piston $dW = (P \times A)dy$

$$dW = PdV$$

where, A is the area of cross section and $Ady = dV$ provides the change in volume.

As in adiabatic process, transformation of heat is not allowed, i.e., $dQ = 0$, so from the first law of thermodynamics, we get

$$dU + PdV = 0 \tag{5.12}$$

As internal energy of the system is utilized to do external work by the gas, so fall in temperature is given by dT and

$$dU = C_V \cdot dT$$

So from equation (5.12), we get

$$C_V \cdot dT + P \cdot dV = 0 \tag{5.13}$$

We know that from the ideal gas equation for one mole of gas

$$PV = RT \tag{5.14}$$

On differentiating equation (5.14), we get

$$P \cdot dV + V \cdot dP = R \cdot dT$$

or

$$dT = \frac{(P \cdot dV + V \cdot dP)}{R} \tag{5.15}$$

Substituting the value of dT from equation (5.15) into equation (5.13),

$$C_V \cdot \frac{(P \cdot dV + V \cdot dP)}{R} + P \cdot dV = 0$$

$$C_V \cdot (P \cdot dV) + C_V \cdot (V \cdot dP) + R(P \cdot dV) = 0$$

$$(C_V + R) \cdot (P \cdot dV) + C_V \cdot (V \cdot dP) = 0$$

But, we know that $C_P - C_V = R$

So,

$$C_P \cdot (P \cdot dV) + C_V \cdot (V \cdot dP) = 0 \tag{5.16}$$

On dividing equation (5.16) by $C_V \cdot PV$, we get

$$\frac{C_P \cdot (P \cdot dV)}{C_P \cdot PV} + \frac{C_V \cdot (V \cdot dP)}{C_P \cdot PV} = 0$$

$$\frac{dV}{V} + \frac{C_V \cdot dP}{C_P \cdot P} = 0$$

But,

$$\frac{C_P}{C_V} = \gamma$$

So,

$$\frac{dV}{V} + \frac{1}{\gamma}\frac{dP}{P} = 0$$

$$\gamma \cdot \frac{dV}{V} + \frac{dP}{P} = 0 \tag{5.17}$$

On integrating equation (5.17), we get

$$\gamma \,.\log_e V + \log_e P = \text{constant}$$

$$\log_e PV^{\gamma} = \text{constant}$$

$$PV^{\gamma} = \text{constant} = C \tag{5.18}$$

Equation (5.18) is known as Poisson's law and gives the relation between pressure and volume in an adiabatic process.

5. Cyclic Process

In a case of cyclic process, the first law of thermodynamics can be written as

$$\oint dQ = \oint dU + \oint dW$$

In a cyclic process, the system is revived to the initial value after completing one cycle.

As we know that the internal energy is a function of state

So, $$\oint dU = 0$$

Hence, $$\oint dQ = \oint dW \tag{5.19}$$

or Total heat supplied = Net work obtained

Therefore, we can say that in a thermodynamic process, work can be obtained only by energy. This heat can be represented in *PV* diagram of figure 5.8 by area *acbda*.

6. Change in Phase

A change of phase occurs in isothermal and isobaric process. At the time of phase change, some of the heat flows out of the system or into the system. Consider a vaporization process of m gram of liquid. Let the pressure be *P* and temperature as *T*. The specific volumes of the liquid and vapor phases are represented by v_1 and v_2, respectively.

The work done by the system in vaporization process,

$$W = P \cdot m(v_2 - v_1)$$

Let the temperature be *T*, at this temperature, the latent heat of vaporization is *L*, the amount of heat progress into the system,

$$Q = m \cdot L$$

The difference in the internal energy in this process is given by

$$\Delta U = (Q - W) = m \cdot L - P \cdot m(v_2 - v_1)$$

The difference in the internal energy per unit mass,

$$\frac{\Delta U}{m} = L - P \cdot (v_2 - v_1)$$

7. Joule's Experiment - Free Expansion of Gas

In the Joule's expansion since the pressure $P = 0$ inside B, so the gas does no external work, i.e.,

$$W = 0$$

This expansion is also adiabatic, hence, $Q = 0$.

Therefore, from the first law of thermodynamics, change in the internal energy is also zero, i.e.,

$$\Delta U = 0$$

In the free expansion of gas, as soon as the stop cock or nozzle is opened, a definite pressure change exists and the system of gas is not stable in mechanical equilibrium. Hence, free expansion cannot be made quasistaic. A free expansion of gas is reversible, the gas moving through non-equilibrium state till the closing equilibrium state is reached.

8. Relation Between Specific Heat of Gases

The internal energy of a system of gas can be defined by any two of the state variables P, V, T, etc. If temperature T and volume V are taken as the independent variables,

$$U = f(T, V)$$

Differentiating the above equation, we get

$$dU = \left(\frac{\partial U}{\partial V}\right)_T dV + \left(\frac{\partial U}{\partial T}\right)_V dT \tag{5.20}$$

Let us suppose that a thermodynamical system of an ideal gas is heated by an amount of heat Q, volume of gas increase by dV at a fixed pressure P. Hence, from the first law of thermodynamics

$$dQ = dU + dW = dU + P \cdot dV \tag{5.21}$$

Substituting the value of change in internal energy dU from equation (5.20), we get

$$dQ = \left(\frac{\partial U}{\partial V}\right)_T dV + \left(\frac{\partial U}{\partial T}\right)_V dT + PdV$$

Dividing both sides of the equation by dT,

$$\frac{dQ}{dT} = \left(\frac{\partial U}{\partial V}\right)_T \left(\frac{dV}{dT}\right) + \left(\frac{\partial U}{\partial T}\right)_V + P\left(\frac{dV}{dT}\right)$$

or

$$= \left[\left(\frac{\partial U}{\partial V}\right)_T + P\right]\left(\frac{dV}{dT}\right) + \left(\frac{\partial U}{\partial T}\right)_V \tag{5.22}$$

If an ideal gas is heated at constant volume, we know that

$$\left(\frac{\partial Q}{\partial T}\right)_V = C_V \text{ and } \left(\frac{dV}{dT}\right) = 0$$

where, C_V is specific heat at constant volume. Hence, equation (5.22) can be written as

$$\left(\frac{\partial Q}{\partial T}\right)_V = \left(\frac{\partial U}{\partial T}\right)_V = C_V$$

If an ideal gas is heated at constant pressure, we know that

$$\left(\frac{\partial Q}{\partial T}\right)_P = C_P$$

where, C_P is specific heat at constant volume. Hence, from equation (5.22), we get

$$C_P = \left(\frac{\partial Q}{\partial T}\right)_P = \left[\left(\frac{\partial U}{\partial V}\right)_T + P\right]\left(\frac{dV}{dT}\right) + \left(\frac{\partial U}{\partial T}\right)_V$$

or

$$C_P = \left[\left(\frac{\partial U}{\partial V}\right)_T + P\right]\left(\frac{\partial V}{\partial T}\right)_P + C_V$$

$$C_P - C_V = \left[\left(\frac{\partial U}{\partial V}\right)_T + P\right]\left(\frac{\partial V}{\partial T}\right)_P$$

$$= \left(\frac{\partial U}{\partial V}\right)_T \left(\frac{\partial V}{\partial T}\right)_P + P\left(\frac{\partial V}{\partial T}\right)_P \quad (5.23)$$

In Joule's expansion of the perfect gas on opening the nozzle or stopcock, no work was done and no heat movement took place. Hence,

$$dQ = 0 = dU + 0$$

or

$$dU = 0$$

Therefore, there is no change in internal energy while the volume changes and temperature is kept constant, i.e.,

$$\left(\frac{\partial U}{\partial V}\right)_T = 0 \quad (5.24)$$

From the perfect gas equation, we know that

$$PV = RT$$

So,

$$P\left(\frac{\partial V}{\partial T}\right)_P = R \quad (5.25)$$

Hence, equation (5.25) can be written as

$$C_P - C_V = P\left(\frac{\partial V}{\partial T}\right)_P = R \tag{5.26}$$

The equation (5.26) is known as Mayer's formula or relation for specific heat.
Equation (5.22) can also be written as

$$dQ = \left[\left(\frac{\partial U}{\partial V}\right)_T + P\right]dV + \left(\frac{\partial U}{\partial T}\right)_V dT \tag{5.27}$$

If a process is taken at constant temperature, i.e., in isothermal process, $dT = 0$
So equation (5.27) will take the form

$$(dQ)_T = \left[\left(\frac{\partial U}{\partial V}\right)_T + P\right](dV)_T \tag{5.28}$$

The equation (5.28) expresses the amount of heat energy given to a system in an isothermal reversible process of ideal gas.

5.12 COMPARISON BETWEEN ADIABATIC AND ISOTHERMAL SLOPES

An isothermal process is represented by the equation

$$PV = \text{constant}$$

Differentiating the above equation, we get

$$PdV + VdP = 0$$

or

$$\text{slope of isothermal curve} = \frac{dP}{dV} = -\frac{P}{V} \tag{5.29}$$

An adiabatic process is represented by the expression

$$PV^{\gamma} = \text{constant}$$

Differentiating the above equation, we have

$$P\gamma V^{(\gamma-1)}dV + V^{\lambda}dP = 0$$

or

$$dP = \frac{-P\gamma V^{(\gamma-1)}dV}{V^{\gamma}}$$

So slope of adiabatic curve $= \dfrac{dP}{dV} = -\gamma\left(\dfrac{P}{V}\right)$ (5.30)

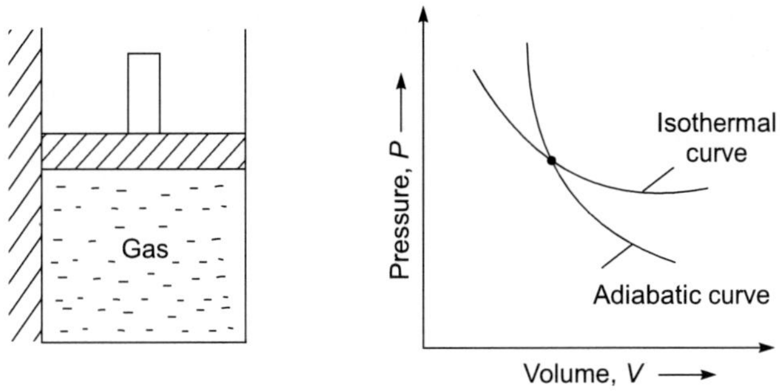

Figure 5.9

From equations (5.29) and (5.30), it is clear that the slope of adiabatic curve is γ times of the slope of the isothermal curve, i.e., the adiabatic curve is sharper than the isothermal curve at the point of intersection of both the curves (Figure 5.9).

5.13 RELATION BETWEEN ELASTICITY OF ADIABATIC AND ISOTHERMAL PROCESSES

Gases do not have their own structures, when pressure is applied on them their volume changes. Hence, we can say that gases exhibit elastic property. Let the initial pressure and volume of the gases be P and V, respectively. If the pressure changes by $(P + dP)$, then volume changes by $(V - dV)$.

Hence, volume elasticity, $\mathcal{E}$, of the gas is defined as the ratio of volume stress and volume strain,

i.e., $$\mathcal{E} = \frac{\text{Volume stress}}{\text{Volume strain}} = \frac{dP}{-(dV/V)} = -V\frac{dP}{dV} \tag{5.31}$$

As the volume of the gases can be changed by adiabatic or isothermal process, so gases have two types of elasticity.

An adiabatic process is represented by expression

$$PV^{\gamma} = \text{constant}$$

Differentiating the above equation, we have

$$P\gamma V^{(\gamma-1)}dV + V^{\lambda}dP = 0$$

or $$\frac{dP}{-\left(\dfrac{dV}{V}\right)} = \gamma P$$

Hence, from definition of elasticity, we get

or $$\mathcal{E}_{\text{adi}} = \gamma P \tag{5.32}$$

An isothermal process is represented by

$$PV = \text{constant}$$

Differentiating above equation, we get

$$PdV + VdP = 0$$

or
$$\frac{dP}{-\left(\frac{dV}{V}\right)} = P$$

So, from the definition of elasticity, we have

$$\varepsilon_{\text{iso}} = P \tag{5.33}$$

Comparing equation (5.32) for elasticity of adiabatic process and (5.33) for elasticity of adiabatic process, we get

$$\varepsilon_{\text{adi}} = \gamma \varepsilon_{\text{iso}}$$

Hence, the elasticity of adiabatic process is γ times the elasticity of isothermal process.

Numerical Examples

1. A gas is contained in a vessel of 10^{-5} mm of Hg pressure at temperature 0°C. 1 gm gas molecule occupies 22.4 litres at N.T.P. If the molecular diameter is 1.99 Å, determine the mean free path of molecules of gas.

   Solution: We know that the number of molecules in 22.4 litres of gas at 760 mm Hg and 0°C or 273°K temperature $= 6.023 \times 10^{23}$
   Hence, the molecules per cm^3 in the vessel at 273°K temperature and 10^{-6} mm of pressure

   $$n = \frac{6.023 \times 10^{23} \times 10^{-5}}{224000 \times 760}$$

   $$n = 3.5399 \times 10^{10} \text{ molecules cm}^{-3}$$

   and $$d = 1.99 \times 10^{-8} \text{ cm}$$

   We know that the mean free path is given by

   $$\lambda = \frac{1}{\sqrt{2}\,(\pi d^2 n)}$$

   So, $$\lambda = \frac{1}{1.44 \times (3.14 \times 1.99 \times 10^{-8} \times 3.5399 \times 10^{10})}$$

   $$= 19.68 \times 10^{3} \text{ cm}$$

2. Determine the coefficient of viscosity for O_2 at N.T.P. Given that: $\bar{v} = 4.545 \times 10^2$ m/sec, $\lambda = 9.45 \times 10^8$ m and $\rho = 1.24$ kg/m^3.

Solution: We know that the mean free in terms of viscosity is given by

$$\lambda = 3\left(\frac{\eta}{\rho \bar{v}}\right)$$

So coefficient of viscosity, $\eta = \frac{1}{3}(\lambda \rho \bar{v})$

$$\eta = \frac{1}{3} \times 9.45 \times 10^8 \times 1.24 \times 4.545 \times 10^2$$

Viscosity per unit velocity gradient,

$$\eta = 1.7867 \times 10^{-5} \text{ N/m}^2$$

3. The root mean square velocity of a gas molecule is 4.49×10^2 m/sec and the density of gas is 1.249 gm/litre. The viscosity of gas per unit gradient at N.T.P. is 1.659×10^{-4} dynes/cm^2. Determine (a) mean free path, (b) frequency of collisions of molecules, and (c) diameter of gas molecules.

Solution:

(a) We know that the mean free path is given by

$$\lambda = 3\left(\frac{\eta}{\rho \bar{v}}\right)$$

$$= \left(\frac{3 \times 1.659 \times 10^{-4}}{1.249 \times 10^{-3} \times 4.49 \times 10^4}\right)$$

$$= 8.962 \times 10^{-8} \text{ m}$$

(b) The number of collisions per second is defined as the frequency of collision and we know that

$$\lambda = \frac{v_{rms}}{N}$$

$$N = \frac{v_{rms}}{\lambda}$$

$$= \frac{4.49 \times 10^2}{8.962 \times 10^{-8}}$$

$$= 4.99 \times 10^9$$

(c) We know that Maxwell's relation of the mean free path

$$\lambda = \frac{1}{\sqrt{2}\,(\pi d^2 n)}$$

$$d^2 = \frac{1}{\sqrt{2}\,(\pi n \lambda)}$$

$$d = \left[\frac{1}{\sqrt{2}\,(\pi n \lambda)}\right]$$

$$= \frac{1}{\left[1.44 \times 3.14 \times \left(\frac{6.023 \times 10^{23}}{22400}\right) \times 8.962 \times 10^{-6}\right]^{1/2}}$$

$$= 2.94 \times 10^{-10} \text{ m}$$

4. The $3.01 \times 10^{25}/\text{m}^3$ molecules of oxygen are present in a container. The diameter of oxygen molecule is 3.61×10^{-8} cm. Determine the frequency of collision of molecules and the mean free path.

Solution: We know that the mean speed of molecules of a gas is given by

$$\bar{v} = \sqrt{\frac{8kT}{m\pi}}$$

$$= \sqrt{\frac{8 \times 1.38 \times 10^{-23} \times 273}{\frac{32}{6.023 \times 10^{26}} \times 3.14}}$$

$$= 4.259 \times 10^2 \text{ m/sec}$$

We know that the collision frequency of molecules of gas

$$= \pi d^2 \bar{v} n$$

$$= 3.14 \times 3.61 \times 10^{-10} \times 3.61 \times 10^{-10} \times 4.259 \times 10^2 \times 3.01 \times 10^{25}$$

$$= 5.199 \times 10^9$$

We know that the mean free path is given by

$$\lambda = \frac{1}{\sqrt{2}\,(\pi d^2 n)}$$

$$= \frac{1}{3.14 \times 3.61 \times 10^{-10} \times 3.61 \times 10^{-10} \times 3.01 \times 10^{25}}$$

$$= 5.821 \times 10^{-6} \text{ cm}$$

5. Determine the frequency of sound wave at which its wavelength becomes half of the mean free path of oxygen molecules at 273°K and one atmospheric pressure. The diameter of oxygen molecule is 3.01×10^{-6} m.

Solution: We know that the mean free path is given by

$$\lambda = \frac{1}{\sqrt{2}\,(\pi d^2 n)}$$

$$= \frac{1}{1.41 \times 3.14 \times 3.01 \times 10^{-6} \times 3.01 \times 10^{-6} \times \frac{6.023 \times 10^{23}}{22400}}$$

$$= 9.369 \times 10^{-8} \text{ m}$$

Let us suppose that the wavelength of sound wave is λ',

$$\lambda' = \frac{\lambda}{2}$$

The frequency of sound wave is given by

$$v = \frac{u}{\lambda'} = \frac{u}{\frac{\lambda}{2}} = \frac{2u}{\lambda}$$

$$v = \frac{2 \times 332}{9.369 \times 10^{-8}}$$

$$= 3.5389 \times 10^{9} \text{ Hz}$$

6. The radius of oxygen molecule is 3.035×10^{13} cm and mass is 5.30×10^{-23} gm. Determine the coefficient of thermal conductivity at 0°C. Given that the specific heat at constant pressure is 20.90×10^{3} J/kilo-mole°K.

Solution: We know that the coefficient of thermal conductivity is given by

$$K = \frac{mvC_V}{3\sqrt{2}\,(\pi d^2)}$$

where, u is the root mean square velocity, given by

$$v = v_{rms} = \sqrt{\frac{3kT}{m}}$$

So we get

$$K = \frac{5.31 \times 10^{-26} \times (\sqrt{3 \times 1.38 \times 10^{23} \times 273}) \times 20.90 \times 10^{3}}{3 \times 1.41 \times 3.14 \times 3.035 \times 10^{13} \times 3.035 \times 10^{13}}$$

$$K = 24.02 \times 10^{-3} \text{ J/ m sec}^{\circ}\text{K}$$

7. Diameter of a molecule is 1.99×10^{-10} cm at a pressure of 1 atm and 292°K temperature. Determine (i) mean free path, and (ii) frequency of collision. Given that the velocity of molecules is 510×10^2 cm/sec.

Solution:

(i) We know that the ideal gas equation

$$PV = RT = NkT$$

$$n = \frac{N}{V} = \frac{P}{k \times T}$$

$$= \frac{1 \times 10^5}{1.38 \times 10^{-23} \times 292}$$

$$= 2.512 \times 10^{25} \text{ molecules/ m}^3$$

The mean free path is given by

$$\lambda = \frac{1}{\sqrt{2}\,(\pi d^2 n)}$$

$$= \frac{1}{1.41 \times 3.14 \times 1.99 \times 10^{-10} \times 1.99 \;\; 10^{-10} \times 2.512 \times 10^{25}}$$

$$= 2.249 \times 10^{-5} \text{ cm}$$

(ii) We know that the collision frequency is defined as

$$\nu = \frac{\overline{v}}{\lambda}$$

$$= \frac{510}{2.249 \times 10^{-7}}$$

$$= 2.269 \times 10^9 \text{ / sec}$$

8. The mean free path of a gas molecule is 2.12×10^{-7} m and the number of molecules are 3×10^{19}/cc. Determine the diameter of molecules.

Solution: The mean free path of a gas molecule is given by

$$\lambda = \frac{1}{\sqrt{2}\,(\pi d^2 n)}$$

$$d = \left[\frac{1}{\sqrt{2}\,(\pi \lambda n)}\right]^{1/2}$$

$$= \left[\frac{1}{1.41 \times 3.14 \times 2.12 \times 10^{-5} \times 3 \times 10^{19}} \right]^{1/2}$$

$$= 2.01 \times 10^{-10} \text{ m}$$

9. Determine the mean free path of nitrogen molecule. The diameter of nitrogen molecules is 3.22×10^{-10} m and the number of molecules per m^3 is 2.7×10^{25}.

Solution: We know that the mean free path is given by

$$\lambda = \frac{1}{\sqrt{2}\,(\pi d^2 n)}$$

$$= \frac{1}{1.41 \times 3.14 \times 3.22 \times 10^{-10} \times 3.22 \times 10^{-10} \times 2.7 \times 10^{25}}$$

$$= 8.01 \times 10^{-10} \text{ cm}$$

10. Determine the mean free path of molecules of nitrogen at 1 atmospheric pressure and 26°C temperature. The diameter of nitrogen molecules is 3.49×10^{-10} m.

Solution: We know that the mean free path in terms of pressure is given by

$$\lambda = \frac{k \times T}{\sqrt{2}\,(\pi d^2 P)}$$

$$= \frac{1.38 \times 10^{23} \times 299}{1.41 \times 3.14 \times (76 \times 10^{-3} \times 13.6 \times 10^{3} \times 9.8) \times 3.49 \times 10^{-10} \times 3.49 \times 10^{-10}}$$

$$= 7.499 \times 10^{-10} \text{ cm}$$

11. The 2.789×10^{19} molecules of benzene are present in per cc. The mean free path of benzene molecules is 2.19×10^{-8} m. Determine the diameter of benzene molecules.

Solution: We know that the mean free path is given by

$$\lambda = \frac{1}{\sqrt{2}\,(\pi d^2 n)}$$

$$d^2 = \frac{1}{\sqrt{2}\,(\pi \lambda n)}$$

$$= \frac{1}{1.41 \times 3.14 \times 2.19 \times 10^{-10} \times 2.789 \times 10^{19}}$$

$$= 36.2501 \times 10^{-20} \text{ m}$$

$$d = 6.04 \times 10^{-10} \text{ m}$$

12. The mean free path of the molecules of a gas is 8.85×10^{-10} cm. The coefficient of viscosity of gas molecules is 1.67×10^{-5} Nsec/m^2 and density of the gas is 1.24 kgm^{-3}. Determine the average speed of gas molecules.

Solution: We know that the mean free path in terms of viscosity is given by

$$\lambda = \frac{3\eta}{\bar{v}\rho}$$

$$\bar{v} = \frac{3\eta}{\lambda\rho}$$

$$\bar{v} = \frac{3 \times 1.67 \times 10^{-5}}{8.85 \times 10^{-10} \times 1.24}$$

$$= 4.49 \text{ cm/sec}$$

13. The number of oxygen molecules present in per cc at N.T.P. is 3.01×10^{19}. The mean free path of oxygen molecules is 8.360×10^{-8} m. Determine the diameter of oxygen molecules and the time between collision of molecules of oxygen gas.

Solution: The mean free path is given by

$$\lambda = \frac{1}{\sqrt{2}\,(\pi d^2 n)}$$

$$d^2 = \frac{1}{\sqrt{2}\,(\pi \lambda n)}$$

$$= \frac{1}{1.41 \times 3.14 \times 8.360 \times 10^{-6} \times 3.01 \times 10^{19}}$$

$$d = \frac{1}{[1.41 \times 3.14 \times 8.360 \times 10^{-6} \times 3.01 \times 10^{19}]^{1/2}}$$

$$= 2.99 \times 10^{-8} \text{ cm}$$

The mean time between two successive collisions is given by

$$t = \frac{\lambda}{\bar{v}}$$

where,

$$\bar{v} = \sqrt{\frac{8kT}{m\pi}}$$

$$= \sqrt{\frac{8 \times 1.38 \times 10^{23} \times 273}{\left(\frac{32}{6.023 \times 10^{23}}\right) \times 3.14}}$$

$$= 4.249 \times 10^2 \text{ m/sec}$$

Hence, the average time between the collisions

$$t = \frac{\lambda}{\overline{v}}$$

$$= \frac{8.360 \times 10^{-6}}{4.249 \times 10^4}$$

$$= 19.67 \times 10^{-11} \text{ sec}$$

14. Determine the coefficient of viscosity of oxygen molecules at 0°C. The radius of oxygen molecule is 1.5×10^{-10} m and mass of oxygen molecule is 5.32×10^{-23} gram.

Solution: We know that the coefficient of viscosity is given by

$$\eta = \frac{1}{3\sqrt{2}}\left(\frac{m\overline{v}}{\pi d^2}\right)$$

The mean velocity is given by

$$\overline{v} = \sqrt{\frac{8kT}{m\pi}}$$

$$= \sqrt{\frac{8 \times 1.38 \times 10^{23} \times 273}{\left(\frac{32}{6.023 \times 10^{23}}\right) \times 3.14}}$$

$$= 4.249 \times 10^2 \text{ m/sec}$$

So coefficient of viscosity

$$\eta = \frac{1}{3 \times 1.41}\left(\frac{5.32 \times 10^{-26} \times 4.249 \times 10^2}{3.14 \times 1.5 \times 10^{-10} \times 1.5 \times 10^{-10}}\right)$$

$$= 1.921 \times 10^{-5} \text{ N/m}^2$$

15. Oxygen and hydrogen gases are kept at similar temperature and pressure. Diameter of oxygen molecule and hydrogen molecule are 3.01 Å and 2.49 Å, respectively. Determine the ratio of coefficient of viscosity of oxygen and hydrogen.

Solution: The coefficient of viscosity is given by

$$\eta = \frac{1}{3}(mn\overline{v}\lambda)$$

where,

$$\lambda = \frac{1}{\sqrt{2}\,(\pi d^2 n)}$$

$$\overline{v} = \sqrt{\frac{8kT}{m\pi}}$$

So, the ratio of coefficient of viscosity of oxygen and hydrogen is given by

$$\frac{\eta_O}{\eta_H} = \sqrt{\frac{m_O}{m_H}} \times \left(\frac{d_H^2}{d_O^2}\right)$$

$$= \sqrt{\frac{32}{2}} \times \left(\frac{2.49}{3.01}\right)^2$$

$$= 4 \times (0.827)^2$$

$$= 4 \times 0.6841 = 2.736$$

$$= 2.736$$

16. The mean free path of gas molecules is 4×10^{-8} m at temperature 400°K and pressure P. Determine the mean free path at (a) P, 800°K, (b) $P/2$, 800°K, and (c) $2P$, 400°K.

Solution: The mean free path is given by

$$\lambda = \frac{1}{\sqrt{2}\,(\pi d^2 n)}$$

$$= \frac{1}{\sqrt{2}\left[\pi d^2 \times \left(\frac{P}{kT}\right)\right]}$$

$$= T \times \frac{k}{\sqrt{2}\,[\pi d^2 \times P]}$$

At 400°K and pressure P, the mean free path is

$$\lambda = 400 \times \frac{k}{\sqrt{2}\,[\pi d^2 \times P]}$$

(a) The mean free path at pressure P and temperature 800°K

$$\lambda = 800 \times \frac{k}{\sqrt{2}\,[\pi d^2 \times P]} = 2\lambda = 2 \times 4 \times 10^{-8}\,m = 8\text{Å}$$

(b) The mean free path at pressure $P/2$ and temperature 800°K

$$\lambda = 800 \times \frac{k}{\sqrt{2}\,[\pi d^2 \times P/2]}$$

$$= 1600 \times \frac{k}{\sqrt{2}\,[\pi d^2 \times P]} = 4\lambda$$

$$= 4 \times 4 \times 10^{-8}\, m = 16\text{Å}$$

(c) The mean free path at pressure $2P$ and temperature 400°K

$$\lambda = 400 \times \frac{k}{\sqrt{2}\left[\pi d^2 \times 2P\right]}$$

$$= 200 \times \frac{k}{\sqrt{2}\,[\pi d^2 \times P]} = \lambda/2$$

$$= \frac{1}{2} \times 4 \times 10^{-8}\, m = 2\text{Å}$$

17. A perfect gas is contained in a vessel at N.T.P. The diameter of the molecule of gas is 3 Å. Determine the mean free path of gas molecule.

Solution: We know that the gas equation for perfect gas is given by

$$PV = RT = NkT$$

$$\frac{N}{V} = \frac{P}{kT}$$

Let us suppose that n is the number of gas molecules present in per m^3 volume,

$$n = \frac{N}{V} = \frac{P}{kT}$$

We know that the mean free path is given by

$$\lambda = \frac{1}{\sqrt{2}\,(\pi d^2 n)}$$

$$= \frac{1}{\sqrt{2}\left[\pi d^2 \times \left(\frac{P}{kT}\right)\right]}$$

$$= T \times \frac{k}{\sqrt{2}\,[\pi d^2 \times P]}$$

$$= \frac{273 \times 1.38 \times 10^{-23}}{[1.41 \times 3.14 \times 3 \times 10^{-10} \times 3 \times 10^{-10} \times (0.76 \times 13.6 \times 10^3 \times 9.8)]}$$

$$= 0.933 \times 10^{-7} \text{ m}$$

Exercise

Short Answer Questions

1. What do you mean by the transport phenomenon?
2. Define the mean free path and mean free time.
3. What do you mean by mean free path of a gas molecule?
4. Find out the expression of the mean free path on the basis of Clausius.
5. What is the effect of pressure and temperature on the mean free path?
6. Discuss the effect of temperature and pressure on coefficient of thermal conductivity.
7. What is the effect of temperature and pressure on coefficient of viscosity?
8. Establish the relation between coefficient of diffusion and coefficient of viscosity.
9. What is the effect of temperature and pressure on coefficient of diffusion?
10. Derive the relation between the coefficient of viscosity and coefficient of thermal conductivity.
11. What do you mean by the frequency of collision?
12. Establish the relation between the coefficient of diffusion and the coefficient of thermal conductivity.

Long Answer Questions

1. Define the mean free path of gas molecules and derive the expression for it. Prove that the mean free path is directly proportional to the absolute temperature and inversely proportional to the pressure of gas.
2. Define the term transport phenomenon. Discuss conductivity, diffusion and viscosity phenomenon by the kinetic theory of gases.
3. What do you mean by viscosity? Obtain the expression for coefficient of diffusion and prove that it is directly proportional to $T^{3/2}$.
4. What is the mean free path? Show that the mean free path of a gas molecule is given by the expression

$$\lambda = \frac{1}{\sqrt{2}\,(\pi d^2 n)}$$

where, d is the diameter of the gas molecule and n is the number of molecules present in per unit volume of gas. Explain the effect of pressure and temperature on it.

5. Define thermal conductivity. Obtain the expression for the coefficient of thermal conductivity on the basis of kinetic theory.
6. Discuss mean free path and derive the Clausius' expression of mean free path of molecule of gases.
7. What do you mean by the phenomenon of diffusion? Obtain the expression for the coefficient of self-diffusion on the basis of kinetic theory of gases.
8. Establish the expression for the coefficient of thermal conductivity and prove that hydrogen molecule has the largest coefficient of thermal conductivity.
9. Obtain the expression for the coefficient of self-diffusion and show that it is inversely proportional to the pressure of gas.
10. Define mean free path, mean free time of molecule of gas, and collision frequency. Find out the variation of mean free path with temperature and time.

Numerical Problems

1. The mean free path of oxygen molecule at 100°C and one atmospheric pressure is 9.960×10^{-6}cm. Determine its radius.

   [**Ans.** 1.694×10^{-8} cm]
2. The diameter of a gas molecule is 3.969×10^{-8} cm. Calculate its coefficient of viscosity at 289°K. [**Ans.** 169 micropose]
3. The mean free path of a molecule of gas is 8.790×10^{-6} cm. If its diameter is 3.19 Å, calculate the value of number of molecules in per unit volume.

   [**Ans.** 2.49×10^{25}]
4. Determine the mean free path of nitrogen present in hydrogen at N.T.P. The cross-sectional area for slow neutron in H_2 is 19.90×10^{-28} m^2.

   [**Ans.** 13.39×10^2 cm]
5. Density of oxygen molecule at N.T.P. is 1.45 kg/m^3. Determine its root mean square velocity at 30°C.
6. The diameter of a molecule of a gas is 2.969×10^{-8} cm. Find out its coefficient of viscosity 300°K.
7. The number of molecules of a gas present in per cm^3 is 2.99×10^{19} and diameter of molecule of gas is 7.279×10^{-9} m. Determine its mean free path.

   [**Ans.** 1.99×10^{-10} m]
8. The mean free path of a gas molecule is 3×10^{-5} cm at temperature 300°K and pressure P. Determine the mean free path of molecules of gas (i) $P/2$, 600°K, (ii) P, 600°K, and (iii) $2P$, 300°K.
9. The diameter of the oxygen molecule is 3.09 Å and the mass of oxygen molecule is 5.31×10^{-23} gm. Calculate the coefficient of viscosity at 0°C.

10. Oxygen and nitrogen gases are kept at similar temperature and pressure. Diameters of oxygen molecule and nitrogen molecule are 3.01 Å and 2.89 Å, respectively. Determine their ratio of coefficient of viscosity.
11. The number of oxygen molecules present in per cc at N.T.P. is 3.34×10^{19}. The mean free path of oxygen molecules is 7.360×10^{-8} m. Determine their diameters and the collision frequency.
12. The mean free path of the molecules of a gas is 8.89×10^{-8} m. The coefficient of the viscosity of gas molecules is 1.69×10^{-5} Nsec/m^2 and density of gas is 1.36 kgm^{-3}. Determine the average speed of gas molecules.
13. Determine the mean free path of molecule of nitrogen at 1 atmospheric pressure and 40°C temperature. The diameter of nitrogen molecule is 3.49×10^{-8} cm.
14. The diameter of molecule of a gas is 2×10^{-8} m at a pressure 1 atms and 300°K temperature. Determine (i) mean free path, and (ii) frequency of collision. Given that the velocity of the molecules is 512 m/sec.
15. The radius of oxygen molecule is 3.136×10^{11} m and the mass is 5.30×10^{-26} kg. Determine the coefficient of thermal conductivity at 0°C. Given that specific heat at constant pressure is 18.90×10^{3} J/kilo-mole°K.

Multiple-Choice Questions

1. Transformation of heat from lower temperature body to higher temperature body:
   (i) Is possible in the presence of external work
   (ii) Is not possible
   (iii) Is possible without external work
   (iv) None of these statements
2. A gas initially at T°C is suddenly compressed to 1/8th of its actual volume. The temperature of gas after compression is 887°C. The initial temperature of gas is
   (i) 145°C (ii) 17°C
   (iii) 34°C (iv) 15°C
3. Zeroth law of thermodynamics states that
   (i) Ideal gas contains molecules
   (ii) If two thermodynamic systems are in thermal equilibrium to a third system, then they must be in thermal equilibrium to each other
   (iii) Absolute zero is not possible
   (iv) None of the above
4. The physical quantity which is important in the first law of thermodynamics
   (i) Pressure (ii) Density
   (iii) Temperature (iv) Energy

5. The quantity kept constant in an adiabatic process is

(i) Density (ii) Temperature

(iii) Pressure (iv) Heat

6. The ratio of two elasticities of a gas is

(i) γ^2 (ii) γ^3

(iii) γ (iv) $\gamma^{1/2}$

7. The work in an adiabatic process is

(i) $W = \frac{R}{(1-\gamma)}[T_2 - T_1]$ (ii) $W = \frac{R}{(1-\gamma)}[P_2 - P_1]$

(iii) $W = \frac{1}{R(1-\gamma)}[P_2 - P_1]$ (iv) $W = \frac{RT_1}{(1-\gamma)}[V_2 - V_1]$

8. Internal energy of a real gas is affected by

(i) Pressure of the gas (ii) Volume of the gas

(iii) Density of the gas (iv) Temperature of the gas

9. A Carnot's heat engine is working between 373 K and 323 K. Efficiency of engine will be

(i) 18.34% (ii) 13.39%

(iii) 18.6% (iv) 21.2%

10. Slope of an adiabatic curve is

(i) $\frac{dP}{dV} = -\left(\frac{P}{\gamma V}\right)$ (ii) $\frac{dP}{dV} = -\gamma\left(\frac{P}{V}\right)$

(iii) $\frac{dP}{dV} = -\frac{1}{\gamma}\left(\frac{P}{V}\right)$ (iv) $\frac{dP}{dV} = -\left(\frac{P}{V}\right)$

11. An adiabatic process is represented by

(i) $PV^{2\gamma} = C$ (ii) $P^2V^{\gamma} = C$

(iii) $\frac{P}{V^{\gamma}} = C$ (iv) $PV^{\gamma} = C$

12. The amount of heat energy given to a system in an isothermal reversible process of an ideal gas is

(i) $(dQ)_T = \left[\left(\frac{\partial U}{\partial V}\right)_T + P\right](dV)_T$ (ii) $(dQ)_T = \left[\left(\frac{\partial U}{\partial V}\right)_T - P\right](dV)_T$

(iii) $(dQ)_T = \left[\left(\frac{\partial U}{\partial V}\right)_T + 2P\right](dV)_T$ (iv) $(dQ)_T = \left[\left(\frac{\partial U}{\partial V}\right)_T + P^2\right](dV)_T$

13. An isochoric process is defined by

(i) $\frac{P_1}{P_2} = \frac{T_1}{T_2}$ (ii) $P_1 P_2 = \frac{T_1}{T_2}$

(iii) $\frac{P_1^2}{P_2} = \frac{T_1}{T_2}$ (iv) $\frac{P_1}{P_2} = \frac{T_1^2}{T_2}$

14. In an isothermal process

(i) $dQ = \frac{dW}{2}$ (ii) $dQ = dW$

(iii) $dQ = 2dW$ (iv) $dQ = dW^3$

15. In a cyclic process, the relation between heat and work is

(i) $\oint dQ = \oint dW$ (ii) $\oint dQ = \oint \frac{dW}{2}$

(iii) $\oint dQ = \oint 2dW$ (iv) $\oint dQ = \oint 3dW$

16. Difference in internal energy per unit mass is

(i) $\frac{\Delta U}{m} = L - P \cdot (v_2 - v_1)$ (ii) $\frac{\Delta U}{m} = L - 2P \cdot (v_2 - v_1)$

(iii) $\frac{\Delta U}{m} = L - P \cdot (v_2 - v_1)^2$ (iv) $\frac{\Delta U}{m} = 2L - P \cdot (v_2 - v_1)$

Answers

1. (i)	2. (ii)	3. (ii)	4. (iv)	5. (iv)	6. (iii)	7. (i)	8. (iii)
9. (ii)	10. (ii)	11. (iv)	12. (i)	13. (i)	14. (ii)	15. (i)	16. (i)

6

Second and Third Laws of Thermodynamics and Entropy

6.1 REVERSIBILITY AND IRREVERSIBILITY OF PROCESS

Reversible Process

An ideal or reversible process is one in which the system and local surroundings both can be restored to their initial state by reversing the direction of process without creating any effect in the rest of universe.

Let us consider a gas contained in a cylinder fitted with a frictionless piston, made of perfectly conducting material and immersed in a water bath tub of fixed temperature. The gas is compressed gradually so that its temperature remains constant throughout the process. The heat generated in compression process is passing out in the water bath tub. Now after compressing the volume of the gas, gas is allowed to slowly expand isothermally. In this process, the heat will flow from enclosing water bath tub and the temperature of the gas remains constant. The same amount of heat will be obtained during the expansion of gas, as was given up in compression. Therefore, the very slow isothermal compression and expansion of gas is a reversible process. The specific conditions for reversibility of any process can be stated as follows:

(i) All the processes involved in the compete cycle must be very slow.
(ii) The temperature and pressure of the gas must not differ remarkably from those of the local surroundings during the complete cycle.
(iii) The functioning part of the machine must be frictionless.
(iv) Loss of energy is not allowed due to conduction, friction, radiation or convection.

The complete reversibility is an ideal case. In real world, there is always some loss of heat due to conduction, friction, radiation or convection. A number of processes may be considered reversible. There are a few examples of the reversible process as follows:

(i) Expansion or stretching of a spring
(ii) Motion of solids without friction

(iii) Slow isothermal expansion of gas
(iv) Slow adiabatic expansion process

Irreversible Process

In so many processes, when reversed, the system and local surroundings do not retrace its initial states without any inputs, and leave noticeable change in both the system and surroundings. These type of processes are known as natural or irreversible processes.

The important factors that provide irreversible process are as follows:

(i) Dissipative effects, such as viscosity, friction, inelastic solid deformation, electrical resistance, eddy formation, etc.
(ii) Free expansion of gas
(iii) Inadequacy of equilibrium in the process
(iv) Mixing of more than one substance

In nature, all changes are irreversible. There are some examples of irreversible processes:

(i) Heat transportation due to conduction
(ii) Unstrained or free expansion process of gases
(iii) Heat transfer due to radiation
(iv) Transfer of electricity

6.2 ACTUAL HEAT ENGINE AND ITS EFFICIENCY

A practical machine that converts heat energy into mechanical work is known as heat engine. Heat engines, in their working, take in heat at higher temperature, convert some part of it into mechanical work, and throw out remaining heat at low temperature.

There are various types of actually existing heat engines. Out of them, the most important are steam power plants, jet engines and internal combustion engines. The energy flowchart of a heat engine is shown in Figure 6.1.

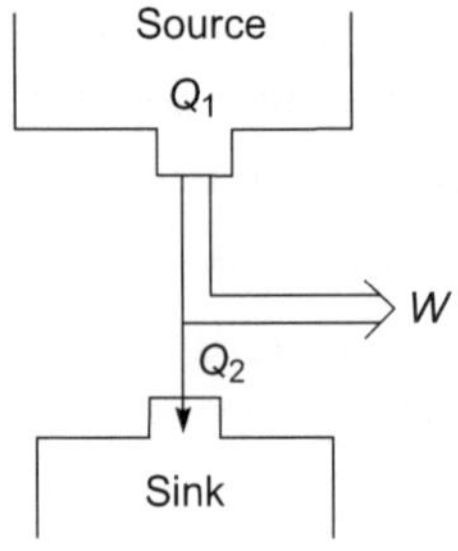

Figure 6.1 Energy flowchart of a heat engine

The essential parts of a heat engine are as follows:

(i) **Heat Reservoir or Source:** A container maintained at fixed high temperature T_1 from which heat engine can take heat, is called source. A source has infinite thermal capacity and heat engine can take any amount of heat from it at constant temperature.

(ii) **Working Substance:** A fluid substance, which performs mechanical work is known as working substance. It undergoes changes in thermodynamical properties such as pressure, volume, temperature, etc. In steam turbines, the working substance is water, in internal-combustion engines, it is the mixture of air and fuel.

(iii) **Cold Reservoir or Sink:** A container maintained at a fixed lower temperature T_2 to which engine can give any amount of heat, is known as sink. A sink has infinite thermal capacity and any amount of heat can be given to it at constant temperature.

In any heat engine, the working substance undergoes change in pressure, temperature and volume and then comes back to the initial state. The complete process through which the working substance goes through from its initial position to final state and then back to its initial state represents one complete cycle.

The thermal efficiency of a heat engine is defined as the ratio of net work done, W, by the engine in one complete cycle to the heat taken from the source. It is denoted by η and mathematically, expressed as follows:

$$\eta = \frac{W}{Q_1}$$

or

$$= \frac{Q_1 - Q_2}{Q_1} = 1 - \frac{Q_2}{Q_1}$$

6.3 CARNOT'S CYCLE AND CARNOT'S HEAT ENGINE

Sadi Carnot, in 1824, conceived a hypothetical reversible heat engine with maximum efficiency. The following assumptions are required for the analysis of Carnot's cycle.

(i) An ideal gas is used as a working substance.

(ii) Two heat reservoirs, source and sink at temperatures T_1 (high temperature) and T_2 (low temperature), respectively are to be considered in the cycle.

(iii) Frictionless cylinder and piston arrangement is to be considered. The cylinder head is constructed in such a way that it can provide a perfect heat insulator or a perfect heat conductor.

(iv) The piston and the walls of cylinder are considered a perfect heat insulator.

The essential parts of Carnot's heat engine are shown in Figure 6.2.

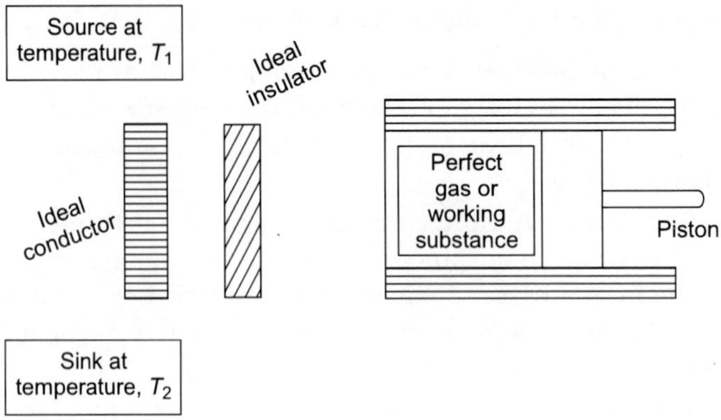

Figure 6.2 Parts of Carnot's heat engine

A continuous supply of work can be obtained by Carnot's cycle. The following sequence of operations is used for different processes in Carnot's cycle (Figure 6.3).

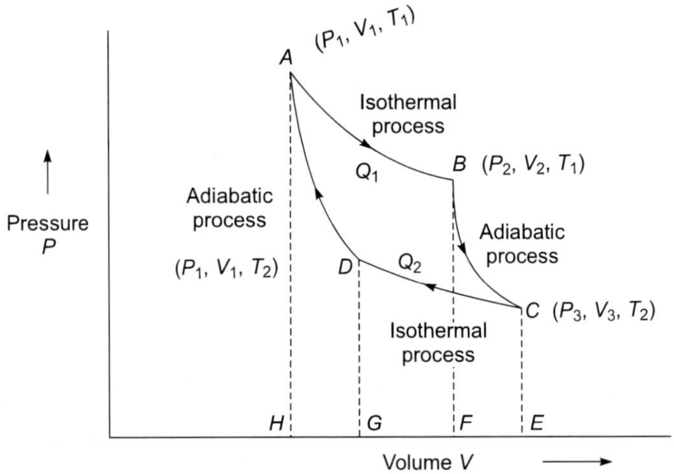

Figure 6.3 Carnot's cycle

1. (*A*-*B*) Isothermal Expansion

The source at temperature T_1 is put in contact with a working substance inside the cylinder through a perfectly conducting base. As heat passes into the cylinder, gas undergoes slow isothermal expansion at a fixed temperature T_1 from state A to B. During this process, the working substance absorbs heat Q_1 from the heat source at temperature T_1 and does work W_1 $(\Delta U = 0)$, which is given by

$$Q_{A-B} = Q_1 = W_1 = RT_1 \log_e \frac{V_2}{V_1} = P_1V_1 \log_e \frac{V_2}{V_1} = \text{area } ABFHA$$

2. (*B-C*) Adiabatic Expansion

Heat source is now removed and the cylinder is placed on an insulator. The gas inside the cylinder is allowed to expand adiabatically and reversibly from state *B* to *C*. Work is done by the working substance at the charge of internal energy and the temperature of gas falls to T_2. This work done is given by

$$W_2 = R\frac{(T_1 - T_2)}{(\gamma - 1)} = \frac{(P_2V_2 - P_3V_3)}{(\gamma - 1)} = \text{area } BCEFB \qquad (Q_{B-C} = 0)$$

3. (*C-D*) Isothermal Compression

The insulator is now removed and the cylinder is placed on a sink at temperature T_2. The piston is moved inwards slowly, so that the gas inside the cylinder is compressed isothermally at constant temperature T_2. The isothermal compression process is represented by *CD*. The gas rejects amount of heat Q_2 to the sink at temperature T_2 and is equal to work done W_3 on the system as $\Delta U = 0$. This work done (W_3) is given by

$$Q_{C-D} = Q_2 = W_3 = -RT_{21}\log_e\frac{V_3}{V_4} = -P_3V_3\log_e\frac{V_3}{V_4} = \text{area } CEGDC$$

Here, negative sign shows that work is done on the working fluid or substance.

4. (*E-F*) Adiabatic Compression

The sink is now removed and the cylinder is again placed on the insulator. The piston is moved inwards and the gas is compressed reversibly and adiabatically, till the working substance returns to its original state *A* by completing the process *DA*. The temperature of the gas rises to T_1 and internal energy of the system increases. The work (W_4) is done on the working substance and is given by

$$W_4 = -R\frac{(T_1 - T_2)}{(\gamma - 1)} = -\frac{(P_1V_1 - P_4V_4)}{(\gamma - 1)} \qquad (Q_{D-A} = 0)$$

Here, negative sign shows that work is done on the working fluid or substance.

Work Done by the Heat Engine in One Cycle

In the Carnot's cycle, the working fluid absorbs heat (Q_1) from the high temperature source and rejects heat (Q_2) to the sink.

So, the net heat absorbed by the working substance in one complete cycle

$$= (Q_1 - Q_2)$$

The net work done by the heat engine in one cycle

$$= (W_1 + W_2 + W_3 + W_4)$$

Since work W_2 and work W_4 are equal and opposite, they cancel each other, and hence, resultant work done

$$(W_1 + W_3) = \text{area } ABCDA$$

In a cyclic process, the first law of thermodynamics can be expressed as

$$\oint dQ = \oint dW$$

or $$(Q_1 - Q_2) = (W_1 + W_3)$$

$$= P_1V_1 \log e \frac{V_2}{V_1} - P_3V_3 \log e \frac{V_3}{V_4}$$

$$= RT_1 \log e \frac{V_2}{V_1} - RT_2 \log e \frac{V_3}{V_4} \tag{6.1}$$

For the adiabatic expansion BC

$$T_1V_2^{(\gamma-1)} = T_2V_3^{(\gamma-1)}$$

or $$\left(\frac{V_2}{V_3}\right) = \left(\frac{T_2}{T_1}\right)^{\frac{1}{(\gamma-1)}} \tag{6.2}$$

Similarly for the adiabatic compression DA

$$T_1V_1^{(\gamma-1)} = T_2V_4^{(\gamma-1)}$$

or $$\left(\frac{V_1}{V_4}\right) = \left(\frac{T_2}{T_1}\right)^{\frac{1}{(\gamma-1)}} \tag{6.3}$$

From equations (6.3) and (6.4), we get

$$\left(\frac{V_2}{V_3}\right) = \left(\frac{V_1}{V_4}\right)$$

or $$\left(\frac{V_2}{V_1}\right) = \left(\frac{V_3}{V_4}\right) \tag{6.4}$$

Substituting this value in equation (6.1), we have

$$(Q_1 - Q_2) = RT_1 \log e \frac{V_2}{V_1} - RT_2 \log e \frac{V_2}{V_1}$$

$$= W = R(T_1 - T_2) \log e \frac{V_2}{V_1} \tag{6.5}$$

In the above cycle, all the four processes are reversible, hence, Carnot's cycle is known as a reversible cycle. The thermal efficiency of the Carnot's heat engine is given by

$$\eta = \frac{\text{Net useful output}}{\text{Input heat}} = \frac{W_{NET}}{Q_1}$$

$$= \frac{W_{NET}}{Q_1} = \frac{(Q_1 - Q_2)}{Q_1}$$

or

$$= \frac{R(T_1 - T_2)\log e \dfrac{V_2}{V_1}}{RT_1 \log e \dfrac{V_2}{V_1}}$$

$$= \frac{(T_1 - T_2)}{T_1} = 1 - \left(\frac{T_2}{T_1}\right) = 1 - \left(\frac{Q_2}{Q_1}\right) \tag{6.6}$$

The following conclusions can be drawn from the efficiency expression:

(i) The thermal efficiency of Carnot's heat engine depends only on the source temperature and sink temperature. It is independent of the working substance. The efficiency gets maximum value, i.e., $\eta = 1$ (100%) at $T_2 = 0°K$, but practically, it is not possible.

(ii) The efficiency is directly proportional to the difference of source and sink temperature. It increases with the increase of this difference.

(iii) The efficiency (η) can be increased either by increasing the temperature of the source or decreasing the temperature of the sink. It will be higher when the temperature difference between the source and sink is higher.

(iv) The efficiency gets minimum or zero value when the temperatures of source and sink are equal, i.e., at $T_1 = T_2$. In this case, the heat engine does not work.

In the Carnot's heat engine, all four processes are reversible, so all internal friction should be zero and also, there will be no friction between the cylinder walls and the piston. The isothermal expansion or compression can be achieved by very slow movement of the piston and also adiabatic expansion and compression can be achieved by very fast movement of the piston. At infinitesimal temperature difference between the source and sink, the absorption and rejection of heat takes place. The Carnot's engine is a perfectly reversible engine. It can also work as a refrigerator when operated in the reverse direction. Carnot's heat engine is an ideal case. Any practical heat engine cannot satisfy all these conditions. Hence, the efficiency of Carnot's engine is higher than the efficiency of all practical engines.

6.4 CARNOT'S HEAT ENGINE AND REFRIGERATOR

Carnot's engine is a reversible engine. When Carnot's cycle is performed in reverse direction, it works as a refrigerator. In the case of heat engine, it absorbs heat amount (Q_1) from the higher temperature (T_1) source, does work (W) and rejects the remaining amount of heat (Q_2) to lower temperature (T_2) sink, as shown in Figure 6.4(a).

When it works as a refrigerator, heat flows from a lower temperature body to a higher temperature body by means of external work done on the working fluid. In this case, all the processes of Carnot's cycle are reversed and the transfer of energy will also be reversed in direction but the magnitude remains the same.

In the case of refrigerator, it absorbs an amount of heat (Q_2) from the lower temperature (T_1) sink. (W) work is done on it by an external source and it rejects the remaining heat (Q_1) to the higher temperature (T_2) source, as shown in Figure 6.4(b).

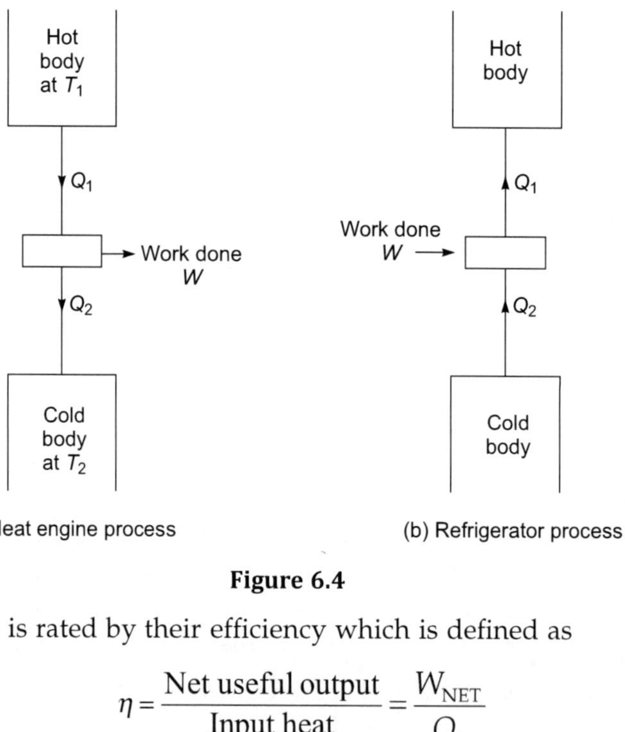

(a) Heat engine process (b) Refrigerator process

Figure 6.4

The heat engine is rated by their efficiency which is defined as

$$\eta = \frac{\text{Net useful output}}{\text{Input heat}} = \frac{W_{\text{NET}}}{Q_1}$$

or

$$= \frac{(T_1 - T_2)}{T_1} = 1 - \left(\frac{T_2}{T_1}\right) = 1 - \left(\frac{Q_2}{Q_1}\right)$$

A refrigerator is rated by its coefficient of performance, which is represented by β and defined as

$$\beta = \frac{\text{Output heat}}{\text{Input work}} = \frac{Q_2}{W}$$

or $$\beta = \frac{T_2}{(T_1 - T_2)} = \frac{Q_2}{(Q_1 - Q_2)} \tag{6.7}$$

The Carnot's cycle was modified by Professor Rankine and he presented a technically acceptable cycle, known as Rankine's cycle. Rankine's cycle is also reversible but it is different from Carnot's cycle.

6.5 SECOND LAW OF THERMODYNAMICS

The first law of thermodynamics is the law of conservation of energy. It says that whatever amount of work is obtained, an equivalent heat is used up or vice versa of it. This law does not tell anything about the direction of flow of heat and the amount of heat which is converted in work. It has no explanation about the flow of heat from a higher temperature body to a lower temperature body. Such type of questions led to the formulation of a new law known as the second law of thermodynamics. This law has been stated by a number of authors in different ways.

Planck's Statement

"It is impossible to design an engine in which, doing work in a complete cycle, will not produce any effect, except increasing the weight and lowering the temperature of heat reservoir."

Lord Kelvin's Statement

"It is impossible to obtain continuous supply of work by cooling a body to a temperature lower than the temperature of its surroundings."

The above two statements are combined to give one equivalent statement, which is known as Kelvin-Planck's statement of the second law of thermodynamics.

Kelvin-Planck's Statement

"It is impossible to design an engine, which does work in a complete cycle and shows no effect other than extracting heat energy from a single heat reservoir and performing an equivalent amount of work."

This statement is applicable to heat engines and tells that no heat engine can be constructed that takes heat (Q) from a high temperature body and converts it into equivalent work. The efficiency of such engine is 100%, i.e., $W = Q$. This is similar to the first law of thermodynamics but against Kelvin's statement.

Hence, a heat engine extracts (Q_1) amount of heat from a higher temperature body, does some work and then, rejects the remaining amount of heat (Q_2) to a lower temperature body, as shown in Figure 6.5.

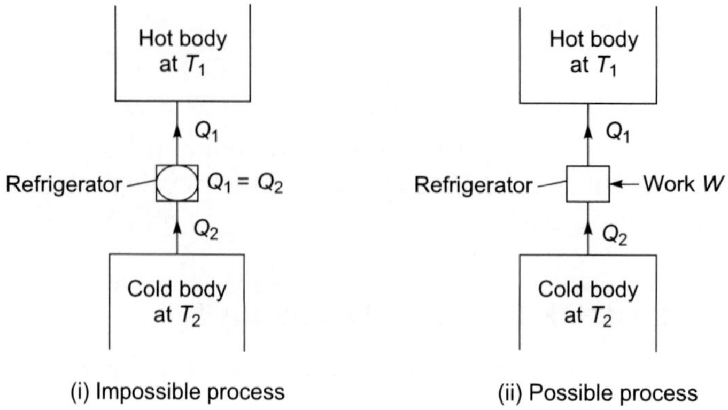

Figure 6.5 Clausius's statement

Clausius's Statement

It is impossible to construct a device working in a complete cycle to transfer heat from a lower temperature body to a higher temperature body without any external source."

This statement is related to the performance of refrigerator or heat pump, working in the reverse direction. The statement tells that heat cannot be transferred from a cold body to hot body without any external means. The arrangement to prove Clausius's statement is shown in Figure 6.6.

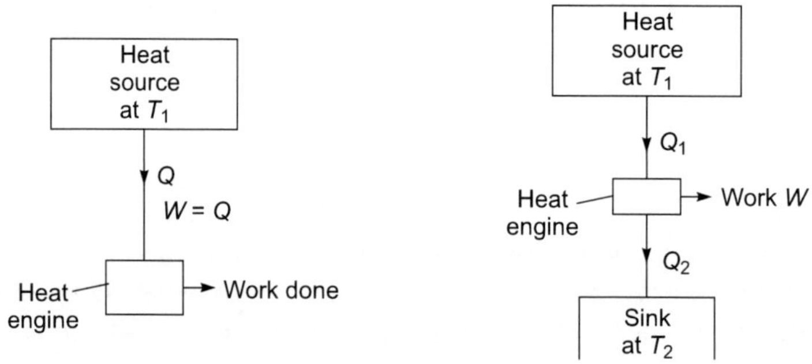

Figure 6.6 Kelvin-Planck's statement representation for heat engine

All statements of the second law of thermodynamics are equivalent, violation of one leads to the violation of others. The second law of thermodynamics determines the direction for the process. It requires different thermodynamic parameters to give a mathematical formulation and description. This important parameter is the entropy of the system.

6.6 CARNOT'S THEOREM

On the basis of the second law of thermodynamics, Carnot's theorem states that:

"No heat engine can be more efficient than a reversible engine operating between the two same temperature reservoirs."

This means that the efficiency of all reversible heat engines working between the two heat reservoirs of same temperatures is the same. It does not depend upon the nature of working substance. Carnot's heat engine has maximum efficiency and no engine can get efficiency more than its.

Proof: Let us consider two engines A_R and A_I working between the same temperature reservoirs at T_1 and T_2. The engine A_R is reversible and engine A_I is irreversible and $T_1 > T_2$. Let the engine A_R be more efficient than engine A_I. Let us assume that in a complete cycle, engine A_R absorbs the amount of heat (Q_1) from the source and rejects the amount of heat (Q_2) to the sink. Similarly, engine A_I absorbs the amount of heat Q_1' from the source and rejects the amount of heat Q_2' to the sink. Suppose both the engines do the same work (W) in a complete cycle. As the engine A_I is more efficient than engine A_R, hence

$$\frac{(Q_1'-Q_2')}{Q_1'} > \frac{(Q_1-Q_2)}{Q_1}$$

or

$$\frac{W}{Q_1'} > \frac{W}{Q_1}$$

$$Q_1' > Q_1$$

and also,

$$(Q_1'-Q_2) = (Q_1-Q_2)$$

or

$$(Q_2-Q_2') = (Q_1-Q_1')$$

Since

$$Q_1' > Q_1 \text{ so } Q_2' > Q_2$$

Let us suppose that both the engines are joined, so that engine A_I drives engine A_R backwards (Figure 6.7). The combination of two makes a self-acting device in which A_R gives external work (W) and A_I absorbs (W) work in the reverse cycle.

For the coupled engines in a complete cycle, net gain of heat by the source is (Q_1-Q_1') and net loss of heat by the sink is (Q_2-Q_2'). External work on the coupled engines is zero, i.e., $W = 0$.

Hence, the coupled engines make a self-operating device, which gives the continuous supply of heat from a cold body to a hot body, without any external means.

This violates the second law of thermodynamics, according to which heat cannot be transferred continuously from a lower temperature body to a higher temperature body without any external source. Hence, our assumptions that A_I is more efficient than A_R is wrong and we can say that a perfect reversible engine has maximum efficiency and no engine can be more efficient than a reversible engine working between the same temperatures, i.e., $\eta_{A_R} \geq \eta_{A_I}$.

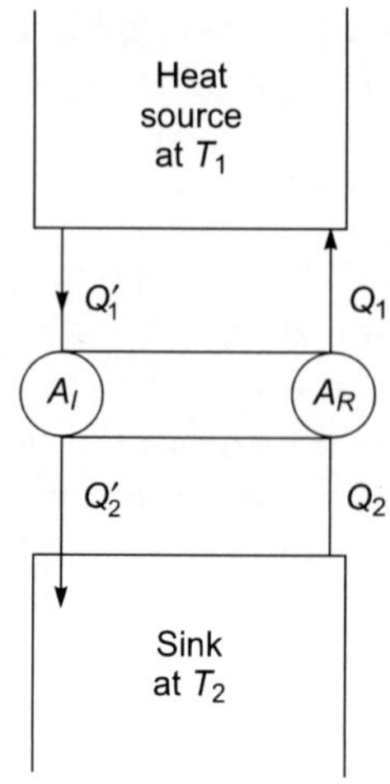

Figure 6.7 Carnot's theorem verification

6.7 ENTROPY OF THE SUBSTANCE

To give the complete description of a working substance in the second law of thermodynamics, a new variable is introduced, which is known as entropy. The entropy is a thermal property which remains constant during adiabatic process.

The variable entropy can be described clearly from *P-V* diagram at different temperatures, as shown in Figure 6.8.

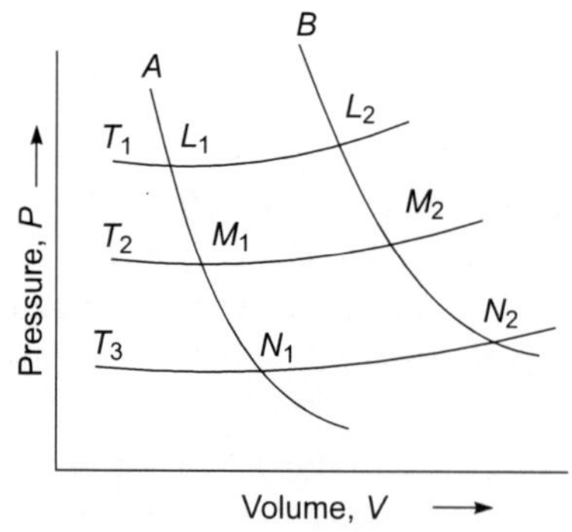

Figure 6.8 *P-V* diagram for entropy concept

Consider a *P-V* diagram for Carnot's reversible cycle consisting of two adiabatics *A* and *B* and three isothermals at temperature T_1, T_2, and T_3 for a perfect gas. In adaiabatics *A* and *B*, volume and temperature change with the change in pressure. Assume that $L_1L_2M_2M_1$ and $M_1M_2N_2N_1$ are Carnot's reversible cycles. Q_1 amount of heat is absorbed in moving from L_1 to L_2 at fixed temperature T_1 and Q_2 heat is rejected at a fixed temperature T_2.

We know that the efficiency of a Carnot's ideal heat engine is given by

$$\eta = 1 - \frac{Q_2}{Q_1} = 1 - \frac{T_2}{T_1}$$

or
$$\frac{Q_2}{Q_1} = \frac{T_2}{T_1}$$

or
$$\frac{Q_1}{T_1} = \frac{Q_2}{T_2} \tag{6.8}$$

Similarly, if we consider the Carnot's cycle $M_1M_2N_2N_1$ bounded by other isothermals and adiabatics, in which Q_2 is the absorbed heat at T_2 temperature and Q_3 is the rejected heat at T_3 temperature. So, we can write

$$\frac{Q_2}{T_2} = \frac{Q_3}{T_3} \tag{6.9}$$

Hence, from equations (6.8) and (6.9), we may write

$$\frac{Q_1}{T_1} = \frac{Q_2}{T_2} = \frac{Q_3}{T_3} = \ldots \text{constant}$$

Hence, if Q is the heat absorbed or rejected at a fixed temperature in going from one adiabatic curve to another, then the above expression may be written as

$$\frac{Q}{T} = \text{constant}$$

If we consider that two adiabatics are very near to each other and the amount of heat, dQ, absorbed or rejected is very small at constant temperature T,

$$\frac{dQ}{T} = \text{constant} \tag{6.10}$$

Then this ratio is defined as the change in entropy and is represented by dS. Hence, equation (6.10) can be written as

$$\frac{dQ}{T} = dS \tag{6.11}$$

Therefore, the change in entropy in going from one state to another is given by

$$\int_{S_{L_1}}^{S_{L_2}} dS = \int_{L_1}^{L_2} \frac{dQ}{T}$$

or
$$S_{L_2} - S_{L_1} = \int_{L_1}^{L_2} \frac{dQ}{T} \tag{6.12}$$

This function S is called entropy of a system and it depends on the thermodynamical variables like pressure, volume, temperature or internal energy. Entropy of a system is an extensive property, as it depends on the mass of the working fluid.

In an adiabatic process, heat cannot enter or be rejected by the system, i.e., $dQ = 0$.

So, $$dS = \frac{dQ}{T} = 0 \tag{6.13}$$

Hence, entropy remains constant in an adiabatic process or in, adiabatic process, change in entropy will be zero. Thus, the adiabatic curve is also known as isoentropic curve.

6.8 ENTROPY AND THE SECOND LAW OF THERMODYNAMICS

In terms of entropy of the system, the second law of thermodynamics may be stated as follows: In nature, all chemical and physical processes occur in such a type that the total sum of entropy of all physical bodies contributing in the process increases. In the special case of a real reversible process, the total sum of entropy remains unchanged or constant.

If there are two states, 1 and 2, which are very close to each other, then the change in entropy in going from state 1 to 2 is given as

$$S_2 - S_1 = dS = \frac{dQ}{T}$$

This above expression of entropy is the mathematical form of the second law of thermodynamics.

6.9 CLAUSIUS THEOREM AND CLAUSIUS INEQUALITY

Let us consider a reversible cycle *LMNO* as shown in Figure 6.9. This closed loop or cycle can be divided into large number of reversible isothermal and adiabatic curves. These cycles are denoted by *a-b-c-d* and *e-f-g-h*, etc.

For the cycle *a-b-c-d*,

The amount of heat absorbed reversibly at temperature T_1 is dQ_1.

The amount of heat rejected reversibly at temperature T_2 is dQ_2.

As heat transferred in a reversible path is equal to the heat transferred in isothermal process, we can write $\frac{dQ_1}{T_1} = \frac{dQ_2}{T_2}$

We can consider the amount of heat supplied as positive and the amount of heat rejected as negative, we get

$$\frac{dQ_1}{T_1} + \frac{dQ_2}{T_2} = 0$$

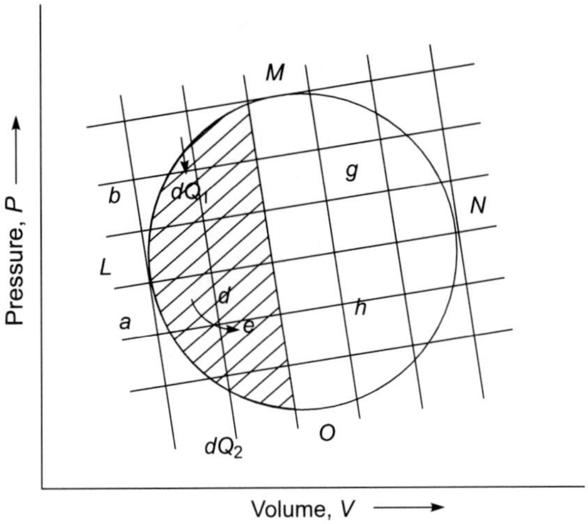

Figure 6.9 A reversible cycle divided into a small number of cycles

For cycle *e-f-g-h*,

$$\frac{dQ_3}{T_3}+\frac{dQ_4}{T_4}=0$$

Hence, for a compete cycle *LMNO*, we get

$$\frac{dQ_1}{T_1}+\frac{dQ_2}{T_2}+\frac{dQ_3}{T_3}+\frac{dQ_4}{T_4}=0$$

or
$$\int_{\text{Re}v}\frac{dQ}{T}=0 \tag{6.14}$$

Hence, for a reversible process, the cyclic integral of the quantity $\frac{dQ}{T}$ is equal to zero. This statement is known as Clausius theorem. This theorem is valid only for a reversible process, for which efficiency is defined by $\eta=1-\frac{Q_2}{Q_1}=1-\frac{T_2}{T_1}$.

Carnot's theorem explains that the efficiency of irreversible heat engine is less than the efficiency of a reversible engine, working between the same range of temperature, i.e., $\eta_{i\,\text{rev}}<\eta_{\text{rev}}$. Therefore,

$$\left(1-\frac{Q_2}{Q_1}\right)<\left(1-\frac{T_2}{T_1}\right)$$

or
$$\left(1+\frac{T_2}{T_1}\right)<\left(1+\frac{Q_2}{Q_1}\right)$$

or
$$\frac{T_2}{T_1} < \frac{Q_2}{Q_1}$$

or
$$\frac{Q_1}{T_1} < \frac{Q_2}{T_2}$$

or
$$\frac{Q_1}{T_1} - \frac{Q_2}{T_2} < 0 \tag{6.15}$$

Equation (6.15) can be written in differential form as

$$\oint \frac{dQ}{T} < 0$$

Equations (6.14) and (6.15) can be combined to write in general form as

$$\oint \frac{dQ}{T} \leq 0 \tag{6.16}$$

Equation (6.16) represents the Clausisus inequality. This give the bases of the reversibility of a cyclic process. Hence, Clausisus theorem of inequality can be written as

If $\oint \frac{dQ}{T} = 0$, the cyclic process is reversible

$\oint \frac{dQ}{T} < 0$, the cyclic process is irreversible or natural

$\oint \frac{dQ}{T} > 0$, this is impossible as it offends the second law of thermodynamics

6.10 THE TEMPERATURE- ENTROPY DIAGRAM

Thermodynamic variation in the state of a matter can be represented by plotting temperature along the vertical axis and entropy along the horizontal axis. This entropy vs temperature diagram is known as *T-S* diagram. The actual amount of heat flow in an infinitesimal reversible process is given by

$$dQ_{\text{Rev}} = TdS$$

The heat transformed in a finite process *p-q* is given by

$$Q_{\text{Rev}} = \int_p^q TdS$$

In a *T-S* diagram, this integral is represented by the area under the path *p-q* as shown in Figure 6.10. If we consider a thin vertical strip in *TS* diagram, the area of the strip is given by

$$TdS = dQ_{Rev}$$

Hence, area in a *T-S* diagram shows the amount of heat similar to work represented by the area in *P-V* diagram. The amount of heat Q_{Rev} is positive in the process *p-q* of the diagram 6.10 and it will be negative for the reverse process, i.e., for *q-p*.

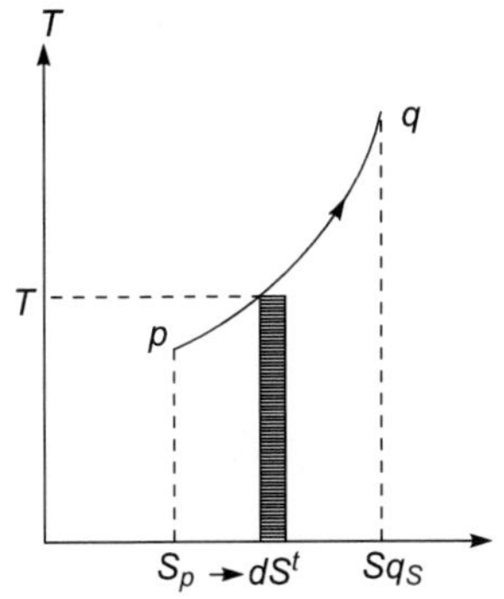

Figure 6.10 The temperature entropy diagram

The *T-S* diagram of a Carnot's cycle is shown in Figure 6.11. Vertical line represents a reversible adiabatic process and horizontal line represents a reversible isothermal process in a *T-S* diagram. The *T-S* plot of Carnot's cycle is a rectangle. Let S_1 and S_2 are the entropies of the working fluid in states *A* and *B*, respectively. Q_1 is the amount of heat absorbed during isothermal expansion process *AB* and Q_2 is the amount of heat rejected in isothermal compression process *CD*. T_1 and T_2 are the constant temperatures at isothermal expansion and isothermal compression, respectively.

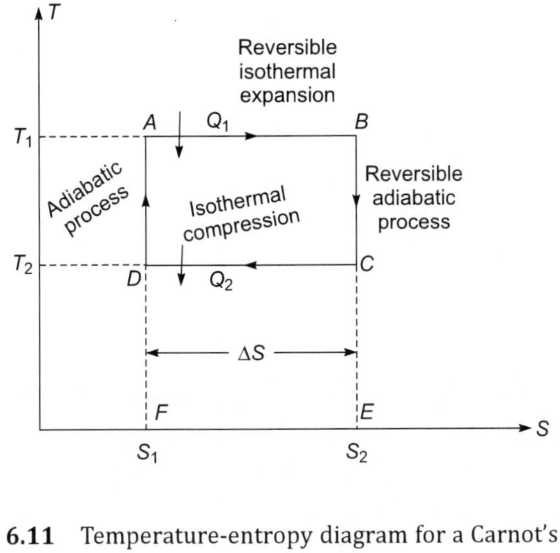

Figure 6.11 Temperature-entropy diagram for a Carnot's engine

Heat absorbed in the isothermal expansion AB is given by

$$Q_1 = T_1 \Delta S = T_1(S_2 - S_1)$$

Heat rejected in the isothermal compression CD is given by

$$Q_2 = T_2 \Delta S = T_2(S_2 - S_1)$$

Heat absorbed in the isothermal expansion AB is given by

$$Q_1 = T_1 \Delta S = T_1(S_2 - S_1)$$

Net work done in the complete process is given by

$$W = (Q_1 - Q_2) = (T_1 - T_2)\Delta S \tag{6.17}$$

Hence, efficiency of the engine is given by

$$\eta = \frac{W}{Q_1} = \frac{Q_1 - Q_2}{Q_1} = \frac{(T_1 - T_2)\Delta S}{T_1 \Delta S}$$

or

$$\eta = 1 - \frac{T_2}{T_1}$$

In equation (6.17), quantity $(Q_1 - Q_2)$ represents the external work done within the cycle and $(T_1 - T_2)\Delta S$ represents the area of the rectangle on T-S plot. Hence, we can say that the area of rectangle in T-S plot is equivalent to the external work in a reversible Carnot's cycle.

6.11 ENTROPY CHANGE IN REVERSIBLE CYCLE

Let us consider a system that executes a complete reversible process, p-q, as in figure (6.10). The entropy change between two state points p and q can be written as

$$S_b - S_a = \int_a^b \frac{dQ}{T} = \left(\frac{Q}{T}\right)_T$$

or

$$dS = \frac{dQ}{T}$$

or

$$dQ = TdS$$

In the T-S diagram of Carnot's cycle (Figure 6.11), A-B represents the amount of heat absorbed in the cycle and C-D represents the amount of heat rejected from the cycle. Hence, area $ABEF$ shows the amount of heat supplied and area $CDFE$ represents the amount of heat rejected. The difference between these two is equivalent to $ABCD$, which gives complete cycle of work and it can also be written as

$$\oint dS = \left(\frac{Q_1}{T_1} - \frac{Q_2}{T_2}\right) = 0$$

Therefore, in a cycle of reversible adiabatic, entropy of the system remains constant or the net change in entropy will always be zero and these processes are known as isoentropic processes. Every cycle of a reversible adiabatic is isoentropic but all isoentropic processes are not reversible adiabatics.

6.12 ENTROPY CHANGE IN IRREVERSIBLE CYCLE

Let us consider a process *a-b* as shown in Figure 6.12. The system comes back to its original state *a*, by a reversible or irreversible cycle, by the path *C*.

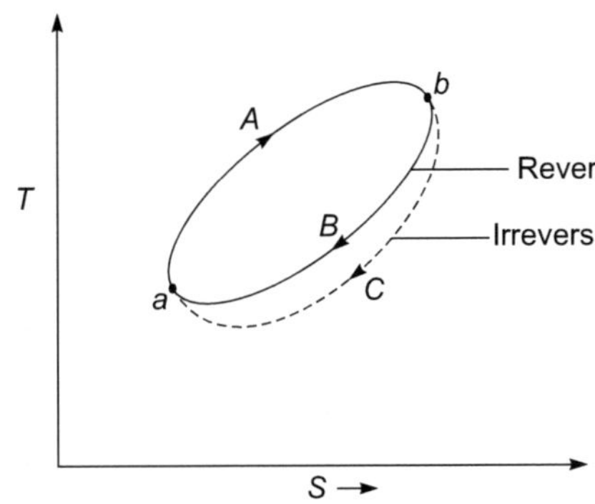

Figure 6.12 Entropy change during an irreversible cycle

In a reversible process *A-B*, we get

$$\oint \frac{dQ}{T} = \int_{aA}^{bA} \frac{dQ}{T} + \int_{bB}^{aB} \frac{dQ}{T} = 0 \tag{6.18}$$

In a reversible process, we may apply the Clausius inequality

$$\oint \frac{dQ}{T} = \int_{aA}^{bA} \frac{dQ}{T} + \int_{bC}^{1C} \frac{dQ}{T} < 0 \tag{6.19}$$

From equation (6.18), we get

$$\int_{aA}^{bA} \frac{dQ}{T} = -\int_{bB}^{aB} \frac{dQ}{T}$$

Substituting this value in equation (6.19) we get

$$-\int_{bB}^{aB} \frac{dQ}{T} + \int_{bC}^{1C} \frac{dQ}{T} < 0$$

or

$$-\int_{bB}^{aB} \frac{dQ}{T} > \int_{bC}^{1C} \frac{dQ}{T} \tag{6.20}$$

For a reversible cycle in path B, we may write

$$\int_{bB}^{aB} \frac{dQ}{T} = \int_{bB}^{aB} dS = \int_{bC}^{aC} dS \tag{6.21}$$

Substituting equation (6.21) in equation (6.20)

$$\int_{bC}^{aC} dS > \int_{bC}^{1C} \frac{dQ}{T}$$

So for any reversible cycle, we get

$$dS > \left(\frac{dQ}{T}\right)$$

and for a reversible cycle, we get

$$dS = \left(\frac{dQ}{T}\right)$$

or in general, we get

$$dS \geq \left(\frac{dQ}{T}\right) \tag{6.22}$$

6.13 THE PRINCIPLE OF INCREASE IN ENTROPY OF A SYSTEM

The entropy of an isolated system either increases or remains constant in all irreversible processes but it will never decrease in such processes. This statement is known as the principle of increase in entropy of a system. All the existing natural processes of the universe are irreversible processes. Hence, entropy of the universe always increases.

As we know that for an isolated system, $dQ = 0$

So, $(dS)_{\text{isolated}} \geq 0$ and

For a reversible isolated process, $(dS)_{\text{isolated}} = 0$

So, $S = \text{constant}$

Hence, in a reversible process, entropy of a system remains unchanged or constant.

For an irreversible process,

$$(dS)_{\text{isolated}} > 0$$

Let us consider a process in which a system at temperature T_1 absorbs the amount of heat dQ from the surroundings at temperature T_2 ($T_2 > T_1$). Consider dW as the work done during this process (Figure 6.13).

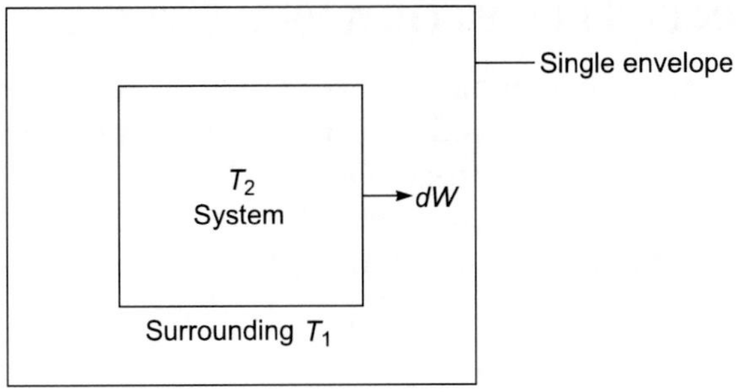

Figure 6.13 Work done in isolated system

For such an isolated system, we may write

$$(dS)_{\text{system}} \geq \frac{dQ}{T_1}$$

As we know that for the surroundings dQ is negative, so

$$(dS)_{\text{surroundings}} = -\frac{dQ}{T_2}$$

Hence, the net change in entropy of the system is given by

$$(dS)_{\text{system}} + (dS)_{\text{surroundings}} \geq \frac{dQ}{T_1} - \frac{dQ}{T_2}$$

or $$(dS)_{\text{system}} + (dS)_{\text{surroundings}} \geq dQ\left(\frac{1}{T_1} - \frac{1}{T_2}\right)$$

As we know that, $T_2 > T_1$, so $\left(\frac{1}{T_1} - \frac{1}{T_2}\right)$ will be positive

So we get,

$$(dS)_{\text{system}} + (dS)_{\text{surroundings}} \geq 0$$

But in the case of $T_1 > T_2$, heat is rejected by the system to the surroundings, so dQ and $\left(\frac{1}{T_1} - \frac{1}{T_2}\right)$ both become negative quantities and will give the same result.

Therefore, for all possible processes of a system in the given surroundings, we get

$$(dS)_{\text{system}} + (dS)_{\text{surrounding}} \geq 0$$

or $$(dS)_{\text{isolated}} \geq 0 \tag{6.23}$$

6.14 CHANGE IN ENTROPY FOR A PERFECT GAS

Let us consider a perfect gas contained in a container occupying the volume V at temperature T and pressure P. Consider that C_P and C_V are specific heat at constant pressure and constant volume, respectively and dT, dV and dP are changes in temperature, volume and pressure, respectively.

From the first law of thermodynamics, we get

$$dQ = dU + dW$$

But for a reversible process, $dW = PdV$, we get

$$(dQ) = dU + PdV \tag{6.24}$$

We know that $$TdS = (dQ)$$

So from equation (6.24), we get

$$TdS = dU + PdV \tag{6.25}$$

As we also know that $$H = U + PV$$

Differentiating the above equation

$$dH = dU + PdV + VdP$$

or $$dU = dH - PdV - VdP \tag{6.26}$$

Substituting this value in equation (6.25), we get

$$TdS = dH - PdV - VdP + PdV$$

$$TdS = dH - VdP \tag{6.27}$$

With the help of equations (6.25) and (6.27), we can determine the change in entropy for different processes. We can calculate the change in entropy ($\Delta S = S_2 - S_1$) of a gas, when it moves from state of T_1, P_1, and V_1 to state of T_2, P_2, and V_2.

(i) Isochoric Process (Constant Volume)

For real gases $$dQ = dU = C_V \cdot dT = TdS$$

So $$dS = \frac{C_V \cdot dT}{T}$$

Integrating the above equation, we get

$$S_2 - S_1 = C_V \log_e\left(\frac{T_2}{T_1}\right) = C_V \log_e\left(\frac{P_2}{P_1}\right) \tag{6.28}$$

(ii) Isobaric Process (Constant Pressure)

From equation (6.27), we can write

$$dQ = TdS = dH \qquad \text{(as } dP = 0\text{)}$$

or $$TdS = C_P dT$$

or $$dS = \frac{C_P \cdot dT}{T}$$

Integrating the above equation, we obtained

$$S_2 - S_1 = C_P \log_e\left(\frac{T_2}{T_1}\right) = C_V \log_e\left(\frac{V_2}{V_1}\right) \tag{6.29}$$

(iii) Isothermal Process (Constant Temperature)

In case of a perfect gas, we can write

$$dQ = dU + PdV$$

or $dQ = PdV$ (as $dU = 0$ for constant temperature)

or $= \dfrac{RT}{V}dV$ (using perfect gas equation, $PV = RT$)

or $TdS = \dfrac{RT}{V}dV$ (using relation $dQ = TdS$)

or $$dS = \frac{R}{V}dV$$

Integrating the above expression, we get

$$S_2 - S_1 = R\log_e\left(\frac{V_2}{V_1}\right) = R\log_e\left(\frac{P_2}{P_1}\right) \tag{6.30}$$

(iv) Polytropic Process for a Real Gas

In case of a polytropic reversible process, we know that

$dQ = dU + PdV$ ($dW = PdV$)

or $TdS = C_V dT + PdV$ ($dU = C_V dT$, and $dQ = TdS$)

$$dS = \frac{C_V dT}{T} + \frac{PdV}{T}$$

$= \dfrac{C_V dT}{T} + \dfrac{RdV}{V}$ (using $PV = RT$)

Integrating the above equation, we get

$$S_2 - S_1 = C_V \log_e\left(\frac{T_2}{T_1}\right) + R\log_e\left(\frac{V_2}{V_1}\right) \tag{6.31}$$

Equation (6.31) represents the change in entropy in terms of volume and temperature.

For a real gas, we can write

$$P_1V_1 = RT_1 \text{ and } P_2V_2 = RT_2$$

or

$$\frac{T_2}{T_1} = \frac{P_2V_2}{P_1V_1}$$

Using the above, equation (6.31) can also be written as

$$S_2 - S_1 = C_V \log_e\left(\frac{P_2V_2}{P_1V_1}\right) + R\log_e\left(\frac{V_2}{V_1}\right)$$

$$= C_V \log_e\left(\frac{V_2}{V_1}\right) + C_V \log_e\left(\frac{P_2}{P_1}\right) + R\log_e\left(\frac{V_2}{V_1}\right)$$

$$= (C_V + R)\log_e\left(\frac{V_2}{V_1}\right) + C_V \log_e\left(\frac{P_2}{P_1}\right)$$

Using the relation $C_P - C_V = R$, we get

$$S_2 - S_1 = C_P \log_e\left(\frac{V_2}{V_1}\right) + C_V \log_e\left(\frac{P_2}{P_1}\right) \tag{6.32}$$

Equation (6.32) represents the change in entropy in terms of volume and pressure.

From the ideal gas equation, we can write,

$$\frac{V_2}{V_1} = \frac{P_1T_2}{P_2T_1}$$

Substituting this value in equation (6.31), we get

$$S_2 - S_1 = C_V \log_e\left(\frac{T_2}{T_1}\right) + R\log_e\left(\frac{P_1T_2}{P_2T_1}\right)$$

$$= C_V \log_e\left(\frac{T_2}{T_1}\right) + R\log_e\left(\frac{P_1}{P_2}\right) + R\log_e\left(\frac{T_2}{T_1}\right)$$

$$= (C_V + R)\log_e\left(\frac{T_2}{T_1}\right) + R\log_e\left(\frac{P_1}{P_2}\right)$$

Using the relation $C_P - C_V = R$, we get

$$S_2 - S_1 = C_P \log_e \left(\frac{T_2}{T_1}\right) + R \log_e \left(\frac{P_1}{P_2}\right) \tag{6.33}$$

Equation (6.33) gives the change in entropy in terms of pressure and temperature.

6.15 THIRD LAW OF THERMODYNAMICS (NERNST'S HEAT LAW)

The third law of thermodynamics was proposed by Nernst in 1906. It states that "the entropy of a thermodynamic substance approaches to zero at absolute zero temperature in the state of equilibrium."

With the help of the third law of thermodynamics, the nature of the system near absolute zero temperature can be explained. The third law of thermodynamics implies that the entropy of a system ceases. In case of many substances, e.g., chemical compounds, amorphous solids, alloys, etc., the entropy is finite and it does not approach to zero at absolute zero temperature.

With the help of this law, the entropy of different substances can be measured. At very low temperatures, the nature of solids can be explained with this law.

6.16 PHYSICAL SIGNIFICANCE OF ENTROPY

The idea of entropy is very difficult, it cannot be demonstrated like volume, temperature and pressure. The entropy is a real physical quantity and has extreme significance in thermodynamics and acquires the following properties:

(a) Entropy defines the state of a working fluid and it is a definite single-valued function.

(b) In every irreversible or natural process, entropy of a system always increases.

(c) Entropy remains constant in adiabatic process.

(d) The second law of thermodynamics can be stated in terms of entropy.

(e) The entropy is a measure of disorderness and this disorderness can be defined in terms of the relation

$$S = k \log W$$

Numerical Examples

1. A reversible heat engine working at the rate of 104.9 kW, has efficiency of 21%. Calculate the amount of heat gain and lost from the working substance.

Solution: Given that: P = 104.9 kW and η = 21%

We know that efficiency of a heat engine is given by

$$\eta = \left(\frac{W}{Q_1}\right)$$

or $$Q_1 = \left(\frac{W}{\eta}\right)$$

$$Q_1 = \frac{104.9 \times 10^3}{0.21} = 499.52 \text{ kJ/sec}$$

Net work in the process, $W = Q_1 - Q_2$

$$104.9 \times 10^3 = 499.52 \times 10^3 - Q_2$$

$$Q_2 = 104.9 \times 10^3 + 499.52 \times 10^3 = 394.62 \text{ kJ/sec}$$

2. 219.9 kJ heat from a sink is absorbed by a reversed engine and gets a mechanical input of 89.9 kJ. Calculate the amount of heat given to the source. Also find out the value of the coefficient of performance for the reversed engine in the case of heat pump and refrigerator.

Solution: We know that the net work in a reversible engine is given by

$$W = Q_1 - Q_2$$

The amount of heat given to source $Q_1 = W + Q_2$

$$Q_1 = 89.9 + 219.9 = 309.8 \text{ kJ}$$

We know that coefficient of performance $\beta = \frac{\text{Output heat}}{\text{Input work}} = \frac{Q_2}{W}$

So for the case of heat engine $\beta_H = \frac{Q_1}{W} = \frac{309.8}{89.9} = 3.446$

and for refrigerator $\beta_R = \frac{Q_2}{W} = \frac{219.9}{89.9} = 2.446$

3. A Carnot's reversible heat engine converts 1/6th of input heat into work. If the sink temperature is reduced by 69.9°K, the efficiency of engine gets two times of its original value. Calculate the temperature of the higher temperature source and the lower temperature sink.

Solution: If T_1 is the temperature of the source and T_2 is the temperature of the sink, the efficiency of Carnot's heat engine is given by

$$\eta = 1 - \left(\frac{T_2}{T_1}\right) = \frac{(T_1 - T_2)}{T_1}$$

Given that $$\frac{(T_1 - T_2)}{T_1} = \frac{1}{6}$$

So, $$(6T_1 - 6T_2) = T_1$$

$$5 \times T_1 = T_2$$

$$T_1 = 0.5T_2$$

Also given that

$$\frac{T_1 - (T_2 - 69.9)}{T_1} = 2\left(\frac{1}{6}\right)$$

$$6T_1 - 6T_2 + 419.4 = 2T_1$$

$$4T_1 - 6T_2 + 419.5 = 0$$

Substituting the value of T_1, we get

$$2T_2 - 6T_2 + 419.5 = 0$$

$$4T_2 = 419.5$$

$$T_2 = 104.85°\text{C}$$

So, $$T_1 = 6 \times 104.85 = 629.1°\text{C}$$

4. The development of a reversible heat engine is claimed by an inverter with the given data:

   (i) power = 74 kW
   (ii) consumption of fuel = 5 kg/hr
   (iii) heating value = 72000 kJ/kg
   (iv) limits of temperature = 900 K to 390 K

   Show that the claim by the inverter is true.

   Solution: Work done for the engine

$$W = 74 \times 3600$$

$$= 266400 \text{ kJ/hr}$$

Amount of heat supplied = fuel mass × heating value of fuel

$$Q = 5 \times 72000 \text{ kJ/ hr}$$

$$= 2360{,}000 \text{ kJ/hr}$$

Hence, the efficiency of the engine claimed by the inverter

$$\eta_1 = \frac{\text{amount of work done}}{\text{supplied heat}}$$

$$= \frac{266400}{360,000} = 0.74 = 74\%$$

Efficiency of the reversible engine is given by

$$\eta = 1 - \left(\frac{T_2}{T_1}\right) = \frac{(T_1 - T_2)}{T_1}$$

$$= \frac{(900 - 390)}{900} = 0.5666 = 56.66\%$$

Since the efficiency of the claimed engine (74%) is greater than the efficiency of the reversible engine (56.6%), so the claim is not valid.

5. A reversible heat engine takes heat from the source at temperature T_1 and rejects the heat to the sink at temperature 770°K. The sink of this reversible engine works as a source of second reversible engine. Second reversible engine rejects heat to another low temperature reservoir at 250°K. Calculate the temperature T_1 (i) when the efficiency of two engines are same, and (ii) when two engines give the same work.

Solution: For engine one, efficiency

$$\eta_1 = 1 - \left(\frac{T_2}{T_1}\right) = \frac{(T_1 - T_2)}{T_1}$$

For engine two, efficiency

$$\eta_2 = 1 - \left(\frac{T_3}{T_2}\right) = \frac{(T_2 - T_3)}{T_2}$$

(i) In this case, $\eta_1 = \eta_2$

So, $$1 - \left(\frac{T_3}{T_2}\right) = 1 - \left(\frac{T_2}{T_1}\right)$$

$$T_2 = \sqrt{T_1 \times T_3}$$

$$T_1 = \frac{(T_2)^2}{T_3} = \frac{770 \times 770}{250} = 2371.6°\text{K}$$

$$= 2371.6°\text{K}$$

(ii) Here, work done is equal, i.e., $W_1 = W_2$

For the engine one,

$$W_1 = Q_1 - Q_2$$

For the second reversible engine,

$$W_2 = Q_2 - Q_3$$

So,
$$Q_1 - Q_2 = Q_2 - Q_3$$

or
$$\frac{Q_1 - Q_2}{Q_2} = \frac{Q_2 - Q_3}{Q_2}$$

or
$$\frac{Q_1}{Q_2} - 1 = 1 - \frac{Q_3}{Q_2}$$

$$\left(\frac{Q_1}{Q_2} + \frac{Q_3}{Q_2}\right) = 2$$

As both the engines are reversible,

So,
$$\left(\frac{Q_1}{Q_2}\right) = \left(\frac{T_1}{T_2}\right) \text{ and } \left(\frac{Q_3}{Q_2}\right) = \left(\frac{T_3}{T_2}\right)$$

Hence,
$$\left(\frac{T_1}{T_2} + \frac{T_3}{T_2}\right) = 2$$

$$(T_1 + T_3) = 2T_2$$

$$T_1 = 2 \times (T_2) - T_3$$

$$= 2 \times 770 - 250$$

$$= 1290°\text{K}$$

6. A heat reservoir at temperature 99.9°C and a pot of 1 kg water at temperature 0.5°C are placed in contact with each other. (i) Determine the change in the entropy of water in reaching the temperature up to 100°C, and (ii) also determine the change in entropy of the universe and the reservoir.

Solution:

(i) We know that the increase in entropy when temperature of m gm of water changes from T_1 to T_2 is given by

$$dS = \int_{T_1}^{T_2} \frac{dQ}{T}$$

$$dS = dQ \log_e \left(\frac{T_2}{T_1}\right)$$

$$= ms \times 2.3026 \times \log_{10}\left(\frac{T_2}{T_1}\right)$$

So change in entropy of 1000 gm of water in changing temperature from 0°C to 100°C will be

$$dS = 1000 \times 1 \times 2.3026 \times \log_{10}\left(\frac{372.9}{273.5}\right)$$

$$= 311.8 \text{ cal/K}$$

(ii) (a) Increase in entropy of the reservoir

$$dS = -\frac{dQ}{T}$$

$$= -\frac{1000 \times 1 \times 99.4}{372.5}$$

$$= -266.84 \text{ cal/K}$$

(b) Entropy change in case of universe

$$dS = (311 - 266.84) \text{ cal/K} = -44.15 \text{ cal/K}$$

Therefore, the increase in entropy of the universe = 44.15 cal/K

7. The temperature of 0.9 g of nitrogen increases from 323°K to 373°K at constant volume. Determine the change in entropy during the process. The specific heat of nitrogen is $C_V = 6.4 \times 10^{-3}$.

Solution: We know that the change in entropy is given by

$$dS = \int_{T_1}^{T_2} \frac{dQ}{T}$$

$$= (0.9 \times 0.00064) \times \int_{323}^{373} \frac{dT}{T}$$

$$= (0.9 \times 0.00064) \times 2.3026 \times \log_{10}\left(\frac{373}{323}\right)$$

$$= (0.9 \times 0.00064) \times 2.3026 \times 0.06245$$

$$= 8.319 \times 10^{-4} \text{ cal/K}$$

8. 10 kg of water at 363°K is changed into steam at the same temperature. Determine the rise in entropy during the process of phase change.

Solution: Heat absorbed by 10 kg of water in the process of phase change is given by

$$dQ = mL$$

$$= 10000 \times 540$$

$$= 5400000 \text{ cal}$$

The increase in entropy, $dS = \dfrac{dQ}{T}$

$$= \frac{5400000}{363}$$

$$= 14876.03 \text{ cal/K}$$

9. 1 gm of gas expands very slowly to 2 times of its volume. Determine the entropy change in terms of *R*.

Solution: We know that work done is given by

$$W = \int_{V_1}^{V_1} PdV$$

$$= \int_{V_1}^{V_1} \frac{RT}{V} dV$$

$$= RT \log_e \frac{V_2}{V_1} = RT \times 2.3026 \times \log_{10} \frac{V_2}{V_1}$$

$$= RT \times 2.3026 \times \log_{10} 2$$

Hence, the rise in entropy is given by

$$dS = \frac{dQ}{T}$$

$$= \frac{1}{T}\left(\frac{W}{J}\right)$$

$$= \frac{1}{T}\left(\frac{RT \times 2.3026 \times \log_{10} 2}{J}\right)$$

$$= \left(\frac{R \times 2.3026 \times 0.3010}{J}\right)$$

$$= 0.69 \frac{R}{J} \text{ cal/K}$$

10. 1 kg of ice is converted into steam. Determine the increase in entropy during the process. Given that, latent heat of ice is 80×10^3 cal/kg, specific heat of water is 1×10^3 cal kg°C and latent heat of steam is 540×10^3 cal/kg.

Solution: Change in entropy during the conversion of 0°C of water into 0°C of ice

$$dS = \frac{dQ}{T}$$

$$= \frac{80}{273} = 0.0292 \text{ cal/K}$$

Change in entropy during the change in m kg of water from 0°C to 100°C

$$dS = \int_{T_1}^{T_2} \frac{dQ}{T}$$

$$= ms \times 2.302 \times \log_{10}\left(\frac{373}{273}\right)$$

For 1 kg of water,

$$dS = 1 \times 1 \times 2.302 \times \log_{10}\left(\frac{373}{273}\right)$$

$$= 0.314 \text{ kcal/K}$$

Change in entropy during the process of conversion of 1 kg of water at 100°C into steam at 100°C

$$dS = \frac{dQ}{T}$$

$$= \frac{540}{373} = 1.446 \text{ cal/K}$$

Hence, the net increase in entropy during the complete process

$$dS = 0.0292 + 0.314 + 1.446 = 1.7892 \text{ cal/K}$$

11. Efficiency of an ideal heat engine changes from 0.16 to 0.33 while increasing the source temperature by 99.9°K. Determine the temperature of the sink.

Solution: The efficiency of an ideal heat engine is given by

$$\eta_1 = 1 - \left(\frac{T_2}{T_1}\right) = 0.16$$

$$\left(\frac{T_2}{T_1}\right) = 1 - 0.16 = 0.84$$

$$T_1 = \frac{T_2}{0.84}$$

and

$$\eta_2 = 1 - \left(\frac{T_2}{T_1 + 99.9}\right) = 0.33$$

$$\left(\frac{T_2}{T_1 + 99.9}\right) = 0.67$$

$$T_2 = 0.67\ (T_1 + 99.9)$$

$$= 0.67\ T_1 + (99.9 \times 0.67)$$

$$= 0.67\ \frac{T_2}{0.84} + (99.9 \times 0.67)$$

$$= 332.95°K$$

Exercise

Short Answer Questions

1. What do you mean by the efficiency of heat engine.
2. Define the reversible and irreversible processes.
3. State the second law of thermodynamics.
4. Prove that Carnot's engine is a more sufficient engine.
5. Write the Kelvin-Planck statement of the second law of thermodynamics.
6. Show that Kelvin and Clausius statements of the second law of thermodynamics are equivalent to each other.
7. State the second law of thermodynamics in two different ways.
8. What do you mean by entropy? Explain its physical significance.
9. Show that for a complete reversible cycle

$$\oint dS = 0$$

10. Prove the principle of increase in entropy.
11. What do you mean by *T-S* diagram? Write its importance.
12. Write the statement of the third law of thermodynamics and give some examples of it.
13. What is the lowest temperature at absolute scale and why?
14. What will be the change in entropy in the free expansion of gas?
15. Define the thermodynamic scale of temperature.
16. Prove that entropy of an irreversible process always increases.
17. Represent the Carnot's reversible cycle on *T-S* diagram and show that the area of *T-S* diagram gives the available energy.
18. Prove that entropy of a reversible adiabatic process remains constant.

19. Prove that the change in entropy $\int_A^B \frac{dQ}{T}$ does not depend on the path of a reversible cycle.
20. What do you mean by the working substance, the source and the sink in a reversible process.

Long Answer Questions

1. Explain Carnot's reversible cycle and derive an expression for the efficiency of heat engine working between temperatures T_1 and T_2.
2. Explain the principle of working of refrigerator. What is the coefficient of performance?
3. Write the statement of Carnot's theorem and show that the efficiency of an ideal engine is given by

$$\eta = 1 - \left(\frac{T_2}{T_1}\right) = 1 - \left(\frac{Q_2}{Q_1}\right)$$

4. Show that the change in entropy of polytropic process for a real gas is given by

$$S_2 - S_1 = C_P \log_e \left(\frac{T_2}{T_1}\right) + R \log_e \left(\frac{P_1}{P_2}\right)$$

5. What do you mean by the thermodynamical scale of temperature? Prove that the ideal gas scale and the thermodynamical scale of temperature are identical and explain the reason.
6. What do you mean by entropy? Explain the *T-S* diagram of entropy.
7. Explain the second law of thermodynamics on the basis of entropy?
8. Explain the physical significance of entropy. Prove that the entropy of the universe increases.
9. Determine the change in entropy of isothermal, isobaric, isochoric and adiabatic processes.
10. Describe the reversible and irreversible processes. Deduce an expression for the Carnot's reversible heat engine and prove that for the same set of two temperatures, the efficiency of all reversible heat engines will be maximum and same.
11. Explain the Carnot's theorem. Obtain the work done in a reversible cycle.
12. What do you mean by the efficiency of an engine? Obtain an expression for the efficiency of heat engine and give the ways to increase efficiency.
13. Define entropy as a property of system on the basis of Classius inequality.
14. Define absolute scale of temperature. Compare the Kelvin ideal gas scale with the thermodynamical scale of temperature.

15. Two identical systems at the temperature T_1 and $T2$ are working as reservoirs for a heat engine. The heat capacity at constant pressure is C_P. If there is no phase change in the process and the pressure of system remains constant, Prove that:
   (i) The resultant temperature of the system is given by

$$T = \sqrt{(T_1 \cdot T_2)}$$

   (ii) The net amount of work in the process is given by

$$C_P \times \left(\sqrt{T_1} - \sqrt{T_2}\right)^2$$

16. Draw and explain the *T-S* diagram and obtain the expression of the efficiency of a reversible engine.
17. Define the entropy of a substance. Show that the entropy change of a reversible process is zero but not same for an irreversible process.
18. Deduce the expression of change in entropy for a process in which pressure, temperature and volume changes.
19. Describe the principle of increase in entropy and also give some examples of it.
20. Determine the efficiency of a reversible engine working between the two temperatures of ice and steam.
21. Give a suitable method to increase the efficiency of a Carnot's ideal heat engine.
22. Write the statement of Carnot's theorem and prove it. Discuss why the temperature below zero Kelvin is impossible.

Numerical Problems

1. A heat engine working at the rate of 105 kW has an efficiency of 23%. Calculate the amount of heat gain and lost from the working substance.
2. A Carnot's reversible heat engine converts 1/5th of input heat into work. If the sink temperature is reduced by 70°K, the efficiency of the engine gets two times of its original value. Calculate the temperature of higher temperature source and lower temperature sink.
3. 230 kJ heat from sink is absorbed by a reversed engine and gets a mechanical input of 70 kJ. Calculate the amount of heat given to the source. Also find out the value of coefficient of performance for the reversed engine in the case of heat pump and refrigerator.
4. A reversible heat engine takes heat from the source at temperature T_1 and rejects heat to the sink at temperature 689°K. The sink of this reversible engine works as a source of the second reversible engine. The second reversible engine rejects heat to another low temperature reservoir at 240°K. Calculate the temperature T_1 (i) when the efficiency of two engines are same, and (ii) when two engines give the same work.

5. The efficiency of a heat engine is 43.97%. If the temperature of the source is 250°C, determine the temperature of sink. **[Ans.** 20°C]
6. Efficiency of an ideal heat engine changes from 1/4th to 1/7th the in increasing the source temperature by 100°K. Determine the temperature of the sink.
7. 2 kg of ice is converted into steam. Determine the increase in entropy during the process. Given that, latent heat of ice is 80×10^3 cal/kg, specific heat of water is 1×10^3 cal kg°C and latent heat of steam is 540×10^3 cal/kg.
8. Tyre of a car has a pressure of 2.98 atm at temperature 300°K. Determine the resultant temperature due to burst in tyre. **[Ans.** 217.98°K]
9. The temperature of 2 g of nitrogen increases from 343°K to 393°K at constant volume. Determine the change in entropy during the process. The specific heat of nitrogen is $C_V = 6.4 \times 10^{-3}$.
10. A reversible heat engine takes heat from the source at temperature T_1 and rejects heat to the sink at temperature 670 K. The sink of this reversible engine works as a source of the second reversible engine. The second reversible engine rejects heat to another low temperature reservoir at 200°K. Calculate the temperature T_1 (i) when the efficiency of two engines are same, and (ii) when two engines give the same work.
11. A reversible Carnot's engine is working between two temperatures with a difference is 100°K. The amount of heat absorbed by the source is 745.9 J and the quantity of heat rejected to sink is 545.98 J. Determine the temperature of the sink and the source. **[Ans.** 273°K and 373°K]
12. The efficiency of a Carnot's heat engine is 20%. Heat engine takes 499 cal heat from source at temperature T and rejects 399.9 cal heat to the sink at 47°C. Determine the temperature of the sink. **[Ans.** T = 127°C]

Multiple-Choice Questions

1. A Carnot's engine is working between 400°K and 300°K. The efficiency of the engine is
   (i) 40% (ii) 25%
   (iii) 60% (iv) 70%
2. In a Carnot's heat engine, the first process is
   (i) Adiabatic compression (ii) Isothermal compression
   (iii) Isothermal expansion (iv) Adiabatic expansion
3. The temperature at which a reversible heat engine can be 100% efficient is
   (i) 14°C (ii) 0°K
   (iii) 0°C (iv) greater than that of sink
4. The theory of the working of a refrigerator is similar to the theory of
   (i) Vapor compression (ii) Oil engine
   (iii) Ice formation (iv) Heat engine

5. A Carnot's engine is working between ice point and steam point. Its efficiency will be

(i) 2 (ii) 26.80%

(iii) 19.50% (iv) 0

6. Which of the following is a correct expression?

(i) $TdS = dH - VdP$ (ii) $TdS = dH + VdP$

(iii) $TdS = -dH + VdP$ (iv) $TdS = -dH - VdP$

7. The equation $dQ = dE + dW$ is applicable for

(i) Only irreversible process

(ii) Reversible or irreversible process

(iii) Only reversible process

(iv) None of above

8. Entropy remains unchanged in

(i) Cyclic process (ii) Isochoric process

(iii) Isobaric process (iv) Isothermal process

9. A machine which converts heat energy into mechanical work is a

(i) Generator (ii) Heat engine

(iii) Refrigerator (iv) Motor

10. The efficiency of Carnot's heat engine is 2/12, when the source is at temperature 390°K. In order to make it 4/12th efficient, the temperature of sink is reduced by

(i) 65°C (ii) 75°C

(iii) 68°C (iv) 89°C

11. Second law of thermodynamics makes the basis of

(i) Temperature (ii) Volume

(iii) Entropy (iv) Pressure

12. Entropy of a system is a measure of

(i) Available kinetic energy (ii) Disorder

(iii) Potential energy (iv) Temperature

13. A heat engine takes 800 kJ/s heat energy at 800°K and liberates 400 kJ at 600°K. The engine works on

(i) Irreversible cycle (ii) Closed cycle

(iii) Reversible cycle (iv) None of the above

14. Pressure remains constant in

(i) Isochoric process (ii) Adiabatic process

(iii) Isobaric process (iv) Isothermal process

15. The amount of total energy of the universe

(i) Is always decreasing (ii) Is always increasing

(iii) Remains constant (iv) Either decreases or increases

16. A system having entropy of two parts as S_1 and S_2 in equilibrium. The net entropy of the system is

(i) $S = S_1 + S_2$ (ii) $S = S_1 - S_2$

(iii) $S = 2(S_1 + S_2)$ (iv) $S = 1/(S_1\ S_2)$

17. The absolute zero entropy is not possible due to the reason

(i) Theoretically, the value of entropy at absolute zero temperature is negative and impossible

(ii) Absolute zero temperature cannot be achieved in finite number of methods

(iii) Both of above

(iv) None of the above

18. The relation between entropy and thermodynamic probability *(W)* is

(i) Thermodynamic probability decreases with entropy

(ii) Thermodynamic probability increases with entropy

(iii) Thermodynamic probability does not change with change in entropy

(iv) None of the above

19. Net entropy change in a irreversible process is

(i) Zero (ii) Positive

(iii) Infinite (iv) Negative

20. An isentropic process can be

(i) Reversible and adiabatic (ii) Irreversible and adiabatic

(iii) Reversible and isothermal (iv) None of above

21. The efficiency of Carnot's engine can be increased more effectively

(i) To increase higher pressure (ii) To decrease higher volume

(iii) To decrease lower temperature (iv) To decrease lower pressure

22. Carnot's cycle has maximum efficiency for

(i) Irreversible engine (ii) Reversible engine

(iii) Petrol engine (iv) Diesel engine

Answers

1. (ii)	2. (iii)	3. (ii)	4. (iv)	5. (ii)	6. (i)	7. (ii)	8. (i)
9. (iii)	10. (i)	11. (iii)	12. (ii)	13. (iii)	14. (iii)	15. (iii)	16. (i)
17. (ii)	18. (ii)	19. (ii)	20. (ii)	21. (iii)	22. (ii)		

7

Thermodynamic Relations and Potentials

7.1 THERMODYNAMIC VARIABLES AND THEIR TYPES

Thermodynamic properties like volume, temperature, pressure, entropy and internal energy can be used to define the thermodynamic state of a matter. When the system changes its state, these properties also change. These are known as thermodynamic variables or microscopic coordinates. There are two types of thermodynamics variables.

(i) Extensive Variables: Extensive variables are macroscopic parameters or variables which are used to describe thermodynamic system in equilibrium. These variables depend on the property of the material present in the system. Net value of extensive variable is equal to the sum of its value in different parts of the system. Entropy, internal energy, length, area, mass, volume, magnetization, heat capacity, etc., are extensive variables.

(ii) Intensive Variables: Intensive variables are also macroscopic specifications which are used to describe thermodynamic system in equilibrium and depend on the property of material. These variables have the same value in each part of the system. Temperature, pressure, refractive index, density, viscosity, magnetic induction, specific volume, e.m.f. force, specific tension, etc., are intensive variables.

An extensive variable may be converted into intensive variable by describing a unit amount of matter. For example, volume is an extensive variable but specific volume (V/M) is an intensive variable. Similarly, mass and heat capacity are extensive but density and specific heat are intensive variables.

7.2 MAXWELL'S THERMODYNAMICAL RELATIONS

The thermodynamic state of a system can be described in terms of any pair of four variables; volume, pressure, temperature and entropy of a system. Maxwell establishes six thermodynamic relations in terms of these variables.

We know that from the first law of thermodynamics,

$$dQ = dU + PdV \tag{7.1}$$

and from the second law of thermodynamics,

$$dQ = TdS \tag{7.2}$$

The above two relations can be combined to define a change in internal energy as,

$$dU = TdS - PdV \tag{7.3}$$

Let us consider that internal energy (U), entropy (S) and volume (V) are functions of the two variables q and r, which are independent and can be any two variables out of four V, S, P and T. Hence, we can write,

$$dU = \left(\frac{\partial U}{\partial q}\right)_r dq + \left(\frac{\partial U}{\partial r}\right)_q dr$$

$$dS = \left(\frac{\partial S}{\partial q}\right)_r dq + \left(\frac{\partial S}{\partial r}\right)_q dr$$

$$dV = \left(\frac{\partial V}{\partial q}\right)_r dq + \left(\frac{\partial V}{\partial r}\right)_q dr$$

Substituting these values in equation (7.3), we get

$$\left(\frac{\partial U}{\partial q}\right)_r dq + \left(\frac{\partial U}{\partial r}\right)_q dr = T\left[\left(\frac{\partial S}{\partial q}\right)_r dq + \left(\frac{\partial S}{\partial r}\right)_q dr\right] - P\left[\left(\frac{\partial V}{\partial q}\right)_r dq + \left(\frac{\partial V}{\partial r}\right)_q dr\right]$$

$$= \left[T\left(\frac{\partial S}{\partial q}\right)_r - P\left(\frac{\partial V}{\partial r}\right)_q\right] dq + \left[T\left(\frac{\partial S}{\partial r}\right)_q - P\left(\frac{\partial V}{\partial r}\right)_q\right] dr$$

We can compare the coefficients of dq and dr on both sides of the above equation, we get

$$\left(\frac{\partial U}{\partial q}\right)_r = T\left(\frac{\partial S}{\partial q}\right)_r - P\left(\frac{\partial V}{\partial q}\right)_r \tag{7.4}$$

$$\left(\frac{\partial U}{\partial r}\right)_q = T\left(\frac{\partial S}{\partial r}\right)_q - P\left(\frac{\partial V}{\partial r}\right)_q \tag{7.5}$$

Differentiating equation (7.4) with respect to r and equation (7.5) with respect to q will give

$$\frac{\partial^2 U}{\partial r \partial q} = \left(\frac{\partial T}{\partial r}\right)_q \left(\frac{\partial S}{\partial q}\right)_r + T\left(\frac{\partial^2 S}{\partial r \partial q}\right) - \left(\frac{\partial P}{\partial r}\right)_q \left(\frac{\partial V}{\partial q}\right)_r - P\left(\frac{\partial^2 V}{\partial r \partial q}\right)$$

and $$\frac{\partial^2 U}{\partial q\,\partial r} = \left(\frac{\partial T}{\partial q}\right)_r \left(\frac{\partial S}{\partial r}\right)_q + T\left(\frac{\partial^2 S}{\partial q \partial r}\right) - \left(\frac{\partial P}{\partial q}\right)_r \left(\frac{\partial V}{\partial r}\right)_q - P\left(\frac{\partial^2 V}{\partial q \partial r}\right)$$

As we know that dU is a perfect differential, so by the property of perfect differentiation, we can write

$$\frac{\partial^2 U}{\partial r \partial q} = \frac{\partial^2 U}{\partial q \partial r}$$

So,

$$\left(\frac{\partial T}{\partial r}\right)_q \left(\frac{\partial S}{\partial q}\right)_r + T\left(\frac{\partial^2 S}{\partial r \partial q}\right) - \left(\frac{\partial P}{\partial r}\right)_q \left(\frac{\partial V}{\partial q}\right)_r - P\left(\frac{\partial^2 V}{\partial r \partial q}\right)$$
$$= \left(\frac{\partial T}{\partial q}\right)_r \left(\frac{\partial S}{\partial r}\right)_q + T\left(\frac{\partial^2 S}{\partial q \partial r}\right) - \left(\frac{\partial P}{\partial q}\right)_r \left(\frac{\partial V}{\partial r}\right)_q - P\left(\frac{\partial^2 V}{\partial q \partial r}\right) \quad (7.6)$$

Since we know that dV and dS are perfect differentials, we get

$$\frac{\partial^2 V}{\partial r \partial q} = \frac{\partial^2 V}{\partial q \partial r} \text{ and } \frac{\partial^2 S}{\partial r \partial q} = \frac{\partial^2 S}{\partial q \partial r}$$

Substituting the above property in equation (7.6), we get

$$\left(\frac{\partial T}{\partial r}\right)_q \left(\frac{\partial S}{\partial q}\right)_r - \left(\frac{\partial P}{\partial r}\right)_q \left(\frac{\partial V}{\partial q}\right)_r = \left(\frac{\partial T}{\partial q}\right)_r \left(\frac{\partial S}{\partial r}\right)_q - \left(\frac{\partial P}{\partial q}\right)_r \left(\frac{\partial V}{\partial r}\right)_q \quad (7.7)$$

Equation (6.5) is Maxwell's thermodynamical relation in general form. Substitution of any two of the variables P, V, T, and S in place of r and q provides the Maxwell's relations. If we take one thermal variable (T or S) and one mechanical variable (V or P), there will be four Maxwell's relations.

Maxwell's First Relation

Putting $q = S$ and $r = V$ in equation (7.7), we get

$$\left(\frac{\partial T}{\partial V}\right)_S \left(\frac{\partial S}{\partial S}\right)_V - \left(\frac{\partial P}{\partial V}\right)_S \left(\frac{\partial V}{\partial S}\right)_V = \left(\frac{\partial T}{\partial S}\right)_V \left(\frac{\partial S}{\partial V}\right)_S - \left(\frac{\partial P}{\partial S}\right)_V \left(\frac{\partial V}{\partial V}\right)_S$$

$$\left(\frac{\partial T}{\partial V}\right)_S = -\left(\frac{\partial P}{\partial S}\right)_V \quad (7.8)$$

where, $\left[\left(\frac{\partial S}{\partial S}\right)_V = 1, \left(\frac{\partial V}{\partial V}\right)_S = 1 \text{ and } \left(\frac{\partial S}{\partial V}\right)_S = 0, \left(\frac{\partial V}{\partial S}\right)_V = 0\right]$

Equation (7.8) represents Maxwell's first relation.

Maxwell's Second Relation

Putting $q = T$ and $r = V$ in equation (7.7), we get

$$\left(\frac{\partial T}{\partial V}\right)_T\left(\frac{\partial S}{\partial T}\right)_V - \left(\frac{\partial P}{\partial V}\right)_T\left(\frac{\partial V}{\partial T}\right)_V = \left(\frac{\partial T}{\partial T}\right)_V\left(\frac{\partial S}{\partial V}\right)_T - \left(\frac{\partial P}{\partial T}\right)_V\left(\frac{\partial V}{\partial V}\right)_T$$

$$\left(\frac{\partial S}{\partial V}\right)_T = \left(\frac{\partial P}{\partial T}\right)_V \tag{7.9}$$

where, $\left[\left(\frac{\partial T}{\partial T}\right)_V = 1, \left(\frac{\partial V}{\partial V}\right)_T = 1 \text{ and} \left(\frac{\partial T}{\partial V}\right)_T = 0, \left(\frac{\partial V}{\partial T}\right)_V = 0\right]$

Equation (7.9) represents Maxwell's second relation.

Maxwell's Third Relation

Putting $q = S$ and $r = P$ in equation (7.7), we get

$$\left(\frac{\partial T}{\partial P}\right)_S\left(\frac{\partial S}{\partial S}\right)_P - \left(\frac{\partial P}{\partial P}\right)_S\left(\frac{\partial V}{\partial S}\right)_P = \left(\frac{\partial T}{\partial S}\right)_P\left(\frac{\partial S}{\partial P}\right)_S - \left(\frac{\partial P}{\partial S}\right)_P\left(\frac{\partial V}{\partial P}\right)_S$$

$$\left(\frac{\partial T}{\partial P}\right)_S = \left(\frac{\partial V}{\partial S}\right)_P \tag{7.10}$$

where, $\left[\left(\frac{\partial S}{\partial S}\right)_P = 1, \left(\frac{\partial P}{\partial P}\right)_T = 1 \text{ and} \left(\frac{\partial S}{\partial P}\right)_S = 0, \left(\frac{\partial P}{\partial S}\right)_P = 0\right]$

Equation (7.10) represents Maxwell's third relation.

Maxwell's Fourth Relation

Putting $q = T$ and $r = P$ in equation (7.7), we get

$$\left(\frac{\partial T}{\partial P}\right)_T\left(\frac{\partial S}{\partial T}\right)_P - \left(\frac{\partial P}{\partial P}\right)_T\left(\frac{\partial V}{\partial T}\right)_P = \left(\frac{\partial T}{\partial T}\right)_P\left(\frac{\partial S}{\partial P}\right)_T - \left(\frac{\partial P}{\partial T}\right)_P\left(\frac{\partial V}{\partial P}\right)_T$$

$$\left(\frac{\partial S}{\partial P}\right)_T = -\left(\frac{\partial V}{\partial T}\right)_P \tag{7.11}$$

where, $\left[\left(\frac{\partial T}{\partial T}\right)_P = 1, \left(\frac{\partial P}{\partial P}\right)_T = 1 \text{ and} \left(\frac{\partial T}{\partial P}\right)_T = 0, \left(\frac{\partial P}{\partial T}\right)_P = 0\right]$

Equation (7.11) represents Maxwell's fourth relation.

Maxwell's Fifth Relation

Putting $q = P$ and $r = V$ in equation (7.7), we get

$$\left(\frac{\partial T}{\partial V}\right)_P\left(\frac{\partial S}{\partial P}\right)_V - \left(\frac{\partial P}{\partial V}\right)_P\left(\frac{\partial V}{\partial P}\right)_V = \left(\frac{\partial T}{\partial P}\right)_V\left(\frac{\partial S}{\partial V}\right)_P - \left(\frac{\partial P}{\partial P}\right)_V\left(\frac{\partial V}{\partial V}\right)_P$$

$$\left(\frac{\partial T}{\partial V}\right)_P\left(\frac{\partial S}{\partial P}\right)_V = \left(\frac{\partial T}{\partial P}\right)_V\left(\frac{\partial S}{\partial V}\right)_P - 1$$

or
$$\left(\frac{\partial T}{\partial P}\right)_V\left(\frac{\partial S}{\partial V}\right)_P - \left(\frac{\partial T}{\partial V}\right)_P\left(\frac{\partial S}{\partial P}\right)_V = 1 \tag{7.12}$$

where,
$$\left[\left(\frac{\partial P}{\partial P}\right)_V = 1, \left(\frac{\partial V}{\partial V}\right)_P = 1 \text{ and } \left(\frac{\partial P}{\partial V}\right)_P = 0, \left(\frac{\partial V}{\partial P}\right)_V = 0\right]$$

Equation (7.12) represents Maxwell's fifth relation in terms of mechanical variable (P, V).

Maxwell's Sixth Relation

Putting $q = T$ and $r = S$ in equation (7.7), we get

$$\left(\frac{\partial T}{\partial S}\right)_T\left(\frac{\partial S}{\partial T}\right)_S - \left(\frac{\partial P}{\partial S}\right)_T\left(\frac{\partial V}{\partial T}\right)_S = \left(\frac{\partial T}{\partial T}\right)_S\left(\frac{\partial S}{\partial S}\right)_T - \left(\frac{\partial P}{\partial T}\right)_S\left(\frac{\partial V}{\partial S}\right)_T$$

$$-\left(\frac{\partial P}{\partial S}\right)_T\left(\frac{\partial V}{\partial T}\right)_S = 1 - \left(\frac{\partial P}{\partial T}\right)_S\left(\frac{\partial V}{\partial S}\right)_T$$

or
$$\left(\frac{\partial P}{\partial T}\right)_S\left(\frac{\partial V}{\partial S}\right)_T - \left(\frac{\partial P}{\partial S}\right)_T\left(\frac{\partial V}{\partial T}\right)_S = 1 \tag{7.13}$$

where,
$$\left[\left(\frac{\partial T}{\partial T}\right)_S = 1, \left(\frac{\partial S}{\partial S}\right)_T = 1 \text{ and } \left(\frac{\partial T}{\partial S}\right)_T = 0, \left(\frac{\partial S}{\partial T}\right)_S = 0\right]$$

Equation (7.13) represents Maxwell's sixth relation in terms of thermodynamic variables (T, S).

Hence, four possible sets of variables P, V, S and T are (S, V), (T, V), (S, P) and (T, P), establishing Maxwell's four thermodynamic relations and the other two sets between mechanical variables (P, V) and thermal variables (T, S), establishing other two thermodynamic relations.

7.3 THERMODYNAMIC POTENTIALS AND THEIR SIGNIFICANCE

Some state functions with their resembling equations of state are used to describe the equilibrium nature of a thermodynamic system in terms of the functions of natural variables (P, V, T, S). There are four such state functions which are known as thermodynamic potentials and represented by U, F, H and G. These thermodynamic potentials can be described in terms of variables P, V, T and S. Maxwell's first four thermodynamic relations can also be derived directly by these potentials.

(i) Internal Energy of System (U): A quantity of self-energy is associated with each system or body, this energy is known as intrinsic or internal energy. This is the total self-energy of the system and is represented by U. We cannot calculate the total internal energy of a system but the change in internal energy while passing from one state to another state can be obtained.

From the first law of thermodynamics,

$$dQ = dU + PdV$$

or

$$dU = dQ - PdV$$

The value of dQ can be obtained by the second law of thermodynamics as $dQ = TdS$

So,

$$dU = TdS - PdV = TdS - dW \quad (7.14)$$

Equation (7.14) provides the change in internal energy while going from one state to another state.

(ii) Helmholtz Free Energy (F): Using the first and second laws of thermodynamics, we get from equation (7.14),

$$dU = TdS - dW$$

As we know that for a system kept at constant temperature T, $TdS = d(TS)$,

So,

$$dU = d(TS) - dW$$

or

$$d(U - TS) = -dW$$

or

$$dF = -dW$$

where,

$$F = (U - TS) \quad (7.15)$$

F is called the Helmholtz work function or Helmholtz free energy in a reversible process.

Now as

$$F = (U - TS)$$

So,

$$dF = dU - TdS - SdT$$

$$= TdS - PdV - TdS - SdT \quad (dU = TdS - PdV)$$

or

$$= -PdV - SdT \quad (7.16)$$

with the help of equation (7.16), we can determine the change in Helmholtz free energy in a process.

In case of reversible isothermal process, we know that $dT = 0$

So from equation (7.16), we get

$$dF = -PdV$$

Hence, Helmholtz free energy represents work done in a reversible isothermal process and it will be negative.

In case of isothermal isochoric process, we know that $dT = 0, dV = 0$

So from equation (7.16), we get

$$dF = 0$$

or $$F = \text{constant}$$

Hence, Helmholtz free energy of a system remains unchanged or constant in a isothermal isochoric process.

(iii) Enthalpy (*H*): Enthalpy is also known as the total or net heat energy and is mathematically expressed as

$$H = U + PV \tag{7.17}$$

We can get change in enthalpy by differentiating the above equation,

$$dH = dU + PdV + VdP$$

or $$= TdS - PdV + PdV + VdP \qquad (\text{as } dU = TdS - PdV)$$

$$= TdS + VdP$$

In case of isobaric adiabatic process, $dP = 0$ and $dQ = TdS = 0$

So, $$dH = 0$$

or $$H = \text{constant}$$

Hence, enthalpy or total heat remains unchanged during isobaric adiabatic process.

In case of a reversible isobaric process, $dP = 0$

So, $$dH = TdS = dQ,$$

Hence, enthalpy is equal to the total amount of heat absorbed during the reversible isobaric process.

(iv) Gibbs Function (*G*): As we know that enthalpy is defined as

$$H = U + PV$$

So, $$dH = dU + PdV + VdP$$

or $$= TdS - PdV + PdV + VdP \qquad (\text{as } dU = TdS - PdV)$$

or $$= TdS + VdP \tag{7.18}$$

For an isothermal isobaric process, $TdS = d(TS)$ and $dP = 0$,

$$dH = d(TS)$$

or $$d(H - TS) = 0$$

or $$dG = 0$$

where, $$G = (H - TS) = \text{constant} \tag{7.19}$$

or $$= (H - TS) = (U + PV - TS) \quad (\text{as } (H = U + PV)$$

or $$= (U - TS) + PV$$

or $$= F + PV \qquad (\text{as } F = U - TS) \tag{7.20}$$

This function G expressed in equations (7.19) and (7.20) is known as Gibbs free energy or Gibbs potential.

7.4 PHYSICAL SIGNIFICANCE OF THERMODYNAMIC POTENTIAL

The potential energy has an important role in mechanics. A stable equilibrium of a mechanical system is obtained with minimum potential energy and the system must work in a direction of lower potential energy.

The internal energy, Helmholtz free energy, enthalpy and Gibbs potential, play same role in thermodynamics as potential energy plays in mechanics. These four functions get a minimum value in different processes and behave like a potential energy and thus known as thermodynamic potentials.

Gibbs free energy get minimum value in an isothermal isobaric process. Enthalpy tends to be minimum in an isobaric adiabatic process. Helmholtz free energy gets minimum value in an isothermal isochoric process.

7.5 DERIVATION OF MAXWELL'S RELATIONS FROM THERMODYNAMIC POTENTIAL

We know that there are four thermodynamic potentials U, F, H and G, which are defined in terms of the thermodynamic variables P, V, S and T. The value of these variables can also be found out by thermodynamic potentials. We can also deduce Maxwell's relations from thermodynamic potentials.

(i) Internal Energy and Maxwell's First Relation

From the first and second laws of thermodynamics, we know that the change in internal energy of a system is given by

$$dU = TdS - PdV \tag{7.21}$$

Differentiating the above equation partially with respect to entropy S, we get

$$\left(\frac{\partial U}{\partial S}\right)_V = T \tag{7.22}$$

and differentiating equation (7.21) partially with respect to potential energy V, we get

$$\left(\frac{\partial U}{\partial V}\right)_T = -P \tag{7.23}$$

Again differentiating equation (7.22) with respect to V and equation (7.23) with respect to S, respectively, we get

$$\left(\frac{\partial^2 U}{\partial V \partial S}\right) = \left(\frac{\partial T}{\partial V}\right)_S \text{ and } \left(\frac{\partial^2 U}{\partial S \partial V}\right) = -\left(\frac{\partial P}{\partial S}\right)_V$$

As we know that dU is a perfect differential, so

$$\left(\frac{\partial^2 U}{\partial S \partial V}\right) = \left(\frac{\partial^2 U}{\partial V \partial S}\right)$$

or

$$\left(\frac{\partial T}{\partial V}\right)_S = -\left(\frac{\partial P}{\partial S}\right)_V \tag{7.24}$$

Equation (7.24) is Maxwell's first thermodynamic relation.

(ii) Thermodynamic Potential (*F*) and Maxwell's Second Relation

We know that the Helmholtz free energy is expressed by the relation

$$F = U - TS$$

Change in Helmholtz free energy will be

$$dF = (dU - TdS - SdT)$$

$$= (TdS - PdV - TdS - SdT) \qquad \text{as } (dU = TdS - PdV)$$

or

$$= (-PdV - SdT) \tag{7.25}$$

Differentiating equation (7.25) partially with respect to temperature T, we obtain

$$\left(\frac{\partial F}{\partial T}\right)_V = -S \tag{7.26}$$

Differentiating equation (7.25) partially with respect to volume V, we get

$$\left(\frac{\partial F}{\partial V}\right)_T = -P \tag{7.27}$$

Again differentiating equation (7.26) with respect to V and equation (7.27) with respect to T, respectively, we get

$$\left(\frac{\partial^2 F}{\partial V \partial T}\right) = -\left(\frac{\partial S}{\partial V}\right)_T \text{ and } \left(\frac{\partial^2 U}{\partial T \partial V}\right) = -\left(\frac{\partial P}{\partial T}\right)_V$$

As we know that dF is a perfect differential, so

$$\left(\frac{\partial^2 F}{\partial V \partial T}\right) = \left(\frac{\partial^2 F}{\partial T \partial V}\right)$$

or

$$\left(\frac{\partial S}{\partial V}\right)_T = \left(\frac{\partial P}{\partial T}\right)_V \tag{7.28}$$

Equation (7.28) is Maxwell's second thermodynamic relation.

(iii) Enthalpy (*H*) and Maxwell's Third Relation

We know that enthalpy is expressed by the relation

$$H = U + PV$$

So, change in enthalpy will be

$$dH = (dU + PdV + VdP)$$

$$= (TdS - PdV + PdV + VdP) \qquad \text{as } (dU = TdS - PdV)$$

or

$$= (TdS + VdP) \tag{7.29}$$

Differentiating equation (7.29) partially with respect to entropy S, we obtain

$$\left(\frac{\partial H}{\partial S}\right)_P = T \tag{7.30}$$

Differentiating equation (7.29) partially with respect to volume P, we get

$$\left(\frac{\partial H}{\partial P}\right)_S = V \tag{7.31}$$

Again differentiating equation (7.30) with respect to P and equation (7.31) with respect to S, respectively, we get

$$\left(\frac{\partial^2 H}{\partial P \partial S}\right) = \left(\frac{\partial T}{\partial P}\right)_S \text{ and } \left(\frac{\partial^2 H}{\partial S \partial P}\right) = \left(\frac{\partial V}{\partial S}\right)_P$$

As we know that dH is a perfect differential, so

$$\left(\frac{\partial^2 H}{\partial P \partial S}\right) = \left(\frac{\partial^2 H}{\partial S \partial P}\right)$$

or $$\left(\frac{\partial T}{\partial P}\right)_S = \left(\frac{\partial V}{\partial S}\right)_P \tag{7.32}$$

Equation (7.32) is Maxwell's third thermodynamic relation.

(iv) Gibbs Potential (*G*) and Maxwell's Fourth Relation

We know that the Gibbs potential is expressed by the relation

$$G = H - TS$$

So, change in *G* will be

$$dG = (dH + TdS - SdT)$$

$$= (TdS + VdP + TdS - SdT) \quad \text{as } (dH = TdS + VdP)$$

or $$= (VdP - SdT) \tag{7.33}$$

Differentiating equation (7.33) partially with respect to entropy *P*, we obtain

$$\left(\frac{\partial G}{\partial P}\right)_T = V \tag{7.34}$$

Differentiating equation (7.33) partially with respect to volume *T*, we get

$$\left(\frac{\partial G}{\partial T}\right)_P = -S \tag{7.35}$$

Again differentiating equation (7.34) with respect to *T* and equation (7.35) with respect to *P*, respectively, we get

$$\left(\frac{\partial^2 G}{\partial T \partial P}\right) = \left(\frac{\partial V}{\partial T}\right)_P \quad \text{and} \quad \left(\frac{\partial^2 G}{\partial P \partial T}\right) = -\left(\frac{\partial S}{\partial P}\right)_T$$

As we know that *dH* is a perfect differential, so

$$\left(\frac{\partial^2 G}{\partial T \partial P}\right) = \left(\frac{\partial^2 H}{\partial P \partial T}\right)$$

or $$\left(\frac{\partial V}{\partial T}\right)_P = -\left(\frac{\partial S}{\partial P}\right)_T \tag{7.36}$$

Equation (7.36) is Maxwell's fourth thermodynamic relation.

Hence, with the help of thermodynamic potentials, we can establish Maxwell's thermodynamic relations. The mechanical variables *P* and *V* and thermodynamic variables *S* and *T* can be defined by equations (7.22) – (7.23), (7.26) – (7.27), (7.30) – (7.31), (7.34) – (7.35).

$$\begin{bmatrix} P = -\left(\frac{\partial U}{\partial V}\right)_S = -\left(\frac{\partial F}{\partial V}\right)_T \\ V = \left(\frac{\partial H}{\partial P}\right)_S = \left(\frac{\partial G}{\partial P}\right)_T \\ S = -\left(\frac{\partial G}{\partial T}\right)_P = -\left(\frac{\partial F}{\partial T}\right)_V \\ T = \left(\frac{\partial U}{\partial S}\right)_V = \left(\frac{\partial H}{\partial S}\right)_P \end{bmatrix} \tag{7.37}$$

The equation (7.37) provides the values of mechanical and thermodynamic variables in the form of thermodynamic potentials.

7.6 APPLICATIONS OF THERMODYNAMIC RELATIONS

1. Joule-Thomson Cooling Effect and Joule-Thomson Coefficient

When a gas moves forcibly from high constant pressure region to low constant pressure region, the emerging gas shows change in temperature. This effect is known as Joule-Thomson cooling effect.

In Joule's Thomson porous plug experiment, even though, on both sides of porous plug, a pressure difference occurs, the enthalpy of the gas remains unchanged during the process, i.e.,

$$H = (U + PV) = \text{constant}$$

Difference in enthalpy is given by

$$dH = (dU + PdV + VdP) = 0$$

But from the first and second laws of thermodynamics, we have

$$dQ = (dU + PdV) = TdS$$

So,

$$dH = (TdS + VdP) = 0 \tag{7.38}$$

As we know that entropy is a function of pressure and temperature, i.e., $S = f(P, T)$ and dS is a perfect differential, we get

$$dS = \left(\frac{\partial S}{\partial T}\right)_P dT + \left(\frac{\partial S}{\partial P}\right)_T dP$$

Substituting this value of dS in equation (7.38), we get

$$T\left[\left(\frac{\partial S}{\partial T}\right)_P dT + \left(\frac{\partial S}{\partial P}\right)_T dP\right] + VdP = 0$$

or
$$T\left(\frac{\partial S}{\partial T}\right)_P dT + T\left(\frac{\partial S}{\partial P}\right)_T dP + VdP = 0$$

or
$$C_P dT + \left[T\left(\frac{\partial S}{\partial P}\right)_T + V\right]dP = 0 \tag{7.39}$$

where $C_P = \left(\frac{\partial Q}{\partial T}\right)_P = T\left(\frac{\partial S}{\partial T}\right)_P$ is specific heat at constant pressure.

From Maxwell's fourth thermodynamic relation, we know that

$$\left(\frac{\partial S}{\partial P}\right)_T = -\left(\frac{\partial V}{\partial T}\right)_P$$

Substituting this value in equation (7.39), we get

$$C_P dT + \left[-T\left(\frac{\partial V}{\partial T}\right)_P + V\right]dP = 0$$

or
$$C_P dT = -\left[-\left(\frac{\partial V}{\partial T}\right)_P + V\right]dP$$

or
$$= \left[T\left(\frac{\partial V}{\partial T}\right)_P - V\right]dP$$

or
$$dT = \frac{1}{C_P}\left[T\left(\frac{\partial V}{\partial T}\right)_P - V\right]dP \tag{7.40}$$

$$\mu = \left(\frac{dT}{dP}\right)_H = \frac{1}{C_P}\left[T\left(\frac{\partial V}{\partial T}\right)_P - V\right] \tag{7.41}$$

where, $\mu = \left(\frac{dT}{dP}\right)_H$ is known as Joule-Thomson coefficient and for perfect gas, its value is zero.

2. Clapeyron's Equation for Latent Heat

We know that Maxwell's second thermodynamic relation is

$$\left(\frac{\partial S}{\partial V}\right)_T = \left(\frac{\partial P}{\partial T}\right)_V$$

Multiplying both sides of above equation by T, we get

$$T\cdot\left(\frac{\partial S}{\partial V}\right)_T = T\cdot\left(\frac{\partial P}{\partial T}\right)_V$$

$$\left(\frac{\partial Q}{\partial V}\right)_T = T \cdot \left(\frac{\partial P}{\partial T}\right)_V \tag{7.42}$$

where $T \cdot \partial S = \partial Q$. In equation (7.43), $\left(\frac{\partial Q}{\partial V}\right)_T$ represents the amount of heat absorbed or rejected per unit change in volume at a fixed temperature. This amount of heat absorbed or rejected at a fixed temperature is known as the latent heat (L) due to the change of state.

Hence, if we consider unit mass of matter, $\partial Q = L$ and $\partial V = V_2 - V_1$

Substituting these values in equation (7.42), we get

$$\left(\frac{L}{V_2 - V_1}\right)_T = T \cdot \left(\frac{\partial P}{\partial T}\right)_V$$

or

$$\left(\frac{L}{V_2 - V_1}\right) = T \cdot \left(\frac{dP}{dT}\right)$$

$$\left(\frac{dP}{dT}\right) = \frac{1}{T}\left(\frac{L}{V_2 - V_1}\right) \tag{7.43}$$

Equation (7.44) represents the Claussius-Clayperon's latent heat equation.

3. Specific Heat Equation or Mayer's Formula

We know that the specific heat at constant pressure C_P and specific heat at constant volume C_V is given by

$$C_P = \left(\frac{\partial Q}{\partial T}\right)_P = T\left(\frac{\partial S}{\partial T}\right)_P \tag{7.44}$$

$$C_V = \left(\frac{\partial Q}{\partial T}\right)_V = T\left(\frac{\partial S}{\partial T}\right)_V \tag{7.45}$$

Entropy is a function of temperature and volume and also a perfect differentiation, we have

$$dS = \left(\frac{\partial S}{\partial T}\right)_V dT + \left(\frac{\partial S}{\partial V}\right)_T dV$$

Dividing both sides of the above equation by dT, we get

$$\frac{dS}{dT} = \left(\frac{\partial S}{\partial T}\right)_V + \left(\frac{\partial S}{\partial V}\right)_T \left(\frac{dV}{dT}\right)$$

$$T\left(\frac{\partial S}{\partial T}\right)_P = T\left(\frac{\partial S}{\partial T}\right)_V + T\left(\frac{\partial S}{\partial V}\right)_T \left(\frac{\partial V}{\partial T}\right)_P \tag{7.46}$$

We know that from Maxwell's second relation

$$\left(\frac{\partial S}{\partial V}\right)_T = \left(\frac{\partial P}{\partial T}\right)_V$$

Substituting this relation in equation (7.46), we get

$$T\left(\frac{\partial S}{\partial T}\right)_P = T\left(\frac{\partial S}{\partial T}\right)_V + T\left(\frac{\partial P}{\partial T}\right)_V\left(\frac{\partial V}{\partial T}\right)_P$$

or $$T\left(\frac{\partial S}{\partial T}\right)_P - T\left(\frac{\partial S}{\partial T}\right)_V = T\left(\frac{\partial P}{\partial T}\right)_V\left(\frac{\partial V}{\partial T}\right)_P$$

Using equations (7.45) and (7.46), we get

$$C_P - C_V = T\left(\frac{\partial P}{\partial T}\right)_V\left(\frac{\partial V}{\partial T}\right)_P \tag{7.47}$$

We know that for a perfect gas, gas equation is

$$PV = RT$$

So, $$\left(\frac{\partial V}{\partial T}\right)_P = \frac{R}{P} \text{ and } \left(\frac{\partial P}{\partial T}\right)_V = \frac{R}{V}$$

Substituting these values in equation (7.47), we get

$$C_P - C_V = T \times \frac{R}{V} \times \frac{R}{P} = \frac{R^2 \times T}{PV} = \frac{R^2 \times T}{RT}$$

or $$C_P - C_V = R \tag{7.48}$$

Equation (7.48) gives relation between specific heat at constant pressure and specific heat at constant volume for ideal gases. It is known as specific heat equation for perfect gases or Mayer's Formula.

We know that the real gas equation is given by

$$\left(P + \frac{a}{V^2}\right)(V - b) = RT$$

or $$\left(P + \frac{a}{V^2}\right) = \frac{1}{(V-b)} RT \tag{7.49}$$

Differentiating the above equation with respect to temperature at constant pressure, we obtain

$$0 - \frac{2a}{V^3} \times \left(\frac{\partial V}{\partial T}\right)_P = \frac{1}{(V-b)} R - \frac{1}{(V-b)^2} RT \times \left(\frac{\partial V}{\partial T}\right)_P$$

or $$\left[-\frac{2a}{V^3} + \frac{RT}{(V-b)^2}\right] \times \left(\frac{\partial V}{\partial T}\right)_P = \frac{1}{(V-b)} R$$

or $$\left(\frac{\partial V}{\partial T}\right)_P = \frac{\frac{1}{(V-b)}R}{\left[-\frac{2a}{V^3}+\frac{RT}{(V-b)^2}\right]} \tag{7.50}$$

Differentiating equation (7.49) with respect to T at constant pressure, we get

$$\left(\frac{\partial P}{\partial T}\right)_V = \frac{1}{(V-b)}R \tag{7.51}$$

Substituting the values of $\left(\frac{\partial V}{\partial T}\right)_P$ and $\left(\frac{\partial P}{\partial T}\right)_V$ from equations (7.50) and (7.51), respectively, in equation (7.47), we get

$$C_P - C_V = T\frac{R}{(V-b)} \times \frac{\frac{R}{(V-b)}}{\left[-\frac{2a}{V^3}+\frac{RT}{(V-b)^2}\right]}$$

$$= \frac{T\left[\frac{R}{(V-b)}\right]^2}{\left[-\frac{2a}{V^3}+\frac{RT}{(V-b)^2}\right]}$$

$$= \frac{R}{\left[-\frac{2a}{V^3}+\frac{(V-b)^2}{RT}+1\right]}$$

As b is a correction in volume and is very small in comparison to V, so we can neglect it,

$$C_P - C_V = \frac{R}{\left[-\frac{2a}{V^3}\times\frac{V^2}{RT}+1\right]}$$

$$= \frac{R}{\left[-\frac{2a}{V}\times\frac{1}{RT}+1\right]}$$

$$= R \times \left(1-\frac{2a}{RTV}\right)^{-1}$$

Expanding the above equation binomially and neglecting the higher powers of a (as a is smaller as compared to V), we get

$$C_P - C_V = R \times \left(1+\frac{2a}{RTV}\right) \tag{7.52}$$

Equation (7.52) provides the relation between C_P and C_V for real gases.

4. The *T-dS* Equations

We know that entropy is a function of temperature and volume, i.e., $S = f(T, V)$ and also s is a perfect differentiation, so we can write

$$dS = \left(\frac{\partial S}{\partial T}\right)_V dT + \left(\frac{\partial S}{\partial V}\right)_T dV$$

Multiplying both sides of the above equation by T, we get

$$TdS = T \times \left(\frac{\partial S}{\partial T}\right)_V dT + T \times \left(\frac{\partial S}{\partial V}\right)_T dV \tag{7.53}$$

We know that from Maxwell's second relation that

$$\left(\frac{\partial S}{\partial V}\right)_T = \left(\frac{\partial P}{\partial T}\right)_V$$

and also specific heat at constant volume is

$$C_V = T\left(\frac{\partial S}{\partial T}\right)_V$$

Substituting the above values in equation (7.53), we get

$$T \cdot dS = C_V \cdot dT + T \cdot \left(\frac{\partial P}{\partial T}\right)_V dV \tag{7.54}$$

Equation (7.54) is known as first $T \cdot dS$ equation.
The entropy can also be taken as a function of T and V, i.e., $S = f(T, V)$.
As dS is a perfect differentiation, we can write

$$dS = \left(\frac{\partial S}{\partial T}\right)_P dT + \left(\frac{\partial S}{\partial P}\right)_T dP$$

Multiplying both sides of the above equation by T, we get

$$TdS = T \times \left(\frac{\partial S}{\partial T}\right)_P dT + T \times \left(\frac{\partial S}{\partial P}\right)_T dP \tag{7.55}$$

We know that from Maxwell's fourth relation that

$$\left(\frac{\partial S}{\partial P}\right)_T = -\left(\frac{\partial V}{\partial T}\right)_P$$

and also specific heat at constant pressure,

$$C_P = T\left(\frac{\partial S}{\partial T}\right)_P$$

Substituting the above values in equation (7.55), we get

$$T \cdot dS = C_P \cdot dT + T \cdot \left(\frac{\partial V}{\partial T}\right)_P dP \tag{7.56}$$

Equation (7.56) is known as the second $T \cdot dS$ equation.

7.7 FIRST AND SECOND ORDER PHASE TRANSITIONS

(i) First Order Phase Transition

When a system is consistent in chemical composition and physical structure, it is called in phase.

If the chemical composition of the system is same but physical structure is different, it will be in different phases, for example, solid, liquid and gas phases. Transition from one to another phase is known as first order phase transition. In the first order transition in the changes in phase takes place at constant pressure and temperature and heat is either absorbed or rejected during the process of change phase. In the first order transition, the entropy and volume (density) change occurs. The Gibbs function G is very useful in phase transition process, which takes place at constant pressure and temperature. The Gibbs potential remains constant in both phases, while its first order derivative with respect to pressure and temperature varies discontinuously at the triple point.

Let us consider a container containing a liquid and its saturated vapor, both in equilibrium. For equilibrium of two phases,

$$g_1 = g_2(T, P)$$

If the temperature increases by $T + dT$ and pressure increases by $P + dP$, at equilibrium, the changes in specific Gibbs function of the two phases will be equal, i.e.,

$$dg_1 = dg_2(T + dT, P + dP)$$

At the position of saturation

$$\frac{dg_1}{dT} = \frac{dg_2}{dT} \tag{7.57}$$

We know that for unit mass

$$dg = -SdT + VdP \tag{7.58}$$

or

$$S = -\left(\frac{\partial g}{\partial T}\right)_P \text{ and } V = \left(\frac{\partial g}{\partial P}\right)_T$$

for two different phases

$$S_1 = -\left(\frac{\partial g_1}{\partial T}\right)_P \text{ and } V_1 = \left(\frac{\partial g_1}{\partial P}\right)_T \tag{7.59}$$

$$S_2 = -\left(\frac{\partial g_2}{\partial T}\right)_P \text{ and } V_2 = \left(\frac{\partial g_2}{\partial P}\right)_T \tag{7.60}$$

As $S_2 \neq S_1$ and $V_2 \neq V_1$,

$$\left(\frac{\partial g_1}{\partial T}\right)_P \neq \left(\frac{\partial g_2}{\partial T}\right)_P \text{ and } \left(\frac{\partial g_1}{\partial P}\right)_T \neq \left(\frac{\partial g_2}{\partial P}\right)_T \tag{7.61}$$

It is clear that even though the specific Gibbs function is continuous during the transition of phase, the first order derivative of Gibbs function changes discontinuously in phase transition (Figure 7.1). Therefore, a first order phase transition can be defined as one in which specific Gibbs function changes discontinuously at the triple point with respect to the constant temperature and pressure.

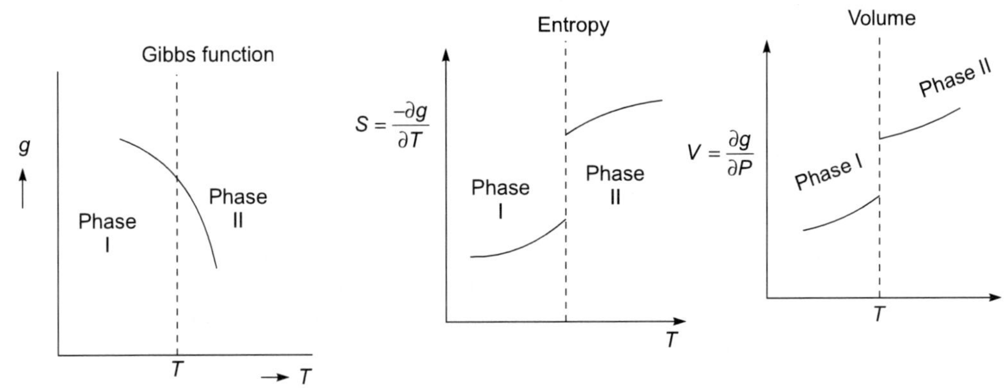

Figure 7.1 First order phase transition

As g is a function of pressure and temperature, so we can write

$$dg = \left(\frac{\partial g}{\partial T}\right)_P dT + \left(\frac{\partial g}{\partial P}\right)_T dP$$

or

$$\left(\frac{\partial g}{\partial T}\right) = \left(\frac{\partial g}{\partial T}\right)_P + \left(\frac{\partial g}{\partial P}\right)_T \left(\frac{\partial P}{\partial T}\right) \tag{7.62}$$

Substituting the values of $\left(\frac{\partial g}{\partial T}\right)_P$ and $\left(\frac{\partial g}{\partial P}\right)_T$ from equations (7.59) and (7.60) to equation (7.62), we get

for state one

$$\left(\frac{dg_1}{dT}\right) = -S_1 + V_1\left(\frac{dP}{dT}\right)$$

and for state two $$\left(\frac{dg_2}{dT}\right) = -S_2 + V_2\left(\frac{dP}{dT}\right)$$

Substituting the above values in equation (7.57), we get

$$-S_1 + V_1\left(\frac{dP}{dT}\right) = -S_2 + V_2\left(\frac{dP}{dT}\right)$$

or $$\left(\frac{dP}{dT}\right) = \frac{S_2 - S_1}{V_2 - V_1}$$

The above equation represents the slope of *P-T* equilibrium diagram in the first order transition.

As we know that $S_2 - S_1 = \left(\frac{dQ}{T}\right) = \frac{L}{T}$

where, *L* is a latent heat of vaporization.

So, $$\left(\frac{dP}{dT}\right) = \frac{L}{T(V_2 - V_1)}$$

The above equation is known as Clausius-Clapeyron latent heat equation.

(ii) Second Order Phase Transition

The phase transition in which heat cannot be transferred during the phase change and also volume and temperature remain unchanged, is known as the second order phase transition. Transition of liquid helium I to liquid helium II is the second order phase transition, in which transition takes place at constant entropy, volume, temperature and pressure. Other examples of the second order phase transitions are transition of ferromagnetic material to paramagnetic material, transition of superconducting metal into a conductor, order-disorder transition in chemical compound and alloys, etc.

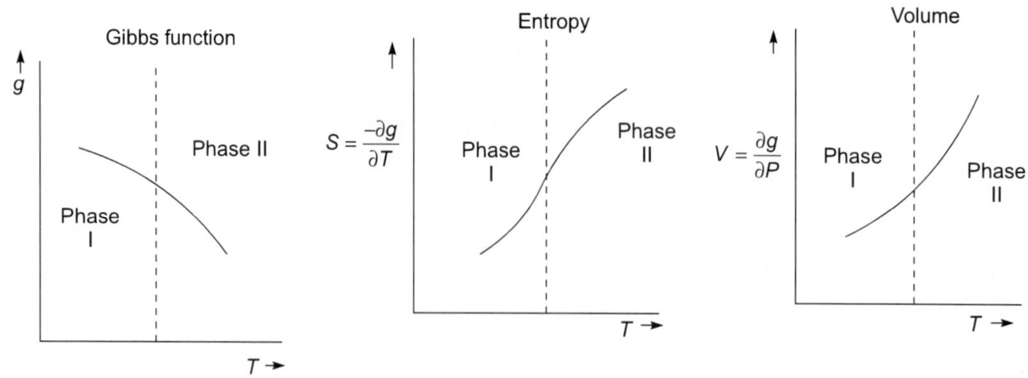

Figure 7.2 Second order phase transition

In the second order phase transition, the Gibbs function and the first order derivatives with respect to T and P at constant pressure and at constant temperature, respectively are continuous. But in this case, the second order derivatives of Gibbs function show discontinuity.

As we know that for a phase transition,

$$g_1 = g_2 \quad \text{or} \quad g_2 - g_1 = 0$$

and
$$S_2 - S_1 = -\left(\frac{\partial g_2}{\partial T}\right)_P + \left(\frac{\partial g_1}{\partial T}\right)_P = 0 \qquad \text{(entropy is constant)}$$

$$V_2 - V_1 = \left(\frac{\partial g_2}{\partial P}\right)_T - \left(\frac{\partial g_1}{\partial P}\right)_T = 0 \qquad \text{(volume is constant)}$$

We know that specific heat at constant pressure is defined as

$$C_P = T\left(\frac{\partial S}{\partial T}\right)_P$$

or
$$\frac{C_P}{T} = \left(\frac{\partial S}{\partial T}\right)_P = -\frac{\partial}{\partial T}\left[\left(\frac{\partial g}{\partial T}\right)_P\right]_P \text{ as } S = \left(\frac{\partial g}{\partial T}\right)_P$$

or
$$\frac{C_P}{T} = \left(\frac{\partial S}{\partial T}\right)_P = -\left(\frac{\partial^2 g}{\partial T^2}\right)_P \tag{7.63}$$

For states one and two, we can write

$$\frac{C_{P_1}}{T} = -\left(\frac{\partial^2 g_1}{\partial T^2}\right)_P \text{ and } \frac{C_{P_2}}{T} = -\left(\frac{\partial^2 g_2}{\partial T^2}\right)_P \tag{7.64}$$

We know that isothermal compressibility (K) is defined as

$$K = -\frac{1}{V}\left(\frac{\partial V}{\partial P}\right)_T$$

or
$$KV = -\left(\frac{\partial V}{\partial P}\right)_T = -\frac{\partial}{\partial P}\left[\left(\frac{\partial g}{\partial P}\right)_T\right]_T = -\left(\frac{\partial^2 g}{\partial P^2}\right)_T \tag{7.65}$$

So,
$$K_1 V = \left(\frac{\partial^2 g_1}{\partial P^2}\right)_T \text{ and } K_2 V = -\left(\frac{\partial^2 g_2}{\partial P^2}\right)_T \tag{7.66}$$

We also know that the volume coefficient of expansion is defined as

$$\alpha = \frac{1}{V}\left(\frac{\partial V}{\partial T}\right)_P$$

or $$\alpha V = \left(\frac{\partial V}{\partial T}\right)_P = \frac{\partial}{\partial T}\left[\left(\frac{\partial g}{\partial P}\right)_T\right]_P = \left(\frac{\partial^2 g}{\partial T \partial P}\right) \tag{7.67}$$

So, $$\alpha_1 V = \left(\frac{\partial^2 g_1}{\partial T \partial P}\right) \text{ and } \alpha_2 V = \left(\frac{\partial^2 g_2}{\partial T \partial P}\right) \tag{7.68}$$

Entropy is a function of temperature and volume, so we can write,

$$dS = \left(\frac{\partial S}{\partial T}\right)_P dT + \left(\frac{\partial S}{\partial P}\right)_T dP$$

From equation (7.64), $\frac{C_P}{T} = \left(\frac{\partial S}{\partial T}\right)_P$ and from Maxwell's fourth relation, $\left(\frac{\partial S}{\partial P}\right)_T = -\left(\frac{\partial V}{\partial T}\right)_P$

So, $$dS = \left(\frac{C_P}{T}\right)dT - \left(\frac{\partial V}{\partial T}\right)_P dP$$

or $$dS = \left(\frac{C_P}{T}\right)dT - V\alpha dP \qquad \text{as } dS = \left(\frac{\partial V}{\partial T}\right)_P = -V\alpha$$

So, $$dS_1 = \left(\frac{C_{P_1}}{T}\right)dT - V\alpha_1 dP \text{ and } dS_2 = \left(\frac{C_{P_2}}{T}\right)dT - V\alpha_2 dP \tag{7.69}$$

For the second order phase transition, entropy of the system remains unchanged, so at constant temperature and pressure, $S_1 = S_2$. If the temperature is changed from T to $T + dT$ and pressure is changed from P to $P + dP$,

$$S_1 + dS_1 = S_2 + dS_2$$

or $$dS_1 = dS_2$$

Hence, from equation (7.69), we get

$$\left(\frac{C_{P_1}}{T}\right)dT - V\alpha_1 dP = \left(\frac{C_{P_2}}{T}\right)dT - V\alpha_2 dP$$

or $$(\alpha_2 - \alpha_1)\, VdP = \left(C_{P_2} - C_{P_1}\right)\frac{dT}{T}$$

or $$\frac{dP}{dT} = \frac{\left(C_{P_2} - C_{P_1}\right)}{(\alpha_2 - \alpha_1)}\frac{1}{TV} \tag{7.70}$$

Volume (V) can be defined as

$$dV = \left(\frac{\partial V}{\partial T}\right)_P dT + \left(\frac{\partial V}{\partial P}\right)_T dP$$

$$dV = V\alpha dT - V\, KdP$$

as $$\left(\frac{\partial V}{\partial T}\right)_P = \alpha \text{ and} \left(\frac{\partial V}{\partial P}\right)_T = -K$$

So, $$dV_1 = V\alpha_1 dT - VK_1 dP \text{ and } dV_2 = V\alpha_2 dT - VK_2 dP$$

For the second order phase transition, $V_1 = V_2$. If the temperature is changed from T to $T + dT$ and pressure is changed from P to $P + dP$,

$$V_1 + dV_1 = V_2 + dV_2 \text{ or } dV_1 = dV_2$$

So, $$V\alpha_1 dT - VK_1 dP = V\alpha_2 dT - VK_2 dP$$

or $$V(K_2 - K_1)dP = (\alpha_2 - \alpha_1)\, VdT$$

or $$\frac{dP}{dT} = \frac{(\alpha_2 - \alpha_1)}{(K_2 - K_1)} \tag{7.71}$$

The condition of equilibrium between the two phases can be described by equations (7.70) and (7.71) and these equations are known as Ehrenfest's equations.

7.8 RELATION BETWEEN ADIABATIC AND ISOTHERMAL ELASTICITY

The bulk modulus or volume elasticity coefficient E is defined by the relation,

$$\mathcal{E} = \frac{\text{Stress}}{\text{Strain}} = -\frac{\partial P}{(\partial V/V)} = -V\frac{\partial P}{\partial V}$$

In an isothermal process, temperature remains constant, so the isothermal elasticity is given by

$$\mathcal{E}_T = -V\left(\frac{\partial P}{\partial V}\right)_T \tag{7.72}$$

In an adiabatic process, entropy remains constant so the adiabatic elasticity is given by

$$\mathcal{E}_S = -\, V\left(\frac{\partial P}{\partial V}\right)_S \tag{7.73}$$

From equations (7.72) and (7.73), the ratio $\mathcal{E}_S$ to $\mathcal{E}_T$ is given by

$$\mathcal{E}_S = \frac{-V\left(\frac{\partial P}{\partial V}\right)_S}{-V\left(\frac{\partial P}{\partial V}\right)_T}$$

$$\frac{\varepsilon_S}{\varepsilon_T} = \frac{\left(\frac{\partial P}{\partial V}\right)_S}{\left(\frac{\partial P}{\partial V}\right)_T} = \frac{\left(\frac{\partial P}{\partial T}\right)_S \left(\frac{\partial T}{\partial V}\right)_S}{\left(\frac{\partial P}{\partial S}\right)_T \left(\frac{\partial S}{\partial V}\right)_T} \tag{7.74}$$

We know that Maxwell's thermodynamic relations are

$$\left(\frac{\partial P}{\partial T}\right)_S = \left(\frac{\partial S}{\partial V}\right)_P, \left(\frac{\partial T}{\partial V}\right)_S = -\left(\frac{\partial P}{\partial S}\right)_V$$

and

$$\left(\frac{\partial P}{\partial S}\right)_T = -\left(\frac{\partial T}{\partial V}\right)_P, \left(\frac{\partial S}{\partial V}\right)_T = \left(\frac{\partial P}{\partial T}\right)_V$$

Substituting these values in equation (7.74), we get

or

$$\frac{\varepsilon_S}{\varepsilon_T} = \frac{-\left(\frac{\partial S}{\partial V}\right)_P \left(\frac{\partial P}{\partial S}\right)_V}{-\left(\frac{\partial T}{\partial V}\right)_P \left(\frac{\partial P}{\partial T}\right)_V}$$

$$= \left[\left(\frac{\partial S}{\partial V}\right)_P \times \left(\frac{\partial V}{\partial T}\right)_P\right]\left[\left(\frac{\partial P}{\partial S}\right)_V \times \left(\frac{\partial T}{\partial P}\right)_V\right]$$

or

$$= \frac{\left(\frac{\partial S}{\partial T}\right)_P}{\left(\frac{\partial S}{\partial T}\right)_V} = \frac{T\left(\frac{\partial S}{\partial T}\right)_P}{T\left(\frac{\partial S}{\partial T}\right)_V}$$

From the second law of thermodynamics, $\partial Q = T\partial S$

or

$$\frac{\varepsilon_S}{\varepsilon_T} = \frac{\left(\frac{\partial Q}{\partial T}\right)_P}{\left(\frac{\partial Q}{\partial T}\right)_V} = \frac{C_P}{C_V} = \gamma$$

or

$$\frac{\varepsilon_S}{\varepsilon_T} = \frac{C_P}{C_V} = \gamma \tag{7.75}$$

where, C_P and C_V are specific heats at constant pressure and volume, respectively and γ is atomicity of the gas. Equation (7.75) shows the relation between the two elasticities.

7.9 JOULE-KELVIN COEFFICIENT FOR IDEAL AND REAL GASES

We know that enthalpy is defined by

$$H = U + PV$$

So, $$dH = dU + PdV + VdP$$

or $$= TdS - PdV + PdV + VdP \qquad \text{as } dU = TdS - PdV$$

or $$= TdS + VdP \tag{7.76}$$

As we know that entropy can be defined as a function of temperature and pressure, we can write

$$dS = \left(\frac{\partial S}{\partial T}\right)_P dT + \left(\frac{\partial S}{\partial P}\right)_T dP$$

Substituting this value in equation (7.77), we get

$$dH = T\left[\left(\frac{\partial S}{\partial T}\right)_P dT + \left(\frac{\partial S}{\partial P}\right)_T dP\right] + VdP$$

$$= \left[T\left(\frac{\partial S}{\partial T}\right)_P\right]dT + \left[T\left(\frac{\partial S}{\partial P}\right)_T + V\right]dP$$

But we know that, $T\left(\frac{\partial S}{\partial T}\right)_P = C_P$

and from Maxwell's fourth relation, $\left(\frac{\partial S}{\partial P}\right)_T = -\left(\frac{\partial V}{\partial T}\right)_P$

So, $$dH = C_P dT + \left[-T\left(\frac{\partial V}{\partial T}\right)_P + V\right]dP$$

or $$C_P dT = dH - \left[-T\left(\frac{\partial V}{\partial T}\right)_P + V\right]dP$$

or $$dT = \frac{dH}{C_P} + \left[T\left(\frac{\partial V}{\partial T}\right)_P - V\right]\frac{dP}{C_P} \tag{7.77}$$

Temperature T can be defined as a function of enthalpy H and pressure P, we can write

$$dT = \left(\frac{\partial T}{\partial H}\right)_P dH + \left(\frac{\partial T}{\partial P}\right)_H dP \tag{7.78}$$

Comparing the coefficient of equations (7.77) and (7.78), we get

$$\left(\frac{\partial T}{\partial H}\right)_P = \left[T\left(\frac{\partial V}{\partial T}\right)_P - V\right]\frac{1}{C_P} \text{ and} \left(\frac{\partial T}{\partial H}\right)_P = \frac{1}{C_P}$$

or
$$\mu = \left(\frac{\partial T}{\partial H}\right)_P = \left[T\left(\frac{\partial V}{\partial T}\right)_P - V\right]\frac{1}{C_P} = \left[\frac{T}{V}\left(\frac{\partial V}{\partial T}\right)_P - 1\right]\frac{V}{C_P} \tag{7.79}$$

where, μ is known as Joule-Kelvin coefficient and volume coefficient, $\alpha = \frac{1}{V}\left(\frac{\partial V}{\partial T}\right)_P$, above equation can be written as

$$\mu = [T\alpha - 1]\,\frac{V}{C_P} \tag{7.80}$$

For real gases, we know that van der Waals equation is

$$\left(P + \frac{a}{V^2}\right)(V - b) = RT$$

or
$$PV - Pb + \frac{a}{V} - \frac{ab}{V^2} = RT$$

Differentiating the above equation at constant pressure, we get

$$PdV - \frac{a}{V^2}dV + 2\frac{ab}{V^3}dV = RdT$$

$$\left[P - \frac{a}{V^2} + 2\frac{ab}{V^3}\right]dV = RdT$$

$$\frac{dV}{dT} = \frac{R}{\left[P - \frac{a}{V^2} + 2\frac{ab}{V^3}\right]}$$

or
$$\left(\frac{\partial V}{\partial T}\right)_P = \frac{R}{\left[\frac{RT}{(V-b)} - \frac{a}{V^2} - \frac{a}{V^2} + 2\frac{ab}{V^3}\right]} \qquad \text{as } P = \frac{RT}{(V-b)} - \frac{a}{V^2}$$

$$\left(\frac{\partial V}{\partial T}\right)_P = \frac{R}{\left[\frac{RT}{(V-b)} - 2\frac{a}{V^3}(V-b)\right]}$$

or
$$\left(\frac{\partial V}{\partial T}\right)_P = \frac{R}{\frac{\left[RTV^3 - 2a\left(V-b\right)^2\right)\right]}{V^3(V-b)}} = -\frac{RV^3(V-b)}{\left[RTV^3 - 2a(V-b)^2\right]}$$

Substituting this value in equation (7.80), we get

$$\mu = \left[\frac{T}{V}\frac{RV^3(V-b)}{\left[RTV^3 - 2a\left(V-b)^2\right)\right]} - 1\right]\frac{V}{C_P}$$

$$= \left[\frac{RTV^2(V-b) - RTV^3 + 2a\left(V-b^2\right)}{\left[RTV^3 - 2a\left(V-b^2\right)\right]}\right]\frac{V}{C_P}$$

$$= \left[\frac{RTV^3 - RTV^2b - RTV^3 + 2a\left(V-b^2\right)}{\left[RTV^3 - 2a(V-b)^2\right]}\right]\frac{V}{C_P}$$

$$= \left[\frac{-RTV^2b + 2a(V-b)^2}{\left[RTV^3 - 2a(V-b)^2\right]}\right]\frac{V}{C_P} \quad (7.81)$$

In equation (7.81), the value of V and C_P is positive, so the value of Joule coefficient μ and temperature of real gas depends on the value of term $[T\alpha - 1]$. There are three possibilities:

(i) In case of $T\alpha > 1$, the temperature of van der Walls gas will increase.
(ii) In case of $T\alpha < 1$, the temperature of van der Walls gas will decrease.
(iii) In case of $T\alpha = 1$, the temperature of van der Walls gas will not change.

The temperature (of a real gas) at which inversion takes place is known as temperature of inversion.

If $T\alpha = 1$, from equation (7.81), we get

$$\mu = 0$$

So from equation (7.81), we get

$$\mu = -RT_iV^2b + 2a\,(V-b)^2 = 0$$

or

$$RT_iV^2b = 2a\,(V-b)^2$$

$$T_i = \frac{2a(V-b)^2}{RV^2b}$$

As we know that b is very small in comparison to volume, so we may neglect it, we get

$$T_i = \frac{2aV^2}{RV^2b}$$

So temperature of inversion,

$$T_i = \frac{2a}{Rb} \quad (7.82)$$

For ideal gases from equation (7.80), we get

$$\mu = \left[T\left(\frac{\partial V}{\partial T}\right)_P - V\right]\frac{1}{C_P}$$

$$= \left[T^2\frac{\partial}{\partial T}\left(\frac{V}{T}\right)_P\right]\frac{1}{C_P}$$

We know that the gas equation for perfect gas is $PV = RT$

So,
$$\mu = \left[T^2\frac{\partial}{\partial T}\left(\frac{R}{P}\right)_P\right]\frac{1}{C_P} = 0 \tag{7.83}$$

Hence, the Joule coefficient for perfect gas is zero, which shows that when a perfect gas passes through porous plug in Joule-Kelvin experiment, its temperature does not show any change.

Numerical Examples

1. If C_1 is the specific heat of a liquid and C_2 is the specific heat of saturated vapor of the same liquid then prove that the Claussius second latent heat equation is given by the relation

$$(C_2 - C_1) = -\frac{L}{T} + \left(\frac{dL}{dT}\right)$$

Solution: If the S_1 and S_2 are entropies of liquid and vapor state of a substance, phase transition of a substance from liquid to vapor state is given by

$$S_2 - S_1 = \frac{L}{T}$$

Differentiating the above equation with respect to T, we get

$$\left(\frac{dS_2}{dT} - \frac{dS_1}{dT}\right) = -\frac{L}{T^2} + \frac{1}{T}\left(\frac{dL}{dT}\right)$$

$$T\left(\frac{dS_2}{dT} - \frac{dS_1}{dT}\right) = -\frac{L}{T} + \left(\frac{dL}{dT}\right)$$

$$\left(T\frac{dS_2}{dT} - T\frac{dS_1}{dT}\right) = -\frac{L}{T} + \left(\frac{dL}{dT}\right)$$

$$(C_2 - C_1) = -\frac{L}{T} + \left(\frac{dL}{dT}\right)$$

The above equation is known as Clausius second latent heat equation.

2. Ice freezes at 271°K under certain pressure. The change in specific volume in the process of freezing of 1 kg water is 90×10^{-6} m^3. Determine the pressure in the atmosphere.

Solution: We know that the freezing point of ice under ordinary pressure is 273°K, so change in freezing point of ice is given by

$$dT = 273 - 271 = 2°\text{K}$$

Given that,

Change in specific volume

$$dV = V_2 - V_1 = 90 \times 10^{-6} \text{ m}^3$$

We know that Claussius-Claypron's latent heat equation is given by

$$\frac{dP}{dT} = \frac{L}{T \cdot (V_2 - V_1)}$$

$$dP = \frac{L \times dT}{T \cdot (V_2 - V_1)}$$

$$= \frac{3.35 \times 10^5 \times 2}{273 \times 90 \times 10^{-6}}$$

$$= 12.704 \times 10^7 \text{ N/m}^2$$

$$= \frac{12.704 \times 10^7}{10^5} = 12.704 \times 10^2 \text{ atm}$$

$$= 1270.4 \text{ atm}$$

where 1 atm = 10^5 N

Hence, pressure at which ice freezed at the temperature 271°K

$$= 1270.4 + 1 = 1271.4 \text{ atm}$$

3. In a vaporization process, the latent heat of vaporization decreases at the rate of 0.54 cal/K. Latent heat of steam is 540 cal. Specific heat of water at 100°C is 1.0. Determine the specific heat of saturated steam.

Solution: We know that the relation between specific heat and latent heat is given by

$$C_2 - C_1 = \left(\frac{dL}{dT} - \frac{L}{T}\right)$$

$$C_2 = C_1 + \left(\frac{dL}{dT} - \frac{L}{T}\right)$$

$$= 1.01 + \left(-0.54 - \frac{540}{373}\right)$$

$$= 0.977 \text{ cal/gm}$$

4. If the pressure over water is increased by 2.711 mm, determine the change in the boiling point of water. Given that, latent heat of steam = 537 cal/gm, and specific volume of steam $V_2 = 1673.9$ cm^3.

Solution: We know that the normal boiling point of water at 1 atm pressure is 373°K

Hence, $T = 373°K$

Latent heat of steam $= 537 \text{ cal/gm} = 537 \times 4.19 \times 10^7 \text{ erg}$

$$V_1 = 1 \text{ cm}^3$$

$$V_2 = 1673.9 \text{ cm}^3$$

We know that Claussius-Claypron's latent heat equation is given by

$$\frac{dP}{dT} = \frac{L}{T \cdot (V_2 - V_1)}$$

So change in temperature

$$dT = \frac{dP \times T \times (V_2 - V_1)}{L}$$

$$= \frac{2.711 \times 13.6 \times 980 \times 373 \times 1673.9}{537 \times 4.2 \times 10^7}$$

$$= 0.99°C$$

5. The following date is given for a metallic copper disc at 27°C,

$$C_P = 24.49 \text{ J/mol°K}$$

$$K = 0.777 \times 10^{-11} \text{ N/m}^2$$

$$\alpha = 50.38 \times 10^{-6}\text{/K}$$

$$V = 7.058 \text{ cm}^3\text{/mol}$$

Obtain the value of specific heat at constant volume.

Solution: We know that the relation

$$(C_P - C_V) = TE\alpha^2 V = T\left(\frac{1}{K}\right)\alpha^2 V$$

$$= \frac{300 \times 50.38 \times 10^{-6} \times 50.38 \times 10^{-6} \times 7.058 \times 10^{-6}}{0.777 \times 10^{-11}}$$

$$= 0.6823 \text{ J/mol°K}$$

$$C_V = 24.49 - 0.6823$$

$$= 23.8077 \text{ J/mol°K}$$

6. In the adiabatic throttling of oxygen, the pressure is reduced by 49.8 atm. If initial temperature is 3000 °K, calculate the decrease in temperature. Given that

$$C_P = 3.49\ R$$

$$R = 8.32 \text{ J/K} - \text{mole}$$

$$a = 13.19 \times 10^{-2} \text{ Nm}^4/\text{mole}^2$$

$$b = 31.19 \times 10^{-6} \text{ m}^3/\text{mole}$$

$$V = 7.058 \text{ cm}^3/\text{mol}$$

Solution: Change in pressure

$$dP = 49.8 \times 0.76 \times 13.60 \times 10^{-2} \times 9.8$$

We know that the drop in temperature in adiabatic throttling is given by

$$dT = \left(\frac{2a}{RT} - b\right) \times \left(\frac{dP}{JMC_P}\right)$$

$$= \left(\frac{2 \times 13.19 \times 10^{-2}}{8.32 \times 300} - 31.19 \times 10^{-6}\right) \times \left(\frac{49.8 \times 0.76 \times 13.60 \times 10^{-2} \times 9.8}{4.2 \times 32 \times 3.49 \times 8.32}\right)$$

$$= 9.678 \times 10^{-2} \text{ °K}$$

7. Determine the values of $\gamma = \left(\frac{C_P}{C_V}\right)$ at 1 atm pressure and 0°C temperature. Given that

$$C_P = 27.9 \text{ J/mol°K}$$

$$\beta = 1.80 \times 10^{6} \text{ °K}$$

$$K = 3.87 \times 10^{-11} \text{ P/atm}$$

$$V = 1.468 \times 10^{-5} \text{ cm}^3/\text{mol}$$

Solution: We know that the relation

$$(C_P - C_V) - T\left(\frac{1}{K}\right)\beta^2 V$$

$$= \frac{273 \times 1.80 \times 10^{-6} \times 1.80 \times 10^{-6} \times 1.468 \times 10^{-5}}{3.87 \times 10^{-11}}$$

$$= 339 \times 10^{-6} \text{ J/mol°K}$$

$$C_P = 27.9 - 3.39 \times 10^{-4}$$

$$C_V = 27.8888 \text{ J/mol°K}$$

So,
$$\gamma = \left(\frac{C_P}{C_V}\right)$$

$$\gamma = \left(\frac{27.9}{27.888}\right) \approx 1$$

8. With the help of Maxwell's thermodynamic relations, show that the adiabatic volume coefficient of expansion and isobaric volume coefficient of expansion have the following relation

$$\frac{\alpha_S}{\alpha_P} = \left(\frac{1}{1-\gamma}\right)$$

Solution: We know that the isobaric volume coefficient is defined by

$$\alpha_P = \frac{1}{V}\left(\frac{\partial V}{\partial T}\right)_P$$

and the adiabatic volume coefficient is defined by

$$\alpha_S = \frac{1}{V}\left(\frac{\partial V}{\partial T}\right)_S$$

So, the ratio of these two is given by

$$\frac{\alpha_S}{\alpha_P} = \frac{\frac{1}{V}\left(\frac{\partial V}{\partial T}\right)_S}{\frac{1}{V}\left(\frac{\partial V}{\partial T}\right)_P}$$

$$= \frac{1}{\left(\frac{\partial V}{\partial T}\right)_P\left(\frac{\partial T}{\partial V}\right)_S}$$

But we know that from Maxwell's third relation,

$$\left(\frac{\partial T}{\partial V}\right)_S = -\left(\frac{\partial P}{\partial S}\right)_V$$

Hence,

$$\frac{\alpha_S}{\alpha_P} = \frac{1}{-\left(\frac{\partial P}{\partial S}\right)_V\left(\frac{\partial V}{\partial T}\right)_P}$$

$$= \frac{1}{-\left[\left(\frac{\partial P}{\partial T}\right)_V \times \left(\frac{\partial T}{\partial S}\right)_V\right]\left(\frac{\partial V}{\partial T}\right)_P}$$

$$= \frac{\left(\frac{\partial S}{\partial T}\right)_V}{-\left(\frac{\partial P}{\partial T}\right)_V \left(\frac{\partial V}{\partial T}\right)_P}$$

$$= \frac{T \times \left(\frac{\partial S}{\partial T}\right)_V}{-T \times \left(\frac{\partial P}{\partial T}\right)_V \left(\frac{\partial V}{\partial T}\right)_P}$$

But we also know that

$$(C_P - C_V) = T\left(\frac{\partial P}{\partial T}\right)_V \left(\frac{\partial V}{\partial T}\right)_P$$

and $$C_V = T \times \left(\frac{\partial S}{\partial T}\right)_V$$

Substituting these values in the expression $\frac{\alpha_S}{\alpha_P}$, we get

$$\frac{\alpha_S}{\alpha_P} = \frac{C_V}{(C_P - C_V)}$$

$$= \frac{C_V / C_V}{-(C_P / C_V - 1)}$$

$$= \frac{1}{-(C_P / C_V + 1)}$$

$$= \frac{1}{(-\gamma + 1)}$$

9. Show that for a ideal gas

$$\left(\frac{\partial U}{\partial V}\right)_T = 0$$

where, U is internal energy of gas, V and T are volume and temperature, respectively.

Solution: We know that Maxwell's first relation is

$$\left(\frac{\partial S}{\partial V}\right)_T = \left(\frac{\partial P}{\partial T}\right)_V$$

But we know that,
$$dS = \frac{dQ}{T}$$

So,
$$\frac{1}{T} \times \left(\frac{\partial Q}{\partial V}\right)_T = \left(\frac{\partial P}{\partial T}\right)_V$$

$$\left(\frac{\partial Q}{\partial V}\right)_T = T \times \left(\frac{\partial P}{\partial T}\right)_V$$

$$\frac{(dU + PdV)}{dV} = T \times \left(\frac{\partial P}{\partial T}\right)_V$$

$$\frac{dU}{dV} + P = T \times \left(\frac{\partial P}{\partial T}\right)_V$$

$$\left(\frac{\partial U}{\partial V}\right)_T + P = T \times \left(\frac{\partial P}{\partial T}\right)_V$$

Ideal gas equation
$$PV = RT$$

So,
$$\left(\frac{\partial P}{\partial T}\right)_V = \frac{R}{V}$$

Substituting this value in the above expression, we get

$$\left(\frac{\partial U}{\partial V}\right)_T + P = T \times \frac{R}{V}$$

$$\left(\frac{\partial U}{\partial V}\right)_T + P = T \times \frac{P}{T}$$

$$\left(\frac{\partial U}{\partial V}\right)_T + P = P$$

$$\left(\frac{\partial U}{\partial V}\right)_T = 0$$

10. Using Maxwell's thermodynamic relations, show that the adiabatic pressure coefficient of expansion and isochoric pressure coefficient of expansion have the following relation

$$\frac{\beta_S}{\beta_P} = \left(\frac{\gamma}{1-\gamma}\right)$$

Solution: We know that the isochoric pressure coefficient is defined by

$$\beta_V = \frac{1}{P}\left(\frac{\partial P}{\partial T}\right)_V$$

and the adiabatic pressure coefficient is defined by

$$\beta_S = \frac{1}{P}\left(\frac{\partial P}{\partial T}\right)_V$$

So the ratio of these two is given by

$$\frac{\beta_S}{\beta_P} = \frac{\frac{1}{P}\left(\frac{\partial P}{\partial T}\right)_S}{\frac{1}{P}\left(\frac{\partial P}{\partial T}\right)_P}$$

$$= \frac{1}{\left(\frac{\partial P}{\partial T}\right)_V \left(\frac{\partial T}{\partial P}\right)_S}$$

But we know that from Maxwell's fourth relation,

$$\left(\frac{\partial T}{\partial P}\right)_S = \left(\frac{\partial V}{\partial S}\right)_P$$

Hence,

$$\frac{\beta_S}{\beta_P} = \frac{1}{\left(\frac{\partial P}{\partial T}\right)_V \left(\frac{\partial V}{\partial S}\right)_P}$$

$$= \frac{1}{\left[\left(\frac{\partial P}{\partial T}\right)_P \times \left(\frac{\partial T}{\partial S}\right)_{PV}\right]\left(\frac{\partial V}{\partial T}\right)_V}$$

$$= \frac{\left(\frac{\partial S}{\partial T}\right)_P}{\left(\frac{\partial P}{\partial T}\right)_V \left(\frac{\partial V}{\partial T}\right)_P}$$

$$= \frac{T \times \left(\frac{\partial S}{\partial T}\right)_P}{T \times \left(\frac{\partial P}{\partial T}\right)_P \left(\frac{\partial V}{\partial T}\right)_V}$$

But we also know that

$$(C_P - C_V) = T\left(\frac{\partial P}{\partial T}\right)_V \left(\frac{\partial V}{\partial T}\right)_P$$

and

$$C_P = T \times \left(\frac{\partial S}{\partial T}\right)_P$$

Substituting these values in the expression $\frac{\beta_S}{\beta_P}$, we get

$$\frac{\beta_S}{\beta_P} = \frac{C_P}{(C_P - C_V)}$$

$$= \frac{C_P/C_V}{(C_P/C_V - 1)}$$

$$= \frac{\gamma}{(\gamma - 1)}$$

Exercise

Short Answer Questions

1. What do you mean by the intensive and extensive variables? Show that an extensive variable may be converted into an intensive variable.
2. Write the Maxwell's thermodynamic relations and explain each term.
3. Using Maxwell's relation, prove that $\frac{1}{\alpha}\left(\frac{\partial T}{\partial P}\right)_S = \frac{TV}{C_P}$.
4. Why Joule-Thomson effect is not applicable on perfect gases?
5. Write the application of Joule-Kelvin effect.
6. What are thermodynamic potentials? Write the four thermodynamical potentials and explain their physical significance.
7. Determine the values of $\left(\frac{\partial U}{\partial V}\right)_T$ and $\left(\frac{\partial U}{\partial P}\right)_V$ for van der Waals gas, where, U is an internal energy of system.
8. Write down Maxwell's four thermodynamic relations. Using these relations, show that

$$\left(\frac{\partial S}{\partial V}\right)_T = \left(\frac{\partial P}{\partial T}\right)_V$$

9. State the first Clausius-Clapeyron equation and prove that

$$\frac{dP}{dT} = \frac{L}{T \cdot (V_2 - V_1)}$$

Here, L is the latent heat, V_1 and V_2 are specific volumes in two states and T is absolute temperature.

10. What is the effect of pressure on (i) melting point of a solid, and (ii) boiling point of a liquid?
11. Write the differences between the internal energy of real gas and ideal gas.
12. What is the physical significance of Maxwell's four thermodynamic relations.
13. Explain the reason why the internal energy of a perfect gas is a temperature-dependent function while the internal energy of van der Waals gas is a temperature- and volume-dependent function?
14. Define four thermodynamic potentials represented by U, F, H and G. Obtain the relationship between four thermodynamic potentials and state variables.
15. What do you mean by the first order transition?
16. Derive Ehrenfest's theorem with the second order phase transition.
17. What are the $T \cdot dS$ equations? Derive the first and second $T \cdot dS$ equation.

Long Answer Questions

1. Write the Maxwell's thermodynamic relations and derive the van der Waals equation by Maxwell's relations.
2. Prove that for homogeneous liquid, the relation between C_P and C_V is given by

$$C_P - C_V = T\left(\frac{\partial P}{\partial T}\right)_V \left(\frac{\partial V}{\partial T}\right)_P$$

where, all the symbols have their usual meaning.

3. Explain the Joule-Thomson cooling effect. Deduce the temperature of inversion using Maxwell's relation.
4. What is Clausius-Clapeyron's latent heat equation? Derive this equation using Maxwell's thermodynamic relations.
5. Define the adiabatic and isothermal elasticities. Prove that the relation between the two is given by the relation

$$\frac{E_S}{E_T} = \frac{C_P}{C_V} S$$

6. Define Joule-Thomson coefficient. Find out its value for a perfect gas and for a real gas.
7. Derive the Clausius-Clapeyron's latent heat equation using Maxwell's thermodynamic relations and discuss the effect of pressure on the boiling and melting points.

8. Write the Maxwell's relation and using those relations show that

$$TdS = C_V dT + T\left(\frac{\partial P}{\partial T}\right)_V dV$$

9. Discuss the first order and second order phase transitions. Prove that

$$\left(\frac{dP}{dT}\right) = \frac{(S_2 - S_1)}{(V_2 - V_1)}$$

where, all the symbols have their usual meaning.

10. Define enthalpy and Gibb's potential. Prove that for simultaneous occurrence of isobaric and isothermal process, the value of Gibb's free energy remains constant.

11. Explain the result of Joule – Kelvin expansion. Prove that Joule – Kelvin coefficient is given by the expression

$$\mu = \frac{1}{C_P}\left[T\left(\frac{\partial V}{\partial T}\right)_P - V\right]$$

and its value for a perfect gas is zero.

12. Define thermodynamic potentials and using these potentials, derive Maxwell' thermodynamic relations.

13. Explain the internal energy and Helmholtz free energy. Show that the relation between internal energy and Helmholtz free energy is given by

$$U = F - T\left(\frac{\partial F}{\partial T}\right)_V$$

14. Define enthalpy of gas. Show that enthalpy remains constant in Joule – Kelvin expansion.

15. What do you mean by higher order phase transition? State and explain Gibb's phase rule with example.

16. Define Gibb's free energy and prove that for process at constant temperature and constant pressure, the value of Gibbs free energy gets minimum value at equilibrium.

17. Write the Maxwell's relations. Explain the process of heat production by adiabatic compression of a substance.

18. Write the differences between the first and second order phase transitions. Derive Clausius-Clapeyron's latent heat equation and show that under some specific conditions, latent heat becomes zero.

19. Define internal energy and show that

$$\left(\frac{\partial U}{\partial V}\right)_T = T\cdot\left(\frac{\partial P}{\partial T}\right)_V - P$$

20. Discuss the Gibbs free energy function and enthalpy. Prove that

$$\left(\frac{\partial H}{\partial T}\right)_P = C_P$$

21. Explain the Joule-Thomson effect and Joule coefficient μ. Using Maxwell's relation, prove that the value of μ for ideal gas is zero and for van der Walls real gas given by

$$\mu = \frac{1}{C_P} \cdot \left(\frac{2a}{RT} - b\right)$$

22. If C_1 is the specific heat of a liquid and C_2 is the specific heat of saturated vapor of the same liquid, then show that the Claussius second latent heat equation is given by the relation

$$(C_2 - C_1) = -\frac{L}{T} + \left(\frac{dL}{dT}\right)$$

23. What do you mean by the thermodynamic potentials? Define Gibbs free energy G and enthalpy H. Prove that the relation between G and H is given by expression

$$G = H + T\left(\frac{\partial G}{\partial T}\right)_P$$

Numerical Problems

1. Ice freezes at 371°K under a certain pressure. The change in specific volume in the process of freezing of 2 kg water is 80×10^{-6} m^3. Determine the pressure in the atmosphere.

2. Benzene boiled at 81.230°K temperature at 80 cm of Hg. Determine the normal boiling point of benzene. Latent heat of vaporization is 380 J/g, density of liquid is 0.89 g/cm^3 and density of vapor is 3.9 g/litre. [**Ans.** 80°C]

3. Determine the values of $\gamma = \left(\frac{C_P}{C_V}\right)$ at 1 atm pressure and 0°C temperature. Given that

$$C_P = 17.9 \text{ J/mol°K}$$

$$\beta = 2.70 \times 10^6 \text{ °K}$$

$$K = 1.67 \times 10^{-11} \text{ P/atm}$$

$$V = 1.429 \times 10^{-5} \text{ cm}^3\text{/mol}$$

4. The following date is given for a metallic disc at 30°C,

$$C_P = 34.40 \text{ J/mol°K}$$

$$K = 0.670 \times 10^{-11} \text{ N/m}^2$$

$$\alpha = 60.40 \times 10^{-6}/\text{K}$$

$$V = 8.32 \text{ cm}^3/\text{mol}$$

Obtain the value of specific heat at constant volume.

5. In a vaporization process, the latent heat of vaporization decreases at the rate of 0.63 cal/K. Latent heat of steam is 540 cal. Specific heat of water at 100°C is 1.0. Determine the specific heat of saturated steam.
6. Determine the change in boiling point of water for 70 cm change of Hg pressure. The specific volume of water steam is 17650 cm^3/gm at 373 °K and 0.76 m Hg pressure. Latent heat is given as 5364 cal/gm.
7. In the evaporation process, 1 gm water occupies 1600 cm^3. The latent heat of steam is given as 511.30 cal/gm. If pressure required to boil water is 787 mm of mercury, determine the boiling temperature of water. [**Ans.** 100.9°C]
8. In adiabatic throttling process, H_2 gas pressure is reduced by 60 atm. If initial temperature of gas is 30°C, determine the fall in temperature during the process.
9. A certain amount of ice is kept in cold storage at the external temperature 27.9°C. Determine the minimum power required to run the motor of refrigerator. Given that specific heat of fusion of ice is 3.29×10^5 J/kg.

   [**Ans.** 372.89 watts]
10. In a process, the latent heat is given as L = 700-0.605 T. Determine the specific heat of steam and show that it is negative.
11. Determine the pressure in atm which is required to decrease the melting point of ice by 2°K. Given: specific volume of water at 273°K = 2 cm^3, specific volume of ice at 273°K = 2.092 cm^3. L = 79.69 cal/g. [**Ans.** 133.19 atm]

Multiple-Choice Questions

1. The relation between C_P and C_V for a perfect gas is
   (i) $C_P - C_V = T^2 E\alpha V$ (ii) $C_P - C_V = TE\alpha V$
   (iii) $C_P - C_V = T^2E^2 \alpha^2 V$ (iv) $C_P - C_V = TE\alpha^2 V$
2. Out of these, thermodynamic potential is
   (i) Pressure (ii) Gibbs function
   (iii) Temperature (iv) Work
3. Helmholtz free energy is defined as
   (i) $F = U + TS$ (ii) $F = H + PV$
   (iii) $F = U - TS$ (iv) $F = H - PV$
4. Example of extensive variable is
   (i) Pressure (ii) Entropy
   (iii) Specific volume (iv) Density

5. First Maxwell's relation is

(i) $\left(\frac{\partial T}{\partial V}\right)_S = -\left(\frac{\partial P}{\partial S}\right)_V$ (ii) $\left(\frac{\partial T}{\partial V}\right)_S = \left(\frac{\partial P}{\partial S}\right)_V$

(iii) $\left(\frac{\partial T}{\partial V}\right)_S = -\frac{1}{T}\left(\frac{\partial P}{\partial S}\right)_V$ (iv) $\left(\frac{\partial T}{\partial V}\right)_S = -T\left(\frac{\partial P}{\partial S}\right)_V$

6. Enthalpy of a system remains constant in

(i) Isochoric isothermal process (ii) Isobaric adiabatic process

(iii) Isochoric adiabatic process (iv) Isentropic adiabatic process

7. Gibb's potential is defined as

(i) $G = F + PV$ (ii) $G = F - PV$

(iii) $G = F + TS$ (iv) $G = F - TS$

8. Specific heat for air at constant pressure C_P is equal to

(i) 0.17 (ii) 0.21

(iii) 0.24 (iv) 1.0

9. The ratio of specific heats for air is equal to

(i) 0.29 (ii) 0.32

(iii) 2.71 (iv) 0.51

10. Change in enthalpy of a system is the heat supplied at

(i) At constant volume (ii) Constant pressure

(iii) Constant temperature (iv) Constant energy

11. Helium and hydrogen at ordinary temperature show

(i) Cooling effect (ii) Heating effect

(iii) Cooling and heating both (iv) None of the above

12. Pulse takes longest time to cook

(i) At Mount Everest (ii) In Jammu

(iii) In Meghalaya (iv) At sea level

13. The value of specific heat of saturated vapor pressure is

(i) Positive (ii) Negative

(iii) Zero (iv) None of the above

14. First order phase transition occurs at

(i) Constant temperature and pressure

(ii) Constant volume and pressure

(iii) Constant energy and pressure

(iv) Constant volume and temperature

Answers

1. (iv) 2. (ii) 3. (iii) 4. (ii) 5. (i) 6. (ii) 7. (iv) 8. (iii)
9. (iii) 10. (ii) 11. (ii) 12. (i) 13. (i) 14. (i)

8

Change of State in Matter

8.1 CHANGE OF STATE IN MATTER

There are three different states of matter, i.e., solid, liquid and gas. A specific state of matter depends on the temperature. Kinetic theory explains that molecules of liquid or gaseous state have higher degree of freedom than solid state molecules. The gaseous molecules are more free to move in comparison to molecules of other two states. A matter state can be changed at a fixed pressure and a given temperature and vice versa. A fixed amount of heat at constant temperature is supplied to convert the solid state into liquid state. This supplied heat is utilized in overcoming the intermolecular forces between the molecules and is known as the melting point. Similarly, a certain amount of heat is supplied at a fixed temperature and at standard atmospheric pressure to change the matter state from liquid to gases. This heat is utilized in increasing the mean distance between the molecules and is known as the boiling point. During a phase transition, the temperature of matter does not show any change, although extra amount of heat is given or taken from the matter.

The excess amount of heat energy is spent in overcoming the latent heat of the matter, so the temperature of matter does not show change until the phase transition process is complete. Figure 8.1 shows the variation in temperature of water when additional heat is given. During transition, the amount of additional heat energy that is given, goes into conquer the latent heat of water, hence, the temperature of water does not change so far the phase change process is complete.

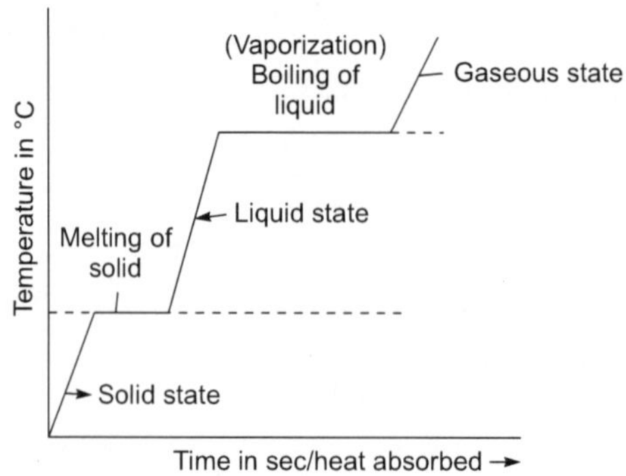

Figure 8.1 Phase changes of a substance

8.2 FUSION LAWS

Fusion of a substance is based on the following laws:

(i) At a fixed temperature and normal pressure, a matter transforms its state from solid to liquid.
(ii) After some duration of the process, no further change occurs in temperature.
(iii) The volume of some specific matter decreases on melting, such as ice, while volume of some matter increases on melting, like ghee, wax, etc.
(iv) To change the state of matter, a definite amount of heat is supplied to 1 gm of matter. This heat is known as latent heat of fusion of that matter. Different matters have different latent heats.
(v) The matter whose volume decreases on melting, shows decrement in its melting point at a high pressure.
(vi) The matter whose volume increases on melting, shows a rise in its melting point with increase in pressure.

8.3 LATENT HEAT OF FUSION OF A SUBSTANCE

Latent heat of a matter or substance is the amount of heat per unit mass required to change the phase of state. The latent heat of fusion specifies the phase change between a solid and liquid and is defined as the quantity of heat needed to transit the state of 1 gm of a matter from solid to liquid, leaving temperature of a substance unaltered (Figure 8.2).

Suppose a small cube of ice is taken in a container and a thermometer is fixed on the pot to measure the temperature of ice. If we heat the container slowly, ice melts but no change appears in the thermometer. After the whole process of melting of ice, the thermometer indicates the rise in temperature of water. Hence, during the process of transition from ice to water, certain amount of heat is utilized in the conversion process of state (ice to water). This amount of heat is known as latent heat of fusion of ice. This amount of heat is hidden and cannot be indicated in thermometer during the transition process. One gm of ice requires 80 cal of heat to change to water. This quantity of heat is latent heat of fusion of ice and is measured in cal/gm.

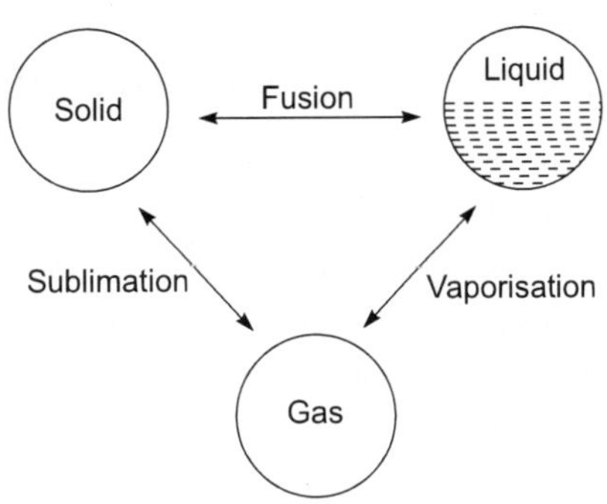

Figure 8.2 Latent heat of fusion

8.4 ICE REGELATION PROCESS

Let us take a big slice of ice and place it above the two weights. Take a thin wire and two weights, each of 5 kg. Fix these weights at the two ends of wire and place the wire on top of the slice, as shown in Figure 8.3.

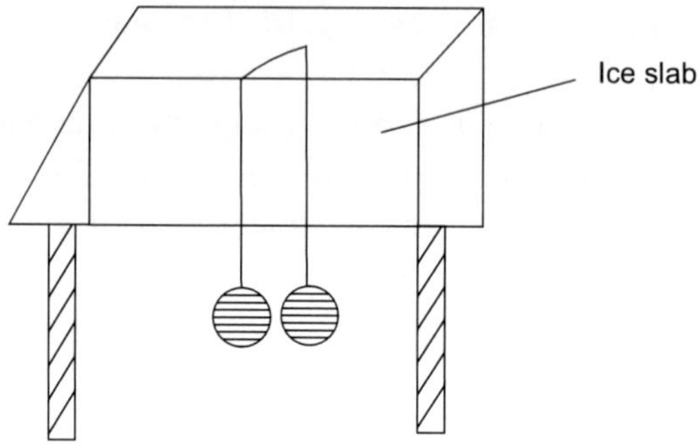

Figure 8.3 Ice regelation process

This wire passes through the ice slice. Due to an increase in pressure below the wire, ice starts to melt at a very low temperature and when the wire crosses the slice, the water collected above the wire freezes once again. Hence, the wire passes through the ice slice but ice slice does not break or split. This refreezing phenomenon is known as regelation. There are some examples of the regelation process as follows:

(i) During the passage of a cart over snow, snow melts by pressure of the wheels. When wheel passes over the snow, water collected over the wheel converts in solid form due to regelation process. That is the reason why wheels are blanketed with snow.

(ii) Pipes of water burst in winters due to the solidification process of water in ice. In this process of conversion of water into ice, pipes burst due to an increase in volume.

(iii) When we skate on snow, due to the increase in pressure, a thin layer of water is formed at the bottom surface of skates. This layer works as a lubricant and skating becomes easy in snow.

8.5 EFFECT OF IMPURITIES ON FREEZING POINT

When some impurities are added to a substance, its freezing point decreases than the actual value. For example, if salt is mixed in ice, ice starts to melt by taking heat energy from salt and the temperature of impure ice decreases. Now salt liquefies and takes some latent heat from the mixture of ice and salt. Again, the temperature decreases. If we take the freezing mixture of ice and salt in the ratio 3 : 1, temperature of the mixture can be lowered to –13°C. Similarly, water can be cooled by dissolving sugar in it.

8.6 PROCESS OF FINDING LATENT HEAT OF FUSION OF ICE

The latent heat of fusion of ice (L) can be determined with the help of a calorimeter. Let us take a calorimeter, lid, stirrer, weighing machine, thermometer, ice pieces, beaker, and blotting paper. First, weigh calorimeter with stirrer and lid but without lid. Take water in a beaker and warm it up to temperature of 10°C above the room temperature (Figure 8.4). Now fill the calorimeter with this warm water. Again weigh calorimeter with stirrer, hot water and lid. Place a thermometer inside the calorimeter and note the temperature of water. This temperature should be 4-5°C more than the room temperature. Take a bloating paper and put some pieces of ice in it and place it inside the calorimeter. The ice chunks are continually added till the temperature is less than the room temperature (up to 4-5°C). Now check if ice is melted completely, then again note the temperature and weigh the whole material.

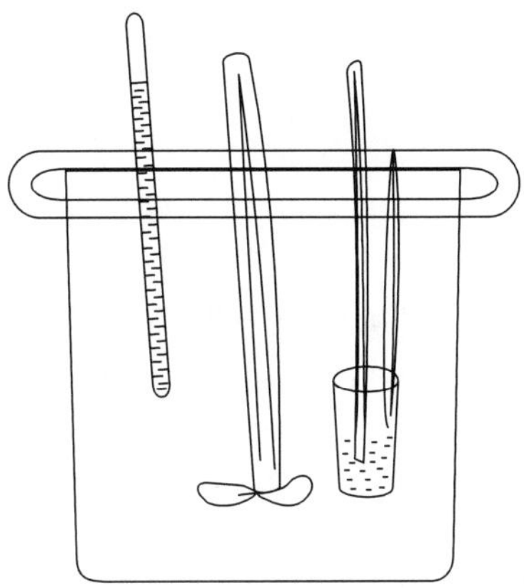

Figure 8.4 Process of fusion of ice

We know that the amount of heat required to melt 1 gm of ice to water at 0°C is known as latent heat of fusion of ice. Let us assume that M gm of ice is mixed in m gm of water at t°C in a calorimeter. Let W be the water equivalent of calorimeter and T is the final temperature.

Amount of heat lost by the calorimeter and water = $(m + W) \times (t - T)$

Quantity of heat gained by ice = $(ML + MT)$

As we know that, quantity of heat gained = quantity of heat lost

So,
$$(ML + MT) = (m + W) \times (t - T)$$

or
$$M(L + T) = (m + W) \times (t - T)$$

or
$$(L + T) = \frac{(m+W)\times(t-T)}{M}$$

or
$$L = \frac{(m+W)\times(t-T)}{M} - T \qquad (8.1)$$

Equation (8.1) gives the relation for determination of latent heat of fusion of ice.

8.7 PROCESS OF FINDING MELTING POINT OF WAX

To determine the melting point of wax, the following setup is designed (Figure 8.5).

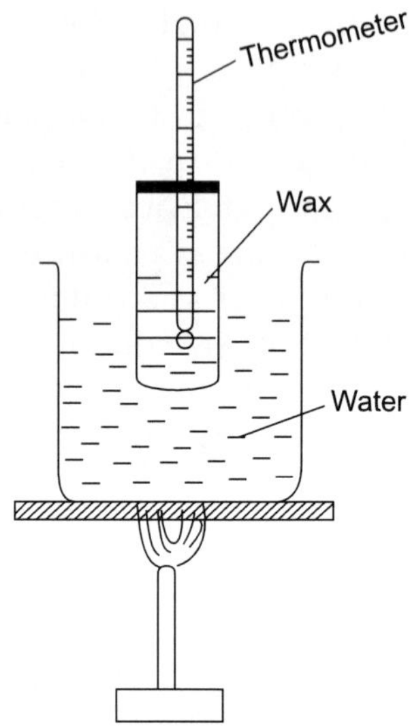

Figure 8.5 Setup for finding melting point of wax

Setup consists of a test tube fitted in a beaker. Wax is taken inside the test tube and water is inside the beaker. A sensitive thermometer is kept in the beaker to measure the temperature. Beaker is placed over the burner and water is allowed to heat up to its boiling point. The wax in the test tube starts to melt and comes in liquid form. Now turn off the burner and allow the wax and water to lower their temperature and cool. Measure the temperature of wax with thermometer in the interval of 1 minute. When the temperature of wax decreases up to 30°C, stop the measurement. Now plot a graph between temperature of wax and time (Figure 8.6).

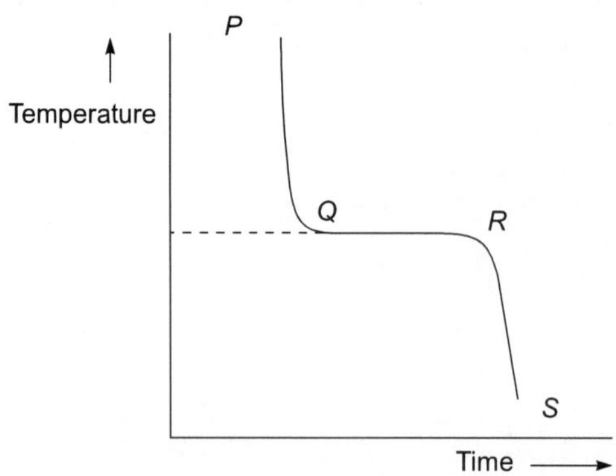

Figure 8.6 Temperature vs time graph to study different state of matter

Time is taken along *X* axis and temperature is taken along *Y* axis. In the graph, a horizontal line *QR* parallel to time axis is obtained. This horizontal line *QR* provides the melting point of wax. At this particular temperature, the liquid wax will again change to solid wax without any change in temperature. The region *PQ* provides the liquid form of wax and the region below *R* provides the solid form of wax.

8.8 VAPORIZATION AND CONDENSATION PROCESS

The process of conversion of a substance from liquid phase to gaseous phase, is known as vaporization and if a substance directly converts from solid phase to gaseous form, then it is known sublimation. There are two possible types of vaporization, one is evaporation and the other is boiling.

Vaporization on the top surface of liquid is termed evaporation. In the evaporation process, the molecules of liquid are excited by absorbing the heat energy from heat source. These excited molecules break the intermolecular bond between each other and convert in the form of gas. For example, the water of a river is evaporating when the molecules at the top surface of water are heated up by sun's heat energy.

Boiling is a different process in which a liquid is heated at a fixed atmospheric pressure. When a substance's internal pressure approaches to atmospheric pressure, the substance starts to boil and molecules in the substance convert in the gaseous form or phase. The temperature at this level is termed the boiling point of the substance. Different substances have different boiling points.

In both types of vaporization process, a liquid converts into gaseous form. In the evaporation process, only the surface of liquid transforms into gas in the presence of low internal vaporization pressure. But in the process of boiling of a substance or liquid at very high vaporization process, the whole liquid is transformed into gas along with the top surface.

In both evaporation and boiling, temperature of liquid gets a fixed value. Such high temperature causes the breakdown of intermolecular attraction between the molecules, and the liquid turns into gaseous phase. If the temperature of liquid starts to reduce and goes back down, then the molecules again turn back into liquid form. This conversion process of a substance from the gas (or vapor) form to liquid phase is known as condensation. For example, steam is converted into water due to the condensation process. Rain is also the condensation process of water vapor present in the form of clouds (Figure 8.7).

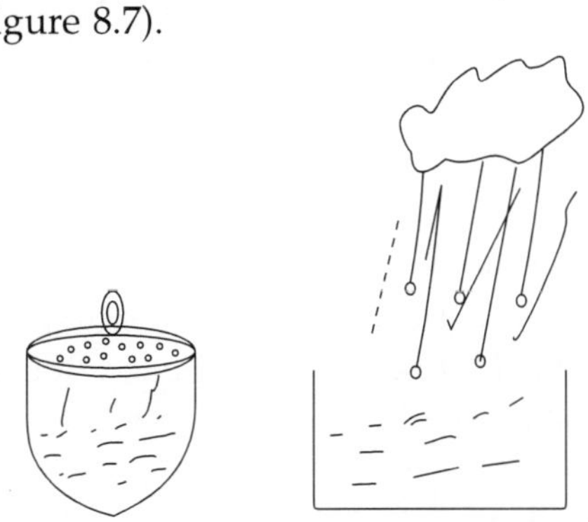

Figure 8.7 Condensation process

8.9 BOILING OR EBULLITION LAWS

There are certain laws to covert a liquid into gaseous form by boiling it:

(i) When a liquid state changes its phase into gaseous state at a particular temperature and at ordinary pressure, it is known as boing point of the substance.
(ii) As the conversion process completes, no further change in temperature is allowed.
(iii) The volume of all liquids increases in the vaporization process.
(iv) A liquid can be boiled at reduced temperature under the presence of low pressure. If the pressure of a liquid increases, then boiling point of liquid will also increase.

8.10 DETERMINATION OF LATENT HEAT OF VAPORIZATION

The amount of heat required to change 1 gm of liquid to vapor state, is known as latent heat of vaporization, with unaltered temperature. Latent heat of vaporization is not same for every substance. Latent heat of vaporization for water is 537 calories per gram at a fixed atmospheric pressure of 76 cm of mercury.

The following setup is used to determine the latent heat of vaporization. The setup consists of a calorimeter, stirrer, lid, boiler, steam trap, wool and thermometer (Figure 8.8).

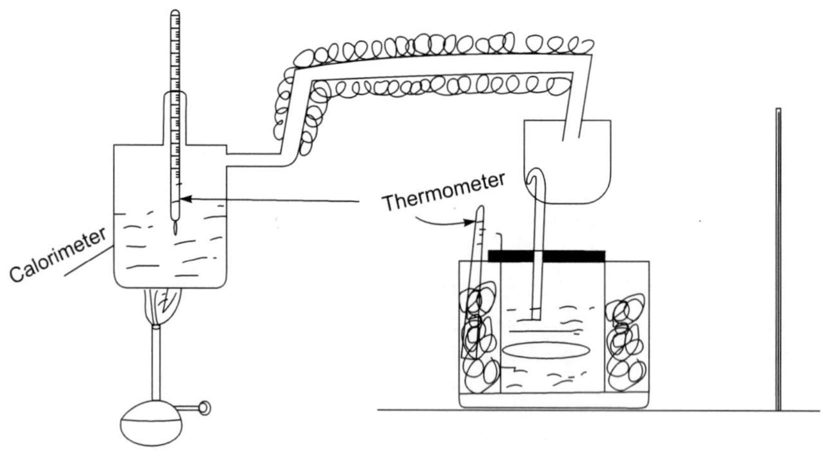

Figure 8.8 Setup to determine latent heat of vaporization

Let us take a calorimeter, weigh the calorimeter with the stirrer and lid. Now fill the calorimeter with two-thirds of water and again weigh the calorimeter with water stirrer and the lid. Start to heat the water in the boiler. Boiler is connected with steam trap and steam container is covered with a cotton wool. As heat starts to come from the boiler and reaches at the end of the tube, temperature is noted down. Measure the temperature of water in calorimeter and wipe the tube ends with a blotting paper. Now place the tube into the calorimeter and allow to pass steam inside it.

Increase the temperature of water by 10°C and take the reading. During the process, continuous stirring is required. Now note down the final temperature and remove the thermometer. Now again weigh the calorimeter.

Let us suppose that m gm of steam is allowed to pass into M gm of water at t_1°C. Water equivalent of calorimeter is w. Let t_2°C and T°C be the temperature of steam and final temperature, respectively.

So amount of heat absorbed by the water and calorimeter = $(M + w) \times (T - t_1)$ and quantity of heat lost by the steam = $mL + m \times (t_2 - T)$.

As we know that the quantity of heat absorbed is equal to heat lost. Hence,

$$(M + w) \times (T - t_1) = mL + m \times (t_2 - T)$$

or $$(M + w) \times (T - t_1) = m \times [L + m(t_2 - T)]$$

or $$\frac{(M+w)\times(T-t_1)}{m} = [L+(t_2-T)]$$

or $$L = \frac{(M+w)\times(T-t_1)}{m} - (t_2 - T)$$

The above expression is used to calculate the latent heat of steam.

8.11 EFFECT OF PRESSURE ON THE BOILING POINT OF A GIVEN LIQUID

The boiling point of a liquid depends on the pressure applied on it. To study this effect of pressure on the boiling point, the following setup is used (Figure 8.9).

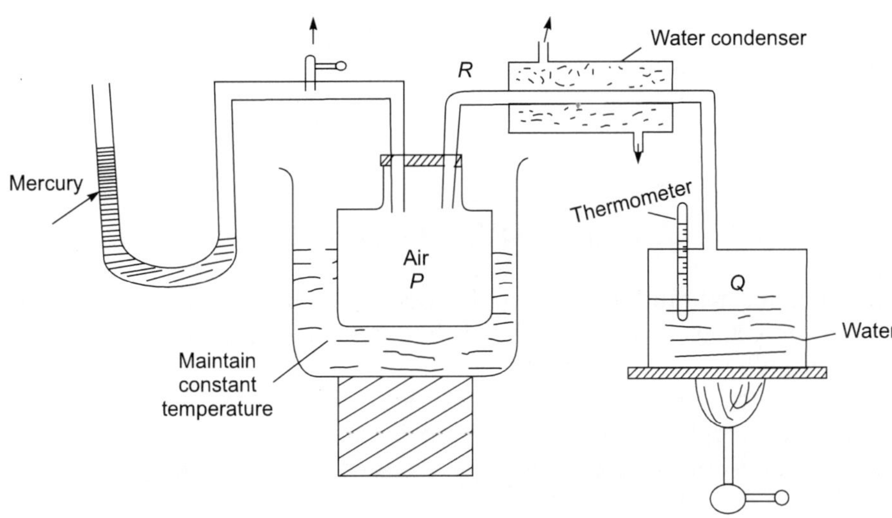

Figure 8.9 Setup to show the effect of pressure on boiling point

A substance changes its state from liquid to vapor at a particular temperature. This temperature is known as the saturation temperature and it increases with increase in pressure. The saturation temperature at one atmosphere pressure is known as the boiling point. In the experimental setup, liquid is contained inside the vessel P and connected with the flask Q by a tube R covered with water condenser. The air pressure over the surface of a liquid is measured with the help of a manometer. The air pressure inside vessel P and flask Q can be changed by using an evacuation pump or compression pump. A stopcock is used between the compression pump and flask Q. This stopcock is open and compression pump is allowed to compress air into Q. Now the stop cock is closed. Pressure of air in P and Q can be measured with manometer, which is higher than the atmospheric pressure. Liquid inside the container P is boiled and its boiling temperature is measured by the thermometer. The temperature of liquid remains unaltered at the boiling point. The steam is not allowed to enter the vessel Q by flowing cold water through the condenser. The steam is allowed to condense and condensed water goes back to the vessel P.

At different pressure, respective boiling point of liquid can be measured. To measure the boiling point at low pressure, vacuum pump is used. After the experiment, it is observed that the boiling point of liquid decreases with decrease in pressure and increases with increase in pressure. The boiling point of water at normal pressure of 76 cm of mercury is 100°C and it is 128°C at pressure 152 cm of mercury (double the normal pressure).

A number of equipment designed on the concept of the principle that the boiling point of a liquid depends on the pressure. Pressure cooker worked on this principle.

The pressure inside the pressure cooker is higher than the normal atmospheric pressure. When a pressure cooker is filled with water up to the normal level and heated by putting it on the stove, vaporization of water starts. The steam produced in the cooker due to heating cannot escape the cooker, so pressure on the top of water surface inside the pressure cooker increases. Hence, the temperature of water or liquid kept inside the cooker is higher than the boiling point (at the normal atmospheric pressure) and food inside the cooker is cooked in a short time.

8.12 EFFECT OF COOLING DUE TO VAPORIZATION

The molecules of a liquid are free and move in a random direction. During their motion, the molecules collide with each other and transfer their energy to each other and any molecule may get energy higher than its own energy. When it happens, the molecule leaves the surface of liquid. In the absence of any additional kinetic energy, the total kinetic energy will show decrement and a fall in temperature. When a liquid evaporates, the rest of the liquid shows a decrease in temperature.

8.13 EXAMPLES OF EVAPORATION

(i) When a person is wearing wet clothes, water evaporation process takes place. During the process of evaporation, latent heat is taken from the body of that person and the person feels cooling.

(ii) In hot season, we swelter. This is due to the evaporation process. We have fine pores in our skin and water comes out of these fine pores. This water evaporates by taking heat from our body and we feel chill.

(iii) We water garden, floor, lawns, etc., during summers. Water evaporation process takes place by using the heat from them and these things become cool.

(iv) Dogs keep usually their tongues out in summers. Evaporation process takes place by taking heat from the tongue and dogs feel chill.

(v) When we place hot liquids in a plate, the surface area of liquid increases. Evaporation process starts and the liquid cools down.

(vi) When we sit in a packed room, we feel uncomfortable due to perspiration. In a packed room, perspiration cannot evaporate and makes us uneasy. As we come out of the room, fast evaporation process of perspiration starts by taking the heat from our body.

(vii) During high fever, the body temperature increases and we keep wet cold cloth on the forehead. When we keep cold cloth on the forehead, rapid evaporation process starts by taking heat from the head and rest of the body and the body temperature falls down.

(viii) During the summers, people use *khas-khas* sheets to keep rooms cool. When water is poured on *khas-khas*, evaporation of water starts by taking heat from the room.

8.14 VAPOR AND GAS STATES

Vapor and gas states are two different stages of a single continuous phenomenon. Both the states can be separated by a specific temperature. This specific temperature is known as the critical temperature. When temperature is below the critical temperature, the substance is in its vapor state and as temperature reaches above the critical temperature, the substance is in the gaseous state.

A gas cannot be liquefied by applying pressure (either small or large) but a vapor can be liquefied by the application of pressure. Hence, to liquefy a gas, it has to be transformed into vapor state by cooling it below the critical temperature. Now, vapor can be liquefied by applying high pressure on it.

Therefore, the critical temperature of a gas or a substance is that particular temperature above which gas cannot be liquefied by the application of pressure. The value of critical temperature is different for different gases. The critical temperature of some gases is given in Table 8.1.

TABLE 8.1 Critical temperature of some important gases

Gases and Substances	Critical Temperature (in°C)
Ammonia	132
Carbon dioxide	31
Hydrogen	–240
Oxygen	–119
Helium	–268
Nitrogen	–147
Argon	–122
Water	374
Hydrogen chloride	51.6
Chlorine	144

8.15 SATURATED AND UNSATURATED VAPORS AND VAPOR PRESSURE

When a liquid form of a substance is taken in a closed container, then the remaining space above the liquid is filled with vapor. If this space cannot contain more vapors, then the vapor is known as saturated vapor and if it can contain more vapors, then it is known as unsaturated vapor. The molecules inside the container are continually leaving and returning in the liquid. At any instant of time, the number of molecules escaping the surface of liquid will be equal to the molecules re-entering the liquid. Hence, equilibrium is set up and the density molecules in air will be maximum. The properties of saturated vapor were studied by Dalton. Figure 8.10(a) shows the state of liquid before saturation has been obtained and Figure 8.10(b) shows the state of saturation (when the number of molecules leaving the surface is equal to the number of molecules re-entering the surface).

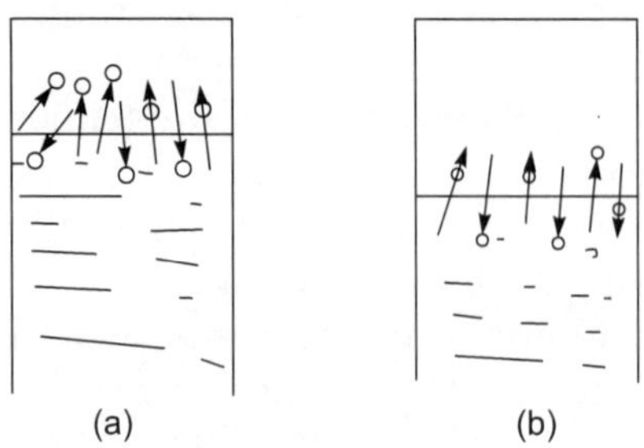

Figure 8.10

The saturated vapor has following properties:

(i) The saturated vapor pressure of a liquid depends on the property of the substance.

(ii) The saturated vapor pressure depends on the temperature of liquid. This pressure increases with increase in temperature and decreases with decrease in temperature.

(iii) The saturated vapors do not obey ordinary gas laws while unsaturated vapors of a liquid follow the gas laws.

(iv) The saturated vapor pressure is independent of volume filled by the vapors.

(v) The saturated vapors of a liquid are not affected by the presence of other liquid vapors and there will be no chemical reactions.

(vi) If we take more than one liquid, then the total pressure applied by the vapors of all liquids is equal to the sum of the individual vapor pressures,

$$P = P_i + P_j + P_k + \dots$$

If we plot a graph between the pressure and temperature of saturated vapor, then pressure of saturated vapor gradually increases with increase in temperature. But the pressure of unsaturated vapor is directly proportional to the temperature.

In the graph (Figure 8.11), region L to M represents the saturated vapor and M to N represents unsaturated vapor.

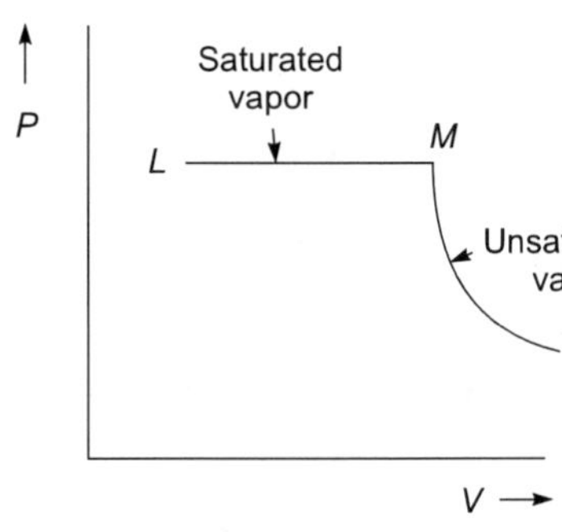

Figure 8.11 Variation of pressure with volume (saturated vapor and unsaturated vapor)

The graph between pressure and volume shows that in the region L to M, pressure increases with decrease in volume, i.e., in this region, vapor obeys Boyle's law. Hence, from L to M, vapor is unsaturated vapor (figure 8.12). At the point M, vapor applied maximum pressure and is saturated vapor. Beyond M, pressure remains constant in increasing volume. M to N region represents the conversion of saturated vapor in the liquid.

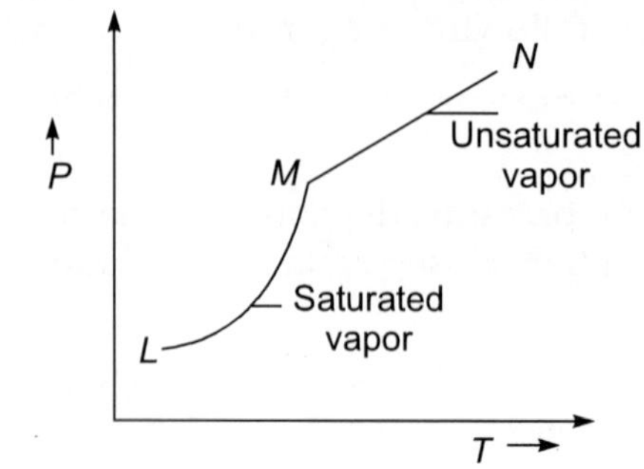

Figure 8.12 Variation of pressure with temperature (saturated vapor and unsaturated vapor)

8.16 DETERMINATION OF VAPOR PRESSURE

The pressure applied by gas to solid or liquid contained in a container at a particular temperature, is known as vapor pressure. The molecules of a liquid, closed in a container, continually go up and come back to liquid. At any instant, the number of molecules leaving the surface of liquid may be equal to the number of molecules entering the liquid and an influential equilibrium will be set up. The pressure exerted by saturated vapor is called saturated vapor pressure of that liquid. As the molecular velocity depends on temperature, so the saturated vapor pressure increases with increase in temperature. Hence, to determine the saturated vapor pressure of a liquid, the temperature must be fixed.

To measure vapor pressure of a liquid, the setup is shown in Figure 8.13. The apparatus consists of a vessel A, which contains the vapor of liquid. A reservoir R is used to reserve mercury and M is a manometer. A stopcock is used to control the volume of gas. When the stopcock is open, the gas enters the tube at a high pressure. With the help of a reservoir, we set the level of the liquid. Let us suppose that the manometer limb B contains gas at pressure P_1 and h is the difference in the level of mercury in manometer limbs. Hence, at temperature T,

$$\text{Vapor pressure of liquid, } P = P_1 + h$$

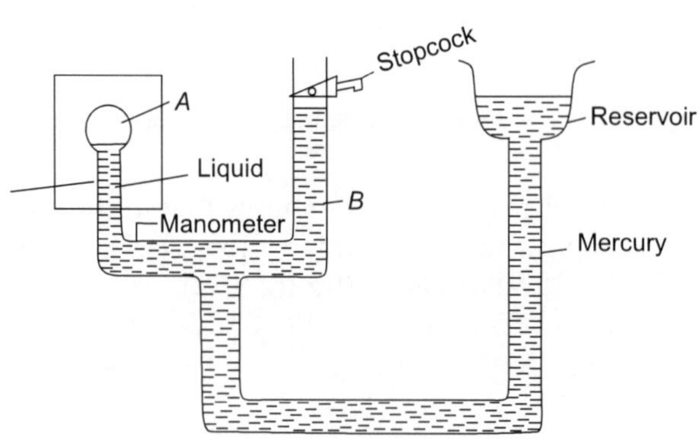

Figure 8.13 Setup to measure vapor pressure

To measure the vapor pressure at low temperature, the reservoir is lifted and the level of mercury in the limb B goes above the stopcock. Now, the stopcock is closed. The vessel B is covered with low temperature bath. The liquid inside the tube moves to bath with the adjustment of reservoir. We can measure vapor pressure directly by the difference of mercury level in the manometer, i.e.,

$$\text{Vapor pressure of liquid, } P = h \qquad \text{(as } P_1 = 0\text{)}$$

In further decrement of temperature, the liquid starts solidifying. At this stage, the temperature and pressure resemble to triple point.

8.17 TRIPLE POINT OF A SUBSTANCE

The triple point of a substance is the point at which solid, liquid and gas, all three phases of a substance exist simultaneously at thermal equilibrium. Hence, the triple point represents equilibrium between three phases of a substance. Let us discuss the three different phases of water and study the behavior of water in P-T diagram.

(i) The melting point or freezing point is the temperature at which a substance freezes at standard atmospheric pressure (760 mm Hg).The melting point of solid decreases with increase in pressure. Ice melts at a temperature lower than absolute zero and at a pressure larger than the ordinary pressure. In P-T diagram, the curve CD is known as fusion line or ice line and it provides the relation between temperature and pressure of melting of solid. In the left side of curve CD, substance will be in solid form and in the right side of it, it will be in liquid form.

(ii) Boiling point of a substance is the temperature at which it boils at normal atmospheric pressure. If water is boiled, then the boiling point of water rises with increase in pressure. In P-T diagram, the curve AB is known as vaporization or steam line. It represents the relation between the pressure and temperature of boiling of water. Above the AB, substance is in liquid form and below AB, substance is in the vapor form. The curve AB shows the equilibrium between the liquid and vapor phase.

(iii) The curve EF is known as sublimation line or hoar frost line. It provides the equilibrium between solid and vapor states of a substance. Below the curve EF, the substance will be in vapor phase and above the curve EF, it will be in solid state.

All these three curves AB, CD and EF should meet at a point O as shown in Figure 8.14(a). If these three curves do not coincide at a point, then in the shaded region, (Figure 8.14(b)), all three phases of a substance exist simultaneously. It is impossible that the matter exists in all three states at the same time. Hence, the three curves represent the three different phases of a substance, which must coincide at a single point known as triple point of that substance.

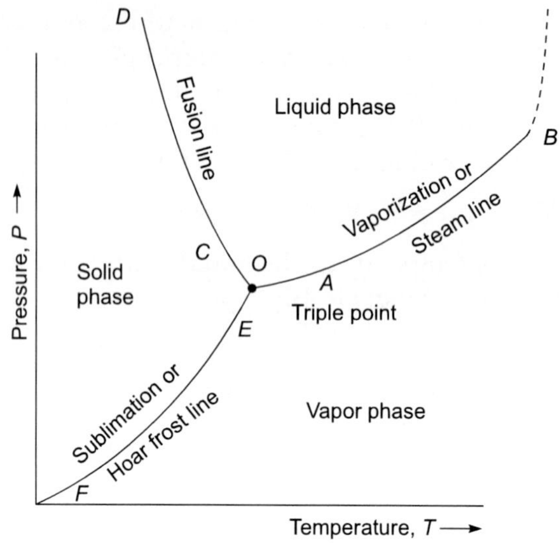

Figure 8.14(a) Triple point

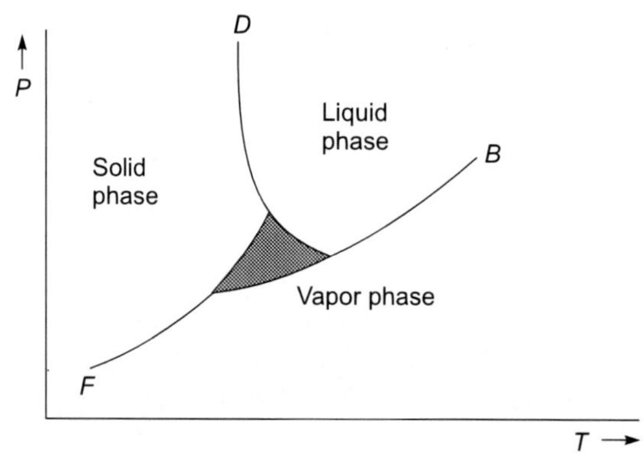

Figure 8.14(b)

The pressure and temperature of different substances are different at triple points. The pressure and temperature of some important substances at triple point are given in the following table.

TABLE 8.2 Triple point data of some important substances

Gases and Substances	Temperature (in°C)	Pressure (atm)
Ammonia	–78	0.06
Carbon dioxide	–57	5.1
Hydrogen	–259	0.071

Contd.

Helium	–271	0.0493
Nitrogen	–219	0.1237
Water	0.01	0.00603
Oxygen	–210	0.0015
Silver	961	0.0001
Copper	1083	0.00000078
Zinc	419	0.05

8.18 GIBBS PHASE RULE

To define the intensive state of a system, some fixed independent intensive properties are required. These minimum intensive properties depend on the complex nature of the system, number of approaches of interaction between the surroundings and system and the number of phases. There are some properties which do not depend on the number of phases and the number of chemical species. These properties are known as degrees of freedom, which is represented by:

$$F = C + 2 - \phi$$

where, C is the components in equilibrium, F is the number of degree of freedom and ϕ is the number of phases present in the heterogeneous system. The above relation establishes the condition of equilibrium between the number of components and phases and is known as Gibbs phase rule.

A component is termed a distinct chemical substance, which is different from other phases. Ice, water and steam are some components that represent water. The Gibbs phase rule also has practical application for the system containing more than one freezing mixture.

Let us consider a pure substance, like water vapor, lying in a single phase, i.e., $\phi = 1$ and $C = 1$. Then the above relation gives $F = 2$. Therefore, to define the state of a pure substance in the presence of a single phase, two independent intensive properties are required. Hence, pressure and temperature each can vary independently, without the affecting other. This stage is represented by the region above curve AB in Figure 8.14(a). In the region DOB, the temperature or pressure can be changed without affecting another and without any change in the phase of substance.

If two phases of water, like liquid and vapor, exist together at equilibrium, i.e., $\phi = 2$ and $C = 1$, so $F = 1$. Therefore, only one intensive property is sufficient to define the state of system. Hence, simultaneously both the properties can not vary independently. If pressure changes, temperature changes itself. This stage is represented by the curve OB in Figure 8.14(a).

Now, if all the three phases of water, i.e., ice (solid), liquid water and steam (vapor) exist together at equilibrium in a system, $\phi = 3$ and $C = 1$, i.e., so $F = 0$. Hence, intensive properties cannot show variation under such conditions. This condition is unique and is known as the triple point. At this stage of a system, all the three phases are present together.

Numerical Examples

1. A calorimeter is containing 200 gm of ice. If one gm of steam is allowed to pass into this calorimeter, then calculate the quantity of ice melted.

Solution: Amount of heat liberated in the conversion of 1 gm of steam into water at 100°C

$$= mL = 1 \times 537 = 537 \text{ calories}$$

Quantity of heat given out in the process of cooling of 1 gm of water from 100°C to 0°C

$$= 1 \times 1 \times 200 = 200 \text{ calories}$$

So, the net amount of heat liberated by 1 gm of steam

$$= 537 + 200 = 737 \text{ calories}$$

Let us suppose that m gm ice is melted,
Heat taken by ice = 80 m calories (as 80 is latent heat of ice)
At equilibrium,
Quantity of heat given out by steam = Quantity of heat taken by ice

$$737 = 80\, m$$

or

$$m = \frac{737}{80} = 9.2125 \text{ gm.}$$

2. A cooling system converts 100 gm of water at 15°C into ice at –40°C in 1 hour. Calculate the amount of heat given out per minute. Given that specific heat of ice = 0.5.

Solution: Quantity of heat given out in cooling water from 15°C temperature to 0°C

$$= ms\, \Delta T = 10 \times 1 \times 15 = 150 \text{ calories}$$

Quantity of heat given out in changing water into ice at 0°C

$$= mL = 100 \times 80 = 8000 \text{ calories}$$

So quantity of heat given out in cooling of ice from 0°C to –40°C

$$= ms\, \Delta T = 100 \times 0.5 \times 40 = 2000 \text{ calories}$$

Hence, the total amount of heat given out in 1 hour

$$= 150 + 8000 + 2000 = 10{,}150 \text{ calories}$$

So amount of heat given out per minute $= \dfrac{10{,}150}{60} = 169.166$ calories/minute

3. How much heat is required to evaporate 1 kg of ice at –20°C? Latent heat of steam is 540 calories per gram and of ice it is 80 calories per gram.

Solution: Quantity of heat required to increase the temperature of ice from –20°C to 0°C

$$ms\,\Delta T = 1000 \times 0.5 \times 20 = 1000 \text{ calories}$$

Quantity of heat required to melt 1 kg of ice at 0°C

$$mL = 1000 \times 80 = 80000 \text{ calories}$$

Quantity of heat required to increase temperature of water from 0°C to 100°C

$$ms\,\Delta T = 1000 \times 1 \times 100 = 10000 \text{ calories}$$

Quantity of heat required to change 1 kg of water into steam at 100°C

$$mL = 1000 \times 480 = 540000 \text{ calories}$$

Hence, net amount of heat required in the whole process

$$= 1000 + 80000 + 100000 + 540000 = 721000 \text{ calories}$$

4. An electric kettle takes 15 minutes to heat a certain amount of water from 0°C to 100°C (in the absence of radiation losses). 80 minutes are required to convert all the water into steam at 100°C. Calculate the latent heat of water steam.

Solution: Let us suppose that m gram of water is converted into steam and latent heat of steam is L

Amount of heat required to increase the water temperature from 0°C to 100°C

$$ms\,\Delta T = m \times 1 \times 100 = 100 \text{ m}$$

So, amount of heat produced by electric kettle in 15 min = $\frac{100\text{ m}}{15}$ calories

and quantity of heat produced by electric kettle in 80 min = $\frac{(100\text{ m} \times 80)}{15}$ calories

Amount of heat required by m gm of water to convert into steam = mL

So, $$mL = \frac{(100\text{ m} \times 80)}{15}$$

or $$L = 533.33 \text{ calories per gram}$$

Exercise

Short Answer Questions

1. What do you mean by the change of state in a system?
2. Write the differences between a gas and vapor of a substance.
3. What are the differences between saturated and unsaturated vapors?
4. What do you mean by the triple point?
5. What is Gibbs phase rule?
6. What do you understand by phase, component and degree of freedom?
7. What are the applications of boiling point of a liquid?

Long Answer Questions

1. What is triple point? Prove that the vaporization line, fusion line and sublimation line meet at the triple point.
2. Discuss the Gibbs phase rule.
3. Describe the concept of saturated and unsaturated vapor and explain their behavior by *P-V* and *P-T* diagram.
4. Explain the effect of cooling due to vaporization with examples.
5. What do you mean by latent heat of vaporization? Determine the latent heat of steam.
6. Describe the effect of pressure on the boiling point.
7. Determine the latent heat of fusion of ice by calorimeter.
8. What do you mean by the melting point? Determine the melting point of wax.
9. What do you mean by the change of state of a substance? Explain the three different phases of a substance by temperature vs time graph.

Numerical Problems

1. 1000 kilogram of ice at –10°C is placed in a bunker to cool vegetables. The ice has taken 24 hours to melt into water at 10°C. Find out the average rate of cooling (obtained by ice).

   Given that: specific heat of water = 4.1868 kJ/kg/°K, specific heat of ice = 1.94 kJ/kg/°K and latent heat of fusion of ice at 0°C = 335 kJ/kg. [**Ans.** 4.586 kW]
2. 52 gm of ice is added with 100 gm of water at 40°C in a container. Calculate the remaining amount of ice and the final amount of water.

   [**Ans.** 2 gm of ice and 150 gm of water at 0°C]
3. How much heat is required to evaporate 2 kg of ice at –15°C? Latent heat of steam is 540 calories per gram and ice is 80 calories per gram.

4. Calculate the result of mixing of 20 gm of water at 20°C with 10 gm of ice at –20°C. Given that specific heat of ice is 0.5 cal/g-K and latent heat of ice is 80 cal/gm.
5. A cooling system converts 200 gm of water at 20°C into ice at –20°C in 2 hours. Calculate the amount of heat given out per minute. Given that specific heat of ice = 0.5.
6. The vapor pressure of a substance at 0°C is 4.60 mm and it is at 1°C is 4.94 mm. The lowering of melting point per atmosphere is 0.0072°C. Find out the temperature for the triple point of water, assuming the absence of air.

   [**Ans.** 0.007156°C]
7. A calorimeter is containing 400 gm of ice. If two gm of steam is allowed to pass into this calorimeter, then calculate the quantity of ice melted.
8. Determine the result of mixing of 400 gm of melting ice with 200 gm of copper chips at 400°C.
9. An electric kettle takes 20 minutes to heat a certain amount of water from 0°C to 100°C (in the absence of radiation losses). 80 minutes are required to convert all the water into steam at 101°C. Calculate the latent heat of water steam.

Multiple-Choice Questions

1. Liquid can be converted to vapor at
   (i) A temperature equal to the boiling point of a liquid
   (ii) A temperature lower than the boiling point of a liquid
   (iii) A temperature higher than the boiling point of a liquid
   (iv) None of the above
2. Conversion of a matter from solid state to liquid state is called
   (i) Boiling (ii) Melting
   (iii) Solidifying (iv) Freezing
3. If a solid is heated, then its particles start to vibrate fast by achieving
   (i) Pressure (ii) Energy
   (iii) Temperature (iv) Volume
4. Which solid undergoes sublimation process?
   (i) Iodine (ii) Potassium chloride
   (iii) Calcium chloride (iv) Sodium chloride
5. A solid has the following property
   (i) Definite shape (ii) Low density
   (iii) High density (iv) Less intermolecular force
6. An example of diffusion in liquid is
   (i) Spreading of ink in water (ii) Smell of perfume
   (iii) Smell of food (iv) None of the above

7. Particles of a matter move far away from each other and start to vibrate strongly by breaking force of attraction among them on

(i) Melting (ii) Solidifying

(iii) Freezing (iv) Boiling

8. Sublimation process is used to purify

(i) Solid (ii) Liquid

(iii) Plasma (iv) Gases

9. During sublimation process

(i) Vapor pressure of a matter in solid state is equal to the partial pressure of the solute in gas state

(ii) Vapor pressure of a matter in plasma state is equal to the partial pressure of the solute in gas state

(iii) Vapor pressure of a matter in liquid state is equal to the pressure in solid state

(iv) Vapor pressure of a matter in gaseous state is equal to the pressure in liquid state

10. Sodium chloride and ammonium chloride mixture can be separated by

(i) Sublimation (ii) Liquefaction

(iii) Vaporization (iv) Solidification

11. Mass of sea water is 52274 kg and volume is 51×10^6 cm^3, its density is

(i) 1024 kg m^{-3} (ii) 1.024 kg m^{-3}

(iii) 102.4 kg m^{-3} (iv) 1026 kg m^{-3}

12. When we hang our cloth on a rope, it dries fast, this is due to

(i) Evaporation of the liquid

(ii) Condensing of the deodorant

(iii) Melting of the liquid

(iv) None of the above

13. Ice changes to water at

(i) 0°C (ii) 150°C

(iii) –4°C (iv) 67°C

14. Specific latent heat of fusion is defined as

(i) $l_f = L_f/m$ (ii) $L_f = ml_f$

(iii) $l_f = L_f\ m^2$ (iv) $l_f = 2L_f/m$

Answers

1. (ii) 2. (ii) 3. (ii) 4. (i) 5. (i) 6. (i) 7. (iv) 8. (i)
9. (i) 10. (i) 11. (i) 12. (i) 13. (i) 14. (i)

9

Transportation of Heat

9.1 INTRODUCTION

Transportation of heat can be achieved by three different processes, i.e., conduction of heat, convection of heat and radiation.

Conduction process transfers heat by direct collision of molecules. In this process, molecules of high temperature vibrate, and due to this vibration and movement they pass the heat energy to the neighboring molecules. The heat energy is transformed from the area of higher kinetic energy to lower energy. The particles which are moving with high speeds are colliding with the particles which are moving with slow speed. Hence, there is a rise in kinetic energy of slow speed particles. Transformation of heat from conduction is the most general form and occurs via physical contact between two bodies, e.g, placing metallic object into an open flame.

The conduction process depends on the following factors: physical properties of matter, material cross section, temperature gradient and length of the path travelled. The direction and rate of heat travel is defined in terms of temperature gradient. Flow of temperature will always be possible from hottest portion to coldest or, higher kinetic energy to lower kinetic energy. Once thermal equilibrium is achieved, further thermal transformation will stop. To construct furnace, cold storage, granite, bricks are used as they have less conductivity.

In the process of convection, heat is transformed from one place to another by the movement of heated particles. When a liquid is heated, the particles move away from the source and carry thermal energy along. The liquid or gas at the hot surface expands and becomes less dense. Trade winds and land and sea breezes develop due to convection. Heating of building by hot water circulation is possible due to convection. A space heater is an example of convection. When a space heater starts to heat, the temperature of air surrounding it near the floor, will increase and starts to expand and reaches to the top of the room.

Radiation is the result of random movements of molecules and atoms in matter or substance. In this process, transformation of heat is possible in the absence of inverting medium. Heat radiation generates directly from the sun without affecting the presence of inverting medium, i.e., thermal radiation generates from the emission of electromagnetic waves. Radiation process occurs via vacuum or any other transparent medium, for example, solid or liquid. The emission of electromagnetic radiation is possible due to the movement of charged nucleons.

The radiated thermal energy of all materials is based on their temperature. The higher the temperature of an object, the emitted radiation will be more. The sun is the best example of thermal radiation that transfers heat energy across the whole solar system. At ordinary room temperature, all objects radiate infrared rays. The wavelength and frequency of the radiated waves depend on the temperature of the material.

9.2 THE THERMAL CONDUCTIVITY COEFFICIENT

Let us consider a cube of side l cm (Figure 9.1). The opposite faces are parallel to each other and are kept at temperatures T_1 and T_2 ($T_1 > T_2$). Through conduction, heat starts to flow in the direction of low temperature.

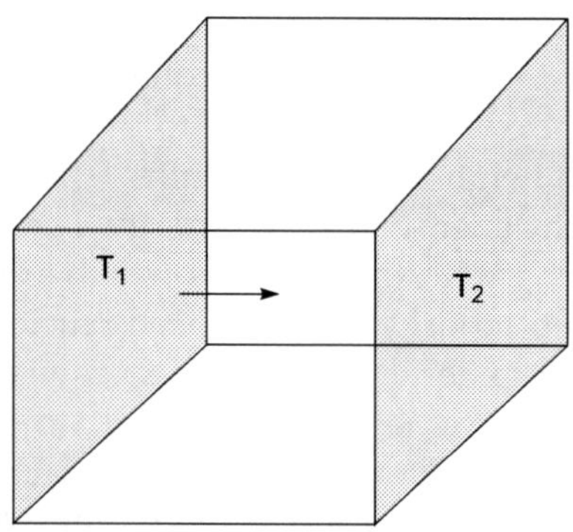

Figure 9.1

The quantity of heat conducted from hotter to colder region across the two opposite faces of cube is proportional to

$Q \propto$ area (A)

$Q \propto$ to the time spent in the flow of heat (t)

$Q \propto$ to the differnce in temperature between two regions $(T_1 - T_2)$

$Q \propto$ inversely proportion to side (l)

Hence, amount of heat absorbed during the process of conduction

$$Q \propto \frac{A(T_1 - T_2)}{l}$$

$$Q = K\frac{A(T_1 - T_2)}{l} \tag{9.1}$$

where K is a constant, which is known as coefficient of thermal conductivity. The coefficient K is a temperature-dependent material parameter. It depends on the physical properties of matter and the pressure applied on the material.

Equation (9.1) can also be written as

$$dQ = KA\left(\frac{dT}{dl}\right)dt \tag{9.2}$$

Here, dQ is the amount of heat conducting in time dt from one face of cube to other face. The quantity $\frac{dT}{dl}$ is known as temperature gradient and represents the state of change in temperature, with respect to distance between two sides. The value of temperature gradient is negative, as value of dT is negative with respect to y.

In equation (9.1), if $A = 1\ \text{cm}^2$, $(T_1 - T_2) = 1°\text{C}$, $t = 1$ sec and $l = 1$ cm, then

$$Q = K$$

Hence, thermal conductivity coefficient is defined as the quantity of heat flowing in the two opposite faces of unit cube in one second, kept at a 1°C temperature difference.

The coefficient of thermal conductivity K is measured in unit of watts per meter degree kelvin (W/mK). Other units of coefficient K are K cal/m/°K/sec. A material with a large value of coefficient K is a good conductor of heat and a material with a small value of K is a good thermal insulator.

9.3 FOURIER EQUATION FOR RECTILINEAR FLOW OF HEAT IN A BAR

Let us consider a rod of uniform cross-sectional area. As the metallic rod is heated at one end, heat starts to flow along the length of rod (Figure 9.2). Let us assume that two planes L and M are perpendicular to the length of rod along the x axis at a distance l and $l + dl$ from the hot end of rod. The temperature at the plane L is T and at the plane M is $\left(T + \frac{dT}{dl}\delta l\right)$.

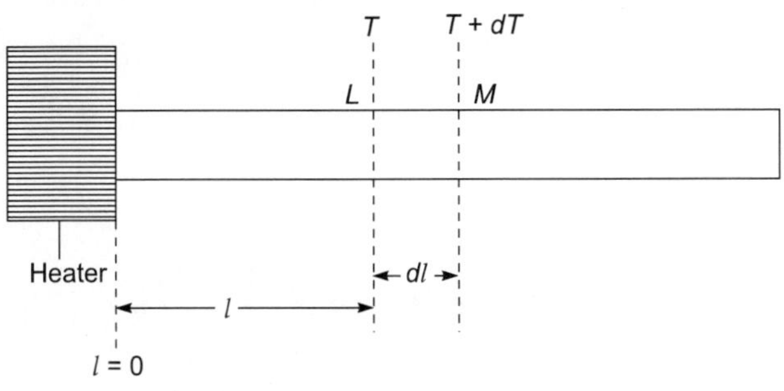

Figure 9.2 Model for rectilinear flow of heat

Hence, the temperature gradient at the plane, $L = \frac{dT}{dl}$

and the temperature gradient at the plane, $M = \frac{d}{dl}\left(T + \frac{dT}{dl}\delta l\right)$

Let Q_1 be the quantity of heat per second flowing through L and Q_2 be the quantity of heat along M. Therefore,

$$Q_1 = -KA\frac{dT}{dl}$$

and

$$Q_2 = -KA\frac{d}{dl}\left(T + \frac{dT}{dl}\delta l\right)$$

Hence, amount of heat gained by section between L and M of the rod

$$Q = Q_1 - Q_2 = -KA\frac{dT}{dl} + KA\frac{d}{dl}\left(T + \frac{dT}{dl}\delta l\right)$$

$$= KA\frac{d^2T}{dl^2}\delta l \tag{9.3}$$

The one part of heat Q is used to increase the temperature of the metallic rod and other part is radiated from the surface. This is the case when steady state is not reached.

Let us suppose that A is the cross-sectional area of the rod, C is the specific heat and dT/dt is rate of rise in the temperature of the rod. Then, the quantity of heat used in increasing the temperature of the metallic rod

= mass × specific heat × rate of rise in temperature

$$= (A \times \delta l \times \rho) \times C \times \frac{dT}{dt}$$

Let E be the emissive power, p be the perimeter, and T be the excess temperature of the bar, then, the amount of heat lost per second = $E \times p \times \delta l \times T$

Hence,

$$Q = (A \times \delta l \times \rho) \times C \times \frac{dT}{dt} + E \times p \times \delta l \times T \tag{9.4}$$

From equations (9.3) and (9.4), we get

$$KA\frac{d^2T}{dl^2}\delta l = (A \times \delta l \times \rho) \times C \times \frac{dT}{dt} + E \times p \times \delta l \times T$$

or

$$\frac{d^2T}{dl^2} = \frac{\rho C}{K} \times \frac{dT}{dt} + \frac{Ep}{KA} \times T \tag{9.5}$$

Equation (9.5) is known as Fourier's differential equation for rectilinear flow of heat along a metallic bar of uniform cross-sectional area.

If the metallic rod is covered by the non-conducting material, heat lost due to radiation is zero, i.e., $E \times p \times \delta l \times T$ is zero. Hence, the total heat absorbed by the metallic rod expands to increase the temperature of the rod. Then equation (9.5) reduces to

$$\frac{d^2T}{dl^2} = \frac{\rho C}{K} \times \frac{dT}{dt} = \frac{1}{d}\frac{dT}{dt} \tag{9.6}$$

where, $\dfrac{K}{\rho C} = d$ = thermal diffusivity of the metallic rod.

Steady State Solution of Fourier's Differential Equation

As the steady state is reached, no heat is required to increase the temperature of the rod. So at the steady state,

$$\frac{dT}{dt} = 0$$

From equation (9.5), we get

$$\frac{d^2T}{dl^2} = \frac{Ep}{KA} \times T$$

Let us assume that $\dfrac{Ep}{KA} = \gamma^2$

So,

$$\frac{d^2T}{dl^2} = \gamma^2 T \tag{9.7}$$

The general solution of equation (9.7) can be written as

$$T = B_1 e^{\gamma l} + B_2 e^{-\gamma l} \tag{9.8}$$

where, B_1 and B_2 are arbitrary constants, which can be determined from the boundary conditions. If the metallic rod is sufficiently long,

(i) For a rod of infinite length.

Let us suppose that T_0 is the excess of temperature of the high temperature end of rod and for low temperature end be zero.

At $l = 0, T = T_0$

Applying this condition in equation (9.8), we get

$$T_0 = B_1 + B_2$$

At $$l = \infty, T = 0$$

So from equation (9.8), we obtain

$$0 = B_1 e^{\infty}$$

Since, $e^{\infty} \neq 0$, so

$$B_1 = 0$$

using $B_1 = 0$, we get $B_2 = T_0$.
Substituting values of B_1 and B_2 in equation (9.8), we get

$$T = T_0 e^{-\gamma l} \tag{9.9}$$

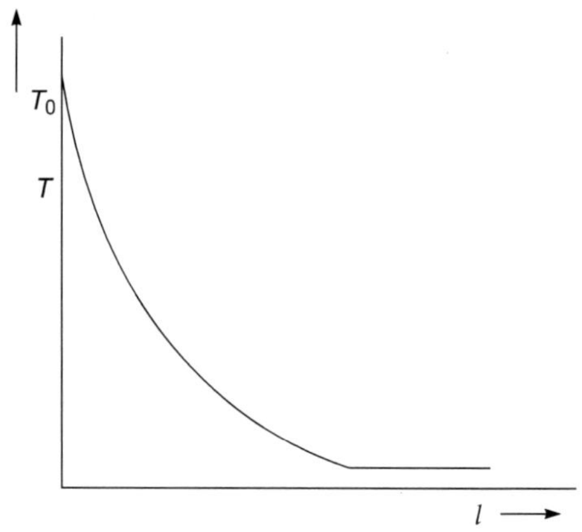

Figure 9.3 Distance vs temperature graph

Equation (9.9) represents an exponential curve (Figure 9.3) for temperature of a point at a distance l from the high temperature end, after the steady state is reached.

(ii) For rod of finite length (radiation loss is zero).
Let us suppose that T is the excess of temperature above the surrounding of the high temperature end of rod and of low temperature end be zero, i.e., $\alpha = 0$
In this case, equation (9.7) reduces to

$$\frac{d^2T}{dl^2} = 0$$

Integrating the above equation, we get

$$\frac{dT}{dl} = D_1 \tag{9.10}$$

where, D_1 is constant of integration and the above equation represents the temperature gradient, along the length of rod, which is constant.

Again integrating equation (9.10), we get

$$T = D_1 l + D_2 \tag{9.11}$$

where, D_2 is another constant of integration.

Let us suppose that the high temperature end of the rod is at T_0 and the low temperature end is at T_L, the boundary conditions are as follows:

At $\quad l = 0,\ T = T_0$

and at $\quad = L_{0,}\ T = T_L$

Apply boundary conditions in equation (9.11), we get

$$T_0 = D_2$$

and

$$T_L = D_1 L + D_2$$

So,

$$D_1 = -\left(\frac{T_0 - T_L}{L}\right)$$

Therefore, the temperature at point l is given by

$$T = -\left(\frac{T_0 - T_L}{L}\right) l + T_0 \tag{9.12}$$

Equation (9.12) shows that the temperature T varies linearly with l (Figure 9.4).

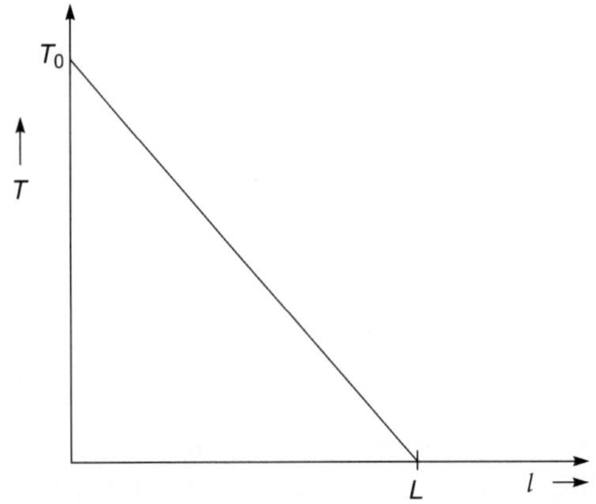

Figure 9.4 Distance vs temperature graph

9.4 PERIODIC FLOW OF HEAT IN AN INSULATED ROD

Assuming that heat is supplied periodically to the rod and not continuously, i.e., one end of a metallic rod is heated and cooled, alternately. As we know that heat waves always flow from higher temperature to lower temperature, so there is formation of two waves for two ends in opposite direction.

Let us consider that loss due to radiation is none. Assuming that the periodic variation of temperature is given by

$$T = T_0 \cos \omega t$$

where, ω is the angular frequency and T_0 is the amplitude along the metallic rod. The variation of temperature, with respect to distance, is a damped sinusoidal curve (Figure 9.5).

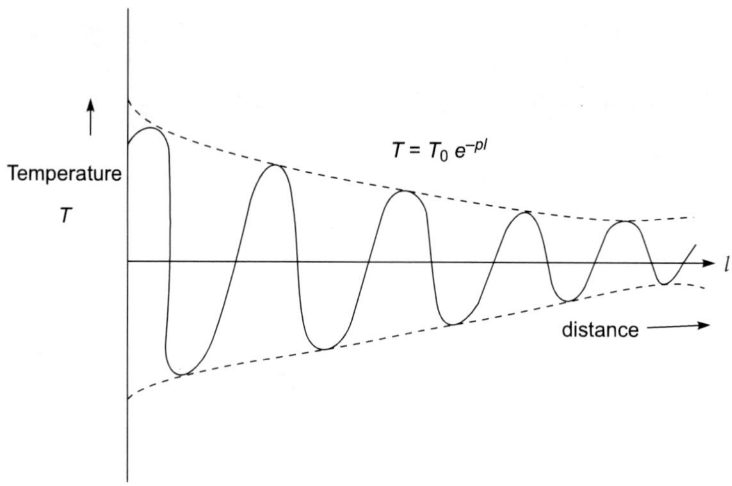

Figure 9.5

Let us consider that heat is flowing along l in x direction. The heat flow along a logged bar is unidirectional and given by Fourier equation as follows:

$$\frac{d^2T}{dl^2} = \frac{1}{d}\frac{dT}{dt} \tag{9.13}$$

where, $\frac{K}{\rho C} = d$ is thermal diffusivity of the metallic rod.

There are two boundary conditions:

At $l = 0$, $T = T_0 \cos \omega t$; and at $x \to \infty$, T will not be ∞.

Let us suppose that the solution of equation (9.13) is given by

$$T = Be^{(\alpha l + i\beta t)} \tag{9.14}$$

where, B, β and α are constants.

First order and second order differentiation of the above equation with respect to T is given by

$$\frac{dT}{dt} = Bi\beta e^{(\alpha l + i\beta t)}$$

and $$\frac{d^2T}{dx^2} = B\alpha^2 e^{(\alpha l + i\beta t)}$$

Substituting the above values in equation (9.13), we get

$$\frac{B}{h} i\beta e^{(\alpha l + i\beta t)} = d\, B\alpha^2 e^{(\alpha l + i\beta t)}$$

or $$i\beta = d\alpha^2$$

or $$\frac{i\beta}{d} = \alpha^2$$

or $$\alpha = \pm\sqrt{\frac{i\beta}{d}}$$

$$\alpha = \pm(1+i)\sqrt{\frac{\beta}{2d}}$$

Substituting these values in equation (9.14), we get

$$T = Be^{\pm(1+i)\sqrt{\frac{\beta}{2d}}\, l + i\beta t}$$

At $x \to \infty$, we can omit +ve sign, so

$$T = Be^{-(1+i)\sqrt{\frac{\beta}{2d}}\, l + i\beta t}$$

$$= Be^{-\sqrt{\frac{\beta}{2d}}\, l + i\left(\beta t - \sqrt{\frac{\beta}{2d}}\, l\right)}$$

$$= Be^{-\sqrt{\frac{\beta}{2d}}\, l} \times e^{i\left(\beta t - \sqrt{\frac{\beta}{2d}}\, l\right)}$$

$$= Be^{-\sqrt{\frac{\beta}{2d}}\, l} \times \left[\cos\left(\beta t - \sqrt{\frac{\beta}{2d}}\, l\right) + i\sin\left(\beta t - \sqrt{\frac{\beta}{2d}}\, l\right)\right] \tag{9.15}$$

At $l = 0$, $\quad T = B(\cos \beta t + i\sin \beta t)$

For small value of time,

$$T = B\cos\beta t$$

But we have taken, at $l = 0$, $T = T_0 \cos \omega t$.

Therefore, $B = T_0$ and constant $\beta = \omega$. Replacing the values of B by T_0 and β by ω in equation (9.15), we get

$$T = T_0 e^{-\sqrt{\frac{\omega}{2d}}\, l} \times \left[\cos\left(\omega t - \sqrt{\frac{\omega}{2d}}\, l \right) \right] \tag{9.16}$$

Let us consider an arbitrary phase constant ϕ, then equation (9.16), can be written as

$$T = T_0 e^{-\sqrt{\frac{\omega}{2d}}\, l} \times \left[\cos\left(\omega t - \sqrt{\frac{\omega}{2d}}\, l - \phi \right) \right] \tag{9.17}$$

Equation (9.17) is a progressive wave equation.

Amplitude of wave $a = T_0 e^{-\sqrt{\frac{\omega}{2d}}\, l} = T_0 e^{-pl}$

where, $p = \sqrt{\frac{\omega}{2d}}$ is known as a damping factor.

Velocity of progressive heat wave $v = \sqrt{2\omega d}$

$$v = \sqrt{2 \frac{2\pi}{T} \frac{K}{\rho C}}$$

$$= \sqrt{\frac{4\pi}{T} \frac{K}{\rho C}}$$

But we know that $$v = \frac{\lambda}{T}$$

So, $$\frac{\lambda}{T} = \sqrt{\frac{4\pi}{T} \frac{K}{\rho C}}$$

or $$K = \frac{\lambda^2 \rho C}{4\pi T} \tag{9.18}$$

Equation (9.18) represents the conductivity of heat wave.

9.5 MEASUREMENT OF THERMAL CONDUCTIVITY BY ANGSTROM'S METHOD

Conductivity of metallic rod was found by Angstrom, using the periodic heat flow method. In this process, one end of metallic rod is enclosed within a container, in which cold water and steam of known temperature are passed for a fixed time (i.e., 10 minutes). As heat is passed in the container for 10 minutes and then cooled for 10 minutes hence, periodic time of heat flow is 20 minutes. This process is repeated for numerous times to achieve the steady state. At steady state, the average temperature at any point remains constant. As steady state is reached, the mean temperature at any two points is noted. Using a thermocouple, the distance between these two points is measured. From this value, the wavelength can be calculated. As values of density and specific heat are known for a particular metal, so the thermal conductivity of the metallic bar can be calculated using equation (9.18).

9.6 INGEN-HAUSZ EXPERIMENT TO COMPARE THE THERMAL CONDUCTIVITIES OF DIFFERENT MATERIALS

Ingen-Hausz experiment is used to compare thermal conductivities of different types of materials.

In this experiment, a hot water or oil container is taken. Rods of different materials are fixed on the base of the container. The size and shape of all rods are same. To maintain an equal emissive power of all rods, outer surfaces of rods are electroplated and are equally polished. The portion of the rods which is outside the container is covered with wax (Figure 9.6). As hot water or oil is put into the container, wax starts melting up to different lengths on different metallic rods.

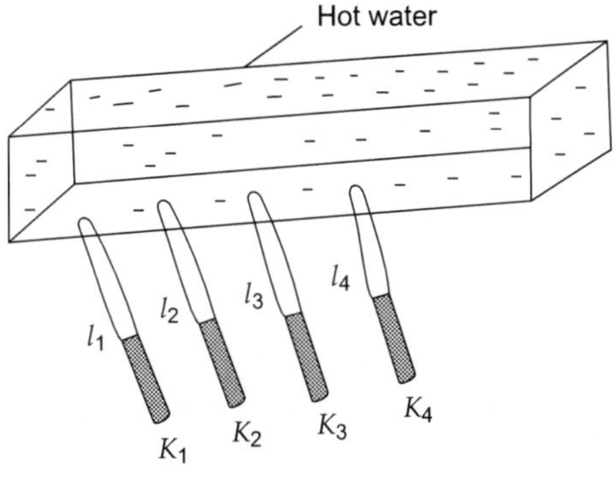

Figure 9.6 Setup for Ingen-Hausz experiment

Let us suppose that the l_1, l_2 and l_3....etc. are lengths of the rods up to which wax has melted. T_0 is the excess of temperature of the hot side of bath and T_m is the excess in temperature of the melting point of coated wax. Both the temperatures are taken above the room temperature.

We know that

$$T = T_0 e^{-\gamma l}$$

For the first metallic rod, temperature is given by

$$T_m = T_0 e^{-\gamma_1 l_1}$$

Temperature for the second metallic rod

$$T_m = T_0 e^{-\gamma_2 l_2}$$

Temperature for the third metallic rod

$$T_m = T_0 e^{-\gamma_3 l_3}$$

Hence, we can write

$$T_m = T_0 e^{-\gamma_1 l_1} = T_0 e^{-\gamma_2 l_2} = T_0 e^{-\gamma_3 l_3}$$

or

$$\gamma_1 l_1 = \gamma_2 l_2 = \gamma_3 l_3 = \ldots \tag{9.19}$$

where,

$$\gamma_1^2 = \frac{Ep}{K_1 A}$$

$$\gamma_2^2 = \frac{Ep}{K_2 A}$$

$$\gamma_3^2 = \frac{Ep}{K_3 A}$$

As values of E, A and p are same for all the metallic rods, so from equation (9.19), we may write

$$l_1 \times \sqrt{\frac{Ep}{K_1 A}} = l_2 \times \sqrt{\frac{Ep}{K_2 A}} = l_3 \times \sqrt{\frac{Ep}{K_3 A}} = \ldots$$

or

$$l_1 \frac{1}{\sqrt{K_1}} = l_2 \frac{1}{\sqrt{K_2}} = l_3 \frac{1}{\sqrt{K_3}} = \text{ constant}$$

or

$$K_1 \frac{1}{l_1^2} = K_2 \frac{1}{l_2^2} = K_3 \frac{1}{l_3^2} = \text{constant}$$

The above equation is used to compare thermal conductivities of any two materials. Thermal conductivity of the material of rod is directly proportional to the square of length of rod up to which wax melts. If the conductivity of one of the rods is available, the thermal conductivity of other rods can be calculated.

9.7 SEARLE'S METHOD TO DETERMINE THE THERMAL CONDUCTIVITY OF ROD

In Searl's method, to determine the thermal conductivity of rod LM, the one end of the rod, L, is covered in a steam container and continuous flow of heat is allowed inside it (Figure 9.7). When steady state is reached, the constant temperatures at different portion T_1, T_2, T_3, T_4 are taken with the thermometer. The mass of water is also recorded at a particular time interval t.

At steady state position, the amount of heat flowing from X to Y is constant and the same amount of heat flows throughout the rod. To avoid conduction and convection of thermal heat to the atmosphere, rod is covered with an insulated material.

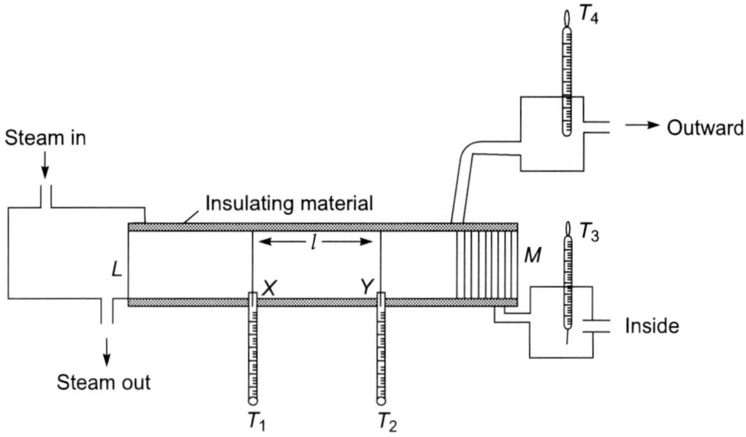

Figure 9.7 Setup for Searle's method

Amount of heat that passes from X to Y in 1 sec $= \dfrac{K \times A \times (T_1 - T_2) \times 1}{l}$ (9.20)

Quantity of heat absorbed by water in one second $= \dfrac{m \times (T_4 - T_3)}{t}$ (9.21)

Comparing equations (9.20) and (9.21), we get

$$\frac{K \times A \times (T_1 - T_2) \times 1}{l}$$

$$= \frac{m \times (T_4 - T_3)}{t}$$

or

$$K = \frac{m \times (T_4 - T_3) \times l}{A \times t \times (T_1 - T_2)}$$

The above equation is used to calculate the conductivity of the material of rod.

9.8 LEE'S METHOD TO DETERMINE THE THERMAL CONDUCTIVITY OF METALS

Thermal conductivity of metals depends on the temperature of the metal and at different temperatures thermal conductivity is different. This concept is used in Lee's method, to determine the thermal conductivity of a metal at different temperatures. Lee's method is performed at low temperatures (Figure 9.8).

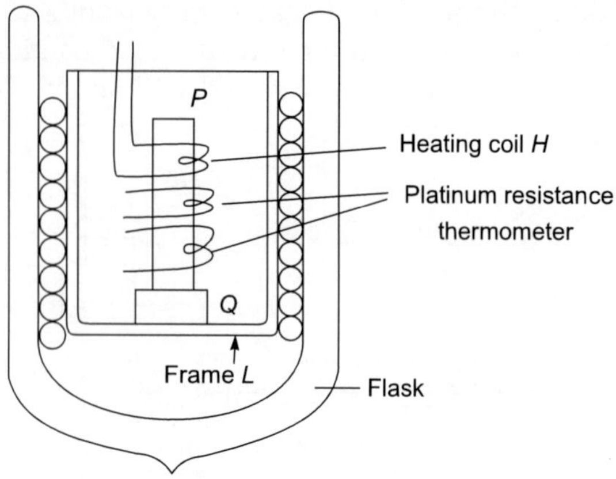

Figure 9.8 Setup for Lee's method for metals

Lee's apparatus consists of rod *PQ*, about 9.9 cm long and 0.4 cm wide (in diameter). This rod *PQ* is fitted inside the frame *L*. T_1 and T_2 are platinum resistance thermometers, used to measure temperature between the two points of rod.

Heat is supplied to the end *P* of the rod by a heating coil *H*. This arrangement is fixed inside a flask. To measure the conductivity of metal, whole setup is immersed in liquid air to maintain its temperature below the measurement temperature. To obtain a desired temperature, liquid air is removed from the whole setup. A current is passed through heating coil to develop temperatures T_1 and T_2. As steady state is reached, thermometers show temperatures T_1 and T_2.

Let us suppose that *I* amperes current is flowing across the heating coil and the potential difference across the coil is *V* volts.

Amount of heat developed in heating coil per second $= V \times I$ joules $= \dfrac{V \times I}{4.2}$ calories

So,
$$V \times I = \frac{K \times A \times (T_1 - T_2)}{l}$$

$$K = \frac{V \times I \times l}{4.2 \times A \times (T_1 - T_2)}$$

The above equation is used to calculate the thermal conductivity of a given metal at a known temperature.

9.9 FORBE'S METHOD TO FIND THERMAL CONDUCTIVITY OF A METAL

Forbe presented a method to determine the conductivity of a conductor. The conductor is taken in the form of long metallic bar, whose one end is heated to get steady state.

Let us consider a section MN of the metallic bar (Figure 9.9). Let us suppose that A is the area of cross section of metallic bar, K is the thermal conductivity of material of the bar, $-\left(\frac{dT}{dx}\right)_M$ and $-\left(\frac{dT}{dx}\right)_N$ are the temperature gradients at M and N, respectively.

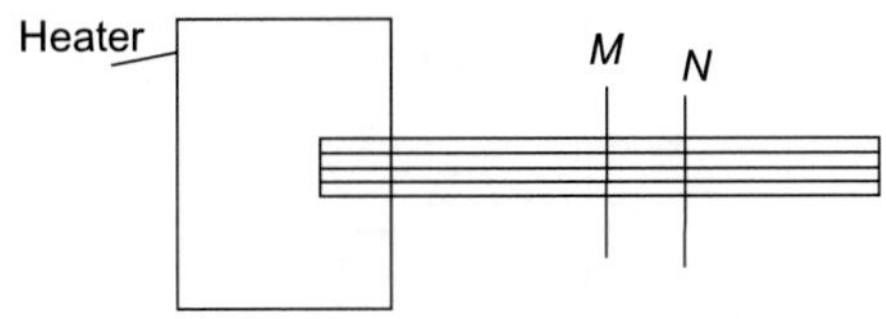

Figure 9.9 Setup for Forbe's method

Rate of flow of heat across the section $M = -K \times A \times \left(\frac{dT}{dx}\right)_M$

and rate of flow of heat across the section $N = -K \times A \times \left(\frac{dT}{dx}\right)_N$

Hence, the amount of excess of heat passing per second across M

$$= -K \times A \times \left[\left(\frac{dT}{dx}\right)_M - \left(\frac{dT}{dx}\right)_N\right] \tag{9.22}$$

As the steady state is reached, this excess amount of heat goes out from the surface of metallic bar between M and N. Let us consider a small element dl of the metallic bar. Let c be the specific heat and ρ be the density of the material and $-\left(\frac{dT}{dt}\right)$ be the rate of cooling of small element, then the amount of heat moving out per second from the element

= mass of element × specific heat × rate of cooling of bar

$$= (A \times dl \times \rho) \times c - \left(\frac{dT}{dt}\right)$$

If we consider a number of strips of small element between M and N, the amounts of total heat moving out per second from all elements

$$= \int_M^N \left(A \times c \times \rho \times \frac{dT}{dt}\right) dl$$

$$= (A \times c \times \rho) \int_M^N \frac{dT}{dt} dl \tag{9.23}$$

Comparing equations (9.22) and (9.23), we get

$$-K\times A\times\left[\left(\frac{dT}{dx}\right)_M-\left(\frac{dT}{dx}\right)_N\right]=(A\times c\times\rho)\int_M^N\frac{dT}{dt}dl$$

or

$$K=\frac{(c\times\rho)\int_M^N\frac{dT}{dt}dl}{\left[\left(\frac{dT}{dx}\right)_M-\left(\frac{dT}{dx}\right)_N\right]} \tag{9.24}$$

To determine the thermal conductivity of the material of bar using equation (9.24), we have to get the values of $\left[\left(\frac{dT}{dx}\right)_M-\left(\frac{dT}{dx}\right)_N\right]$ and the term $\int_M^N\frac{dT}{dt}dl.$

To find out the values of the terms $\left[\left(\frac{dT}{dx}\right)_M-\left(\frac{dT}{dx}\right)_N\right]$ and $\int_M^N\frac{dT}{dt}dl$, two separate experiments are performed. A static experiment is performed for the first term and a dynamical experiment is performed for the second term.

(i) Static experiment to determine the value of the term $\left[\left(\frac{dT}{dx}\right)_M-\left(\frac{dT}{dx}\right)_N\right]$:

In this experiment, a sample of metal is taken in the form of a bar. The bar is kept horizontal and its one end is in a lead bath, whose temperature remains constant (Figure 9.10(a)). To prevent from direct contact of bath with metallic rod, an insulating shield is used.

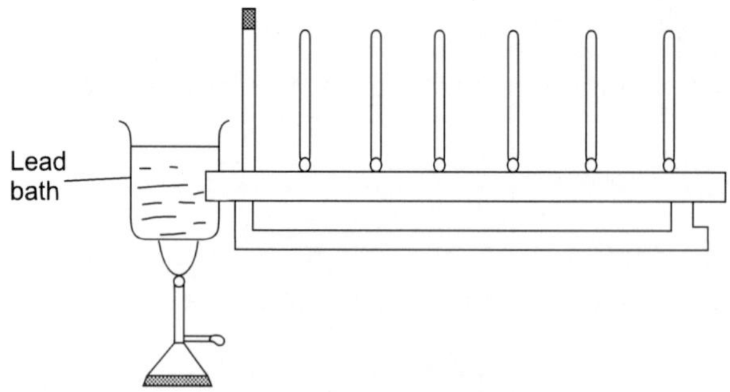

Figure 9.10(a) Forbe's method setup

To measure the temperature at different points of the bar, a number of sensitive thermometers are fitted in the bar. After the steady state, a graph is plotted between the temperature T and distance l from hot end (Figure 9.10(b)).

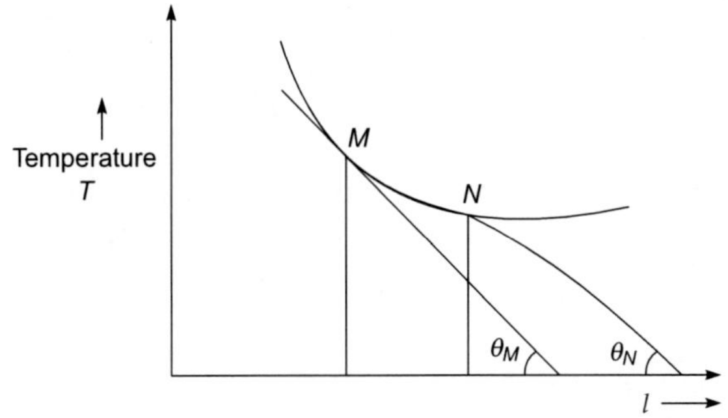

Figure 9.10(b) Distance and temperature graph

To determine the values of $\left(\frac{dT}{dx}\right)_M$ and $\left(\frac{dT}{dx}\right)_N$, two tangents are drawn at points M and N on the curve. Let θ_M and θ_N be the angles which tangents make from the x-axis,

$$\left(\frac{dT}{dx}\right)_M = \tan\theta_M$$

and

$$\left(\frac{dT}{dx}\right)_N = \tan\theta_N$$

So,

$$\left[\left(\frac{dT}{dx}\right)_M - \left(\frac{dT}{dx}\right)_N\right] = \tan\theta_M - \tan\theta_N \tag{9.25}$$

(ii) Dynamic experiment to determine the value of the term $\int_M^N \frac{dT}{dt}dl$:

To perform this experiment, a small piece of actual bar is taken and heated in a heating bath and then, hanged in air by a thread made from a non-conducting material (Figure 9.11). After steady state is reached, its temperature is recorded by the thermometers inserted inside it.

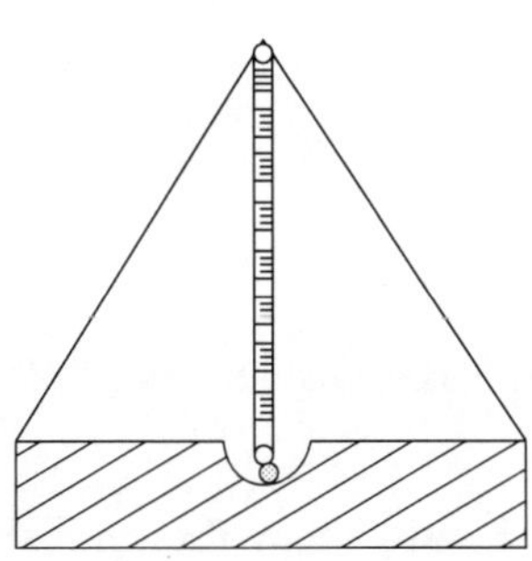

Figure 9.11 Setup for dynamic method

A graph is plotted between temperature T and time t. Tangents are drawn at different points of the graph (Figure 9.12). To measure the value of $\int_M^N \frac{dT}{dt} dl$, slopes of the tangents are determined at various temperatures. For the same temperature, the corresponding values of distance l can be determined from the static method graph (Figure 9.10(b)).

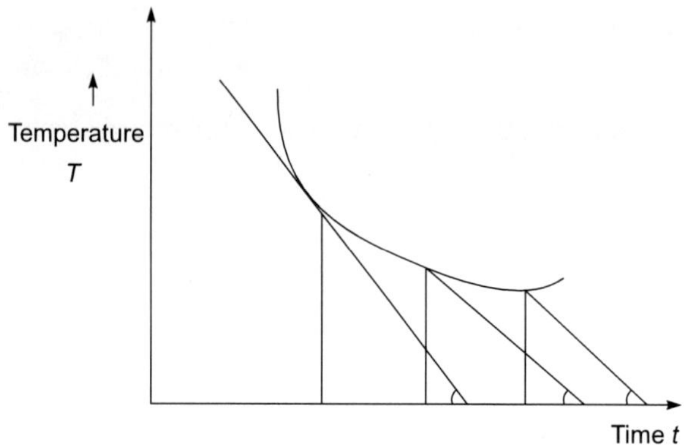

Figure 9.12 Time vs temperature graph

After getting a set of values $\frac{dT}{dt}$, a graph is plotted between $\frac{dT}{dt}$ and distance l (Figure 9.13).

To make a correct measurement, two points M and N are chosen on the curve for two different values of l. Let us consider a small strip of width dl on the graph, the area of strip will be $\frac{dT}{dt} dl$.

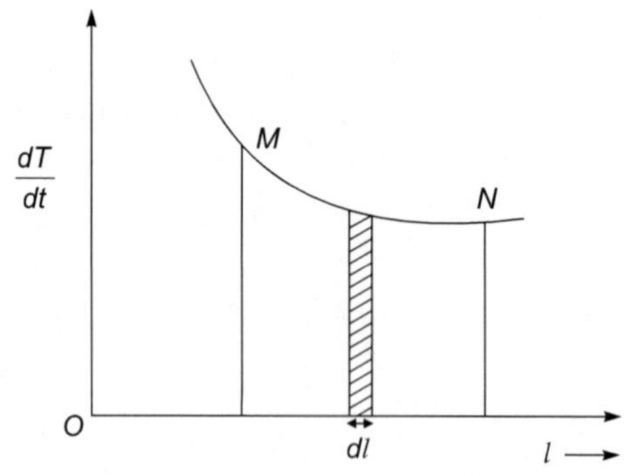

Figure 9.13 Distance vs dT/dt graph

Hence, the integration $\int_M^N \frac{dT}{dt} dl$ will show the total shaded portion under the points M and N. The value of this area can be calculated by plotting a graph.

Therefore, we obtain both values $\left(\frac{dT}{dx}\right)_M - \left(\frac{dT}{dx}\right)_N$ and $\int_M^N \frac{dT}{dt} dl$. If we note down the values of ρ and c from standard table, we can calculate the coefficient of thermal conductivity using equation (9.24).

9.10 RADIAL FLOW OF HEAT BY SPHERICAL SHELL METHOD

To determine the radial flow of heat, a specimen is enclosed between two spheres of radii, R_1 and R_2. Both the spheres are concentric and a heating element is placed between the centers of these cells (Figure 9.14).

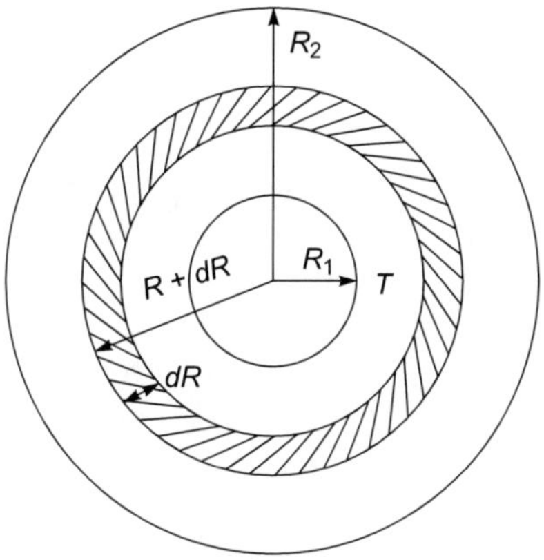

Figure 9.14 Flow of heat in spherical shell method

The conduction of heat will start from the element and it will flow from the inner surface to the outer surface. Let us suppose that T_1 and T_2 are the temperatures of the inner and the outer shell at the steady state.

Let us suppose an imaginary shell of radius R. The thickness of this shell is dR and the temperatures of inner and outer surfaces are T and $T + dT$, respectively.

Flow of heat per second through imaginary shell is given by

$$Q = -KA\left(\frac{dT}{dR}\right)$$

where, $A = 4\pi R^2$

So, $$Q = -K \times 4\pi R^2 \times \left(\frac{dT}{dR}\right)$$

or $$\frac{dR}{R^2} = -\frac{1}{Q} \times K \times 4\pi \times dT$$

Integrating the above equation, we get

$$\int_{R_1}^{R_2} \frac{dR}{R^2} = -\frac{1}{Q} \times K \times 4\pi \times \int_{T_1}^{T_2} dT$$

$$\left(\frac{1}{R_1} - \frac{1}{R_2}\right) = \frac{1}{Q} \times K \times 4\pi \times (T_1 - T_2)$$

or $$K = \frac{Q(R_2 - R_1)}{4\pi \times R_1 R_2 (T_1 - T_2)} \tag{9.26}$$

Equation (9.26) gives the expression for thermal conductivity for radial flow of heat in spherical shell.

9.11 FLOW OF HEAT IN CYLINDRICAL TUBE

There are a number of examples for radial flow of heat in our daily life. Conduction of heat from refrigerator piping and steam pipe are the same case. Let us consider a cylindrical tube of length l. Radii of inner and outer surfaces of the tube are R_1 and R_2, respectively. A hot liquid is poured inside the tube. The temperature of hot liquid inside the tube is T_1 and the temperature of surrounding is T_2 (Figure 9.15). The conduction of heat in the tube will be radially and the thickness of the pipe will be $(R_2 - R_1)$.

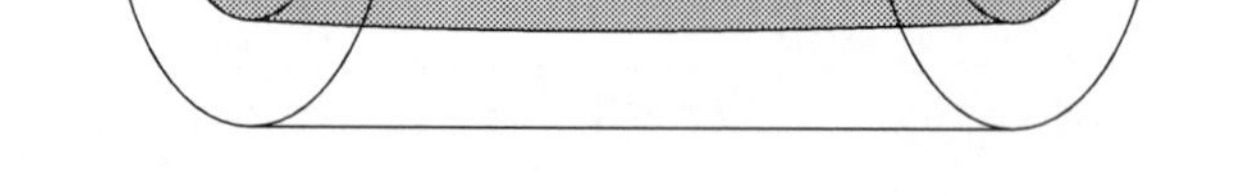

Figure 9.15 Flow of heat in cylindrical tube

To calculate the amount of heat flowing outside, let us consider that cylindrical tube consists of a large number of thin cylinders. These cylinders are co-axial, having radii in the increasing order. Out of these cylinders, let us consider a cylinder of radius R infinitesimal thickness dR. Let us suppose that the drop in temperature across the inner and outer surfaces of cylinder is dT. Let us assume that the temperature gradient across the radius R is $\frac{dT}{dR}$.

The amount of heat conducting per second in the cylinder is given by

$$Q = -KA\left(\frac{dT}{dR}\right)$$

Here, $A = 2\pi Rl$

So,
$$Q = -K \times 2\pi Rl \times \left(\frac{dT}{dR}\right)$$

Negative sign of the above equation represents dependency of temperature with R, i.e., temperature decreases with increase in value of R. The above equation can also be written as

$$Q\frac{dR}{R} = -K \times 2\pi l \times dT$$

Integrating the above equation, we get

$$Q\int_{R_1}^{R_2} \frac{dR}{R} = -K \times 2\pi l \times \int_{T_1}^{T_2} dT$$

$$Q(\log_e R_2 - \log_e R_1) = -K \times 2\pi l \times (T_2 - T_1)$$

$$Q \times \log_e\left(\frac{R_2}{R_1}\right) = K \times 2\pi l \times (T_1 - T_2)$$

$$K = \frac{Q \times \log_e\left(\frac{R_2}{R_1}\right)}{2\pi l \times (T_1 - T_2)}$$

or
$$= \frac{Q \times 2.3026 \times \log_{10}\left(\frac{R_2}{R_1}\right)}{2\pi l \times (T_1 - T_2)} \tag{9.27}$$

Equation (9.27) provides the coefficient of thermal conductivity for radial flow of heat in a cylindrical body.

9.12 FLOW OF HEAT IN COMPOUND WALL

Flow of heat in compound media can be categorized in two parts, one is series combination of bodies and the other one is parallel combination of bodies.

9.12.1 Series Combination of Bodies

Let us assume a compound slab, which is made of two different materials of thickness l_1 and l_2 (Figure 9.16). K_1 and K_2 are the thermal conductivities of two different materials, respectively. Let us suppose that the temperature of the end faces of slab are T_1 and T_2 and temperature of common surface is T.

If $T_1 > T_2$, heat will be conducted from end face 2 to face 1 through the contact surface. At the steady state position, heat will be equal in all sections. Let us suppose that this heat is Q.

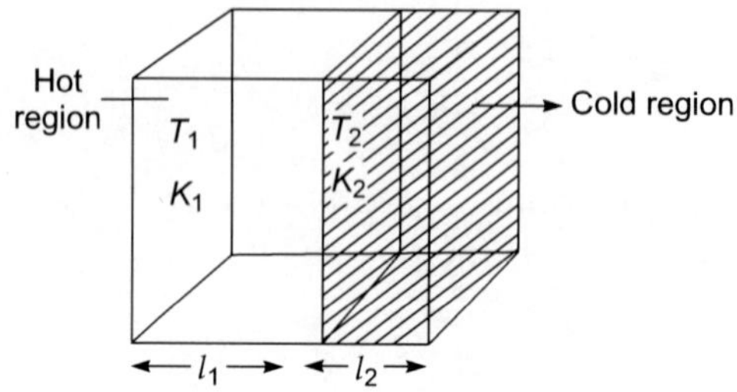

Figure 9.16 Conduction of heat for series combination of bodies

For material 1,

$$Q = K_1 A \frac{(T_1 - T)}{l_1}$$

For material 2,

$$Q = K_2 A \frac{(T - T_2)}{l_2}$$

As amount of heat is equal in each section, after the steady state, so

$$Q = K_1 A \frac{(T_1 - T)}{l_1} = K_2 A \frac{(T - T_2)}{l_2}$$

$$\frac{K_1 A}{l_1} T_1 - \frac{K_1 A}{l_1} T = \frac{K_2 A}{l_2} T - \frac{K_2 A}{l_2} T_2$$

$$A\left(\frac{K_1}{l_1} T_1 + \frac{K_2}{l_2} T_2\right) = A\left(\frac{K_2}{l_2} + \frac{K_1}{l_1}\right) T$$

$$T = \frac{\left(\frac{K_1}{l_1} T_1 + \frac{K_2}{l_2} T_2\right)}{\left(\frac{K_2}{l_2} + \frac{K_1}{l_1}\right)}$$

$$T = \frac{\sum_{j=1}^{p} \left(\frac{K_j}{l_j} T_j\right)}{\sum_{j=1}^{p} \left(\frac{K_j}{l_j}\right)} \tag{9.28}$$

After calculating this value of temperature, we can calculate the value of flow of heat in the compound media in series combination by expression of Q.

9.12.2 Parallel Combination

Let us suppose that two slabs of different materials are arranged in parallel to make a compound wall or slab (Figure 9.17).

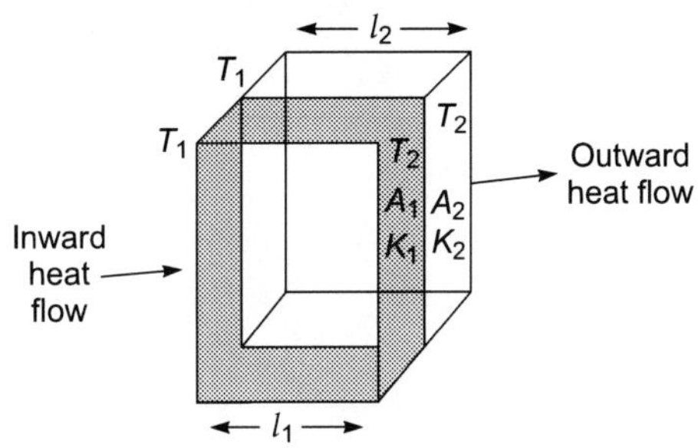

Figure 9.17 Conduction of heat for parallel combination of bodies

Let the thickness of slabs are l_1 and l_2 and thermal conductivities of the two materials be K_1 and K_2. The faces of materials are at temperature T_1 and T_2, respectively and A_1, A_2 are areas of cross section of two materials.

The amount of heat flowing through the first material is in time t and is given by

$$Q_1 = K_1 A_1 \frac{(T_1 - T_2)\times t}{l_1}$$

The amount of heat passing through the second material is in time t and is given by

$$Q_2 = K_2 A_2 \frac{(T_1 - T_2)\times t}{l_2}$$

Hence, the total amount of heat flowing across the compound wall in time t is given by

$$Q_2 = K_1 A_1 \frac{(T_1 - T_2)\times t}{l_1} + K_2 A_2 \frac{(T_1 - T_2)\times t}{l_2}$$

So, the rate of flow of heat will be

$$\frac{Q}{t} = (T_1 - T_2)\left[\frac{K_1 A_1}{l_1} + \frac{K_2 A_2}{l_2}\right]$$

or
$$= (T_1 - T_2)\sum_{j=1}^{n}\frac{K_j A_j}{l_j} \qquad (9.29)$$

Equation (9.29) represents the rate of flow of heat in the compound wall of different materials under uniform temperature conditions.

9.13 THERMAL CONDUCTIVITY OF GLASS TUBE

Let us consider a glass tube, whose radii of inner and outer surfaces are R_1 and R_2, respectively. This glass tube is fixed inside a jacket (Figure 9.18). The whole apparatus is placed in the slanting position so that water is allowed to flow inside the glass tube and it flows from the lower end to the upper end. Two sensitive thermometers are placed to measure the temperature of water entering and leaving the tube. To make good contact of water with the inner walls of tube, a spiral wire is kept inside the tube.

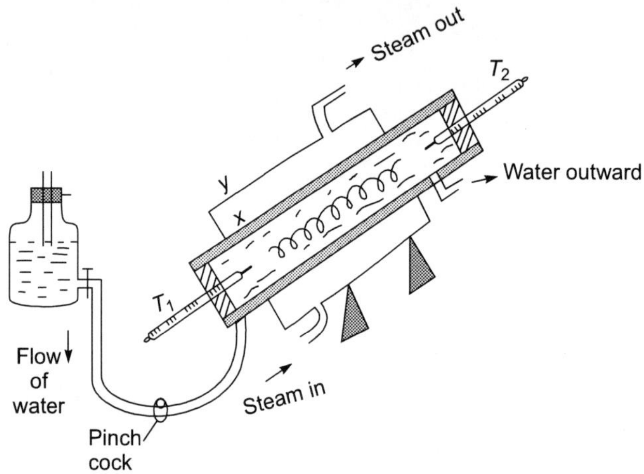

Figure 9.18 Setup for thermal conductivity of glass

To adjust the rate of flow water inside the tube, a pinch-cock is used. Steam is allowed to pass inside the outer jacket and rate of flow of water inside the tube is adjusted such that the temperature difference measured by two thermometers shows sufficient difference. At constant temperature condition at a particular time, water is collected in a flask. Let us suppose that

Temperature of the steam entering the tube = T_1

Temperature of water entering the tube = T_3

Temperature of water moving outside the tube = T_4

Length of the inner tube = 1

Mass of water collected in the flask at t second = m

Amount of heat absorbed by water in t seconds = $m(T_4 - T_3)$

Amount of heat flowing across the tube wall in 1 second

$$Q = \frac{m(T_4 - T_3)}{t}$$

Mean temperature of water inside the glass tube

$$T_2 = \frac{(T_3 + T_4)}{2}$$

But we know that
$$K = \frac{Q \times 2.3026 \times \log_{10}\left(\frac{R_2}{R_1}\right)}{2\pi l \times (T_1 - T_2)}$$

Substituting the values of Q and T_2 in the above expression, we get

$$K = \frac{\frac{m(T_4 - T_3)}{t} \times 2.3026 \times \log_{10}\left(\frac{R_2}{R_1}\right)}{2\pi l \times \left(T_1 - \frac{(T_3 + T_4)}{2}\right)}$$

$$= \frac{m(T_4 - T_3) \times 2.3026 \times \log_{10}\left(\frac{R_2}{R_1}\right)}{2\pi l \times \left(T_1 - \frac{(T_3 + T_4)}{2}\right) \times t} \tag{9.30}$$

From the above expression, we can calculate the value of thermal conductivity of glass.

9.14 THERMAL CONDUCTIVITY OF RUBBER TUBE

The thermal conductivity of rubber tube can be determined by using the theory of cylindrical flow of heat. In this experiment, a fixed quantity of water is taken inside a calorimeter. A rubber tube is taken, whose inner and outer radii are R_1 and R_2, respectively (Figure 9.19).

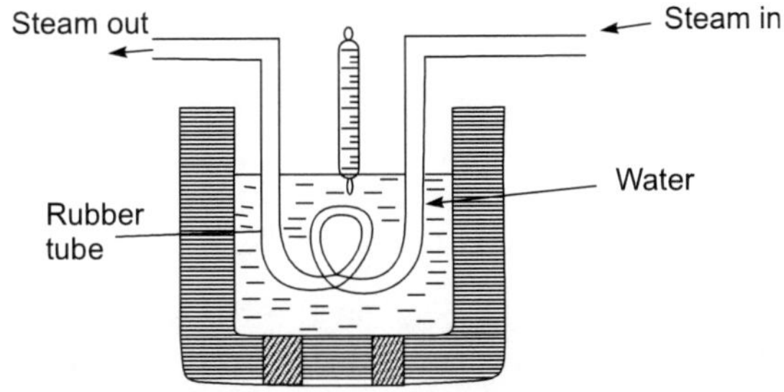

Figure 9.19 Setup for thermal conductivity of rubber

Some portion of this rubber tube is immersed in a water bath tub. A thermometer is placed in the water tub to measure the initial temperature of water. Steam is allowed to pass for a time t inside the rubber tube. Let us assume that,

The initial temperature of water = T_1

The final temperature of water = T_4

Temperature of steam = T_1

The mean temperature of the outer layer of rubber tube

$$T_2 = \frac{(T_3 + T_4)}{2}$$

Mass of water = m

Increase in temperature = $(T_4 - T_3)$

Water equivalent of the calorimeter = w

Heat absorbed by the water = $(m + w)$ $(T_4 = T_3)$

Amount of heat flowing across the rubber tube in 1 second

$$Q = \frac{(m+w)(T_4 - T_3)}{t}$$

But we know that
$$K = \frac{Q \times 2.3026 \times \log_{10}\left(\frac{R_2}{R_1}\right)}{2\pi l \times (T_1 - T_2)}$$

Substituting the values of Q and T_2 in the above expression, we get

$$K = \frac{\frac{(m+w)(T_4 - T_3)}{t} \times 2.3026 \times \log_{10}\left(\frac{R_2}{R_1}\right)}{2\pi l \times \left(T_1 - \frac{(T_3 + T_4)}{2}\right)}$$

$$= \frac{(m+w)(T_4 - T_3) \times 2.3026 \times \log_{10}\left(\frac{R_2}{R_1}\right)}{2\pi l \times \left(T_1 - \frac{(T_3 + T_4)}{2}\right) \times t} \quad (9.31)$$

Equation (9.31) is used to determine the value of thermal conductivity of rubber.

9.15 AUGMENTATION OF ICE IN A POND

Let us consider a thick layer of ice of l cm on the top surface of the water pond. Let us assume that a dl thickness ice is formed in dt time. Assume that the temperature of water below ice is 0°C and the temperature of air at the top surface of ice is $-\theta$°C (Figure 9.20).

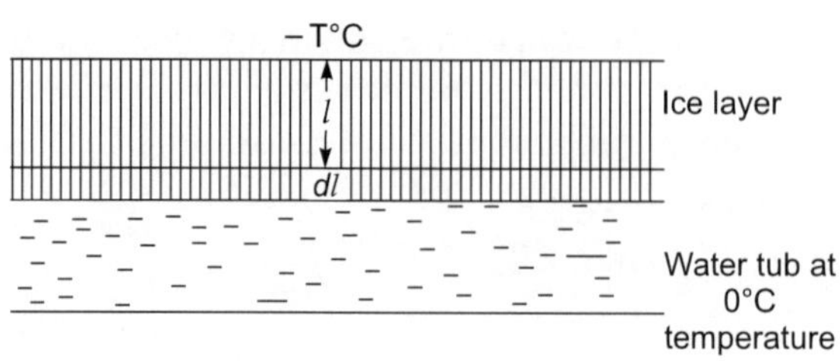

Figure 9.20 Ice accretion process

Let A be the area of the water pond, ρ be the density of ice and L be the latent heat of ice fusion.

Mass of ice formed at the top surface = $Adl\rho$

Loss of heat by water = $Adl\rho \times L$

This loss of heat is conducted towards the thickness layer of ice in upward direction.

Amount of heat conducted = $KA\frac{T}{l}dt$ in calories

Comparing the above two expression of heat, we get

$$KA\frac{T}{l}dt = Adl\rho \times L$$

$$\frac{dl}{dt} = \frac{KT}{\rho \times L \times l}$$

Here, $\frac{dl}{dt}$ shows the rate of development of ice of thickness, dl.

$$dt = \frac{\rho \times L \times l}{KT}dl$$

Net time spent in increasing the thickness of layer by l

$$\int dt = \frac{\rho \times L}{KT}\int l \times dl$$

$$t = \frac{\rho \times L}{KT}\frac{l^2}{2} + B$$

where, B is constant of integration.

At $t = 0$, l = 0,

So, value of constant B is zero

Hence, $$t = \frac{\rho \times L}{KT}\frac{l^2}{2} \qquad (9.32)$$

The above expression is used to measure time in ice accretion process.

9.16 WIEDEMANN-FRANZ LAW FOR THERMAL AND ELECTRICAL CONDUCTIVITY

Wiedemann and Franz, in 1853, gave a law for electrical and thermal conductivities of metal. This law states:

"For all metals, the ratios of thermal and electrical conductivities are same at the same temperature. This ratio has direct proportionality with absolute temperature of the metal."

Let us suppose that σ and K are electrical and thermal conductivities of a particular metal at an absolute temperature T,

$$\frac{K}{\sigma} \propto T$$

or

$$\frac{K}{\sigma \times T} = \text{constant} = L$$

where, L is proportionally constant and known as Lorentz number. Its value remains constant for all metals. Hence, the value of ratio $\frac{K}{\sigma}$ is same for all metals and it is a temperature-dependent function. This well-known law is named as Weidemann-Franz law. Therefore, we can say that a good thermal conductor will be a good electrical conductor. Widemann-Franz law is applicable for a large number of metals, whose temperature lies between 100°C to –100°C.

At low temperatures, the electrical and thermal conductivities of a metal increase with decrease of temperature. This increment in electrical conductivity of metal is higher and at absolute zero, its value approaches to infinity. This stage is similar to superconducting state of a metal. The value of $\frac{K}{\sigma}$ decreases at low temperature and approaches to zero at absolute zero temperature. Free electrons are responsible for thermal and electrical conductivities in metal which is explained by Drude. The thermal conductivity of metals increases with the increase of mean particle velocity, due to increase of the transport of energy in forward direction. While the electrical conductivity of metals decrease with increase in the number of particles. Drude derived a relation for the ratio of thermal and electrical conductivities of metal, which is given by the relation

$$\frac{K}{\sigma \times T} = \frac{3}{q^2}\left(\frac{R^2}{J \times N^2}\right) = \text{constant}$$

where, N is the number of molecules present in 1 gm molecules. R is the universal gas constant, J is Joule's constant and q is the charge of the electron.

TABLE 9.1 Values of $\left(\frac{K}{\sigma \times T}\right) \times 10^8$ for different metals at different temperatures

Metal	0°C	18°C	–50°C	–100°C
Silver	2.32	2.32	2.35	2.29
Aluminum	2.08	2.13	1.97	1.81
Zinc	2.45	2.44	2.41	2.40
Copper	2.30	2.33	2.25	2.17
Lead	2.53	2.52	2.54	2.55
Iron	2.99	2.97	2.93	2.98

9.17 LEE'S METHOD FOR CONDUCTIVITY OF BAD CONDUCTORS

Lee's method is used to determine the thermal conductivity of poor conductors, like rubber, wood, cork, glass, etc. In this method, a specimen conductor is taken in the form of two discs, D_1 and D_2, whose diameters are in order of a few cm and thicknesses are in order of a few mm. The discs D_1 and D_2 are fitted between copper plates, C_1, C_2 and C_3, C_4, respectively. A heating coil H is placed between the disc D_1 and D_2. Four thermocouples are used to measure the temperatures (Figure 9.21).

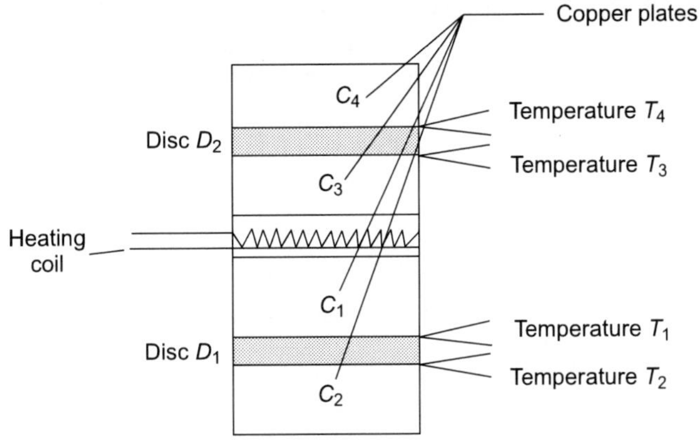

Figure 9.21

To make good thermal contact of disc with copper plates, glycerin coating is used.

When steady current I is allowed to pass through the coil H, E potential difference develops across the coil H. As steady state is achieved, the temperature is recorded in thermocouples. Let us assume that the temperatures in thermocouples are T_1, T_2, T_3 and T_4. The heat generated in the heating coil at steady state is radiated from the surface of copper plates, C_3 and C_4.

Let us suppose that l_1 and l_2 are thicknesses of the discs D_1 and D_2, respectively.

Amount of heat created by the coil in 1 second $= \dfrac{E \times I}{4.2}$ cal

Amount of heat passing through discs D_1 and D_2 in 1 second

$$= \frac{K(T_1 - T_2)A}{l_1} + \frac{K(T_3 - T_4)A}{l_1}$$

Comparing the above two equations of heat, we get

$$\frac{E \times I}{4.2} = \frac{K(T_1 - T_2)A}{l_1} + \frac{K(T_3 - T_4)A}{l_1} \tag{9.33}$$

Equation (9.33) is used to calculate the coefficient of thermal conductivity of bad conductors.

9.18 LEE'S DISC METHOD FOR CONDUCTIVITY OF BAD CONDUCTORS

Lee's disc method is used to determine the thermal conductivity coefficient of bad conductors like cardboard, glass, etc.

Lee's disc setup consists of a deep hollow cylindrical steam chamber *L*. A metallic disc made of poor conductor is placed between the steam chamber and two metallic discs of good conductors. Inlet and outlet tubes are placed in the chamber for steam (Figure 9.22). Radial holes are made to insert the thermometers inside it. Lee's disc method works on the principle that thermal conductivity is the property of a material and when steam is allowed to pass through the cylindrical chamber, a steady state is reached. At the steady state position, amount of heat conducted by the bad conductor is equal to the amount of heat radiated by the Lee's disc.

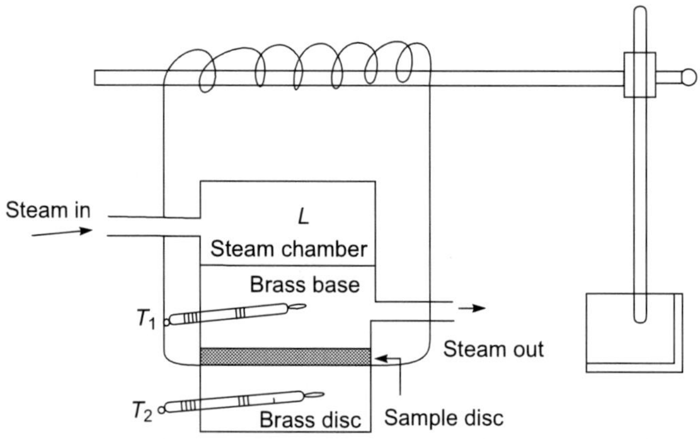

Figure 9.22 Setup for Lee's disc method

Let us suppose that the thickness of poor conductor disc is l and radius is r. As the heat is allowed to enter the steam chamber, a steady state or equilibrium is achieved very soon. In this case, the loss of heat by the lower disc in convection process is the same as the amount of heat passed through the disc made by poorly conducting material.

Let us assume that the temperature of upper disc is T_2 and that of the lower disc is T_1. After recording temperatures T_1 and T_2, the poor conductor is removed from the setup and the lower metallic disc starts heating up the upper metallic disc at temperature T_2. Now, the upper metallic disc and steam chamber are also removed and in place of these, a disc made of a good insulator is placed. The metal disc is then allowed to cool through temperature $T_1 < T_2$ and towards room temperature T_0. After cooling the metal, disc temperature is recorded and a cooling curve is plotted (Figure 9.23).

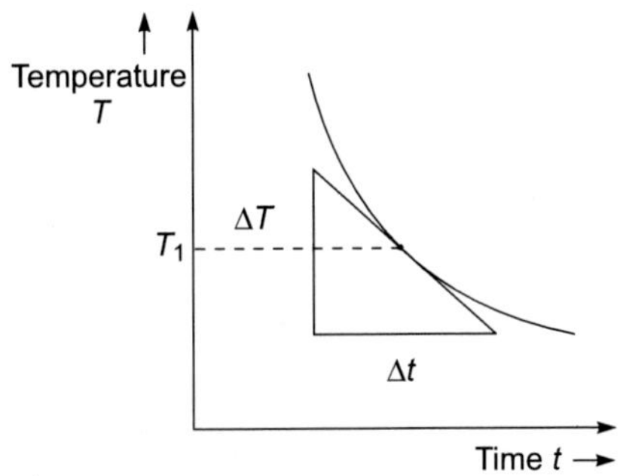

Figure 9.23

The slope of the cooling curve at temperature T_1 is given by $= \dfrac{\Delta T}{\Delta t}$.

At the position of steady state, heat transfer Q by conduction is given by

$$Q = \frac{KA(T_2 - T_1)}{l}$$

where, K is the coefficient of thermal conductivity of the sample insulator, A is the area of cross section of insulated disc, $(T_2 - T_1)$ is the temperature difference between the two ends of the sample, and l is the thickness of the bad conductor.

At the steady state, the temperatures T_1 and T_2 are constant. The rate of conduction through the brass disc must be equal to the rate of loss of heat due to cooling from the bottom surface of the same disc through air convection process. To determine the rate of heat loss, one can measure cooling time of brass disc at the steady state temperature T_1.

Let us suppose that the rate of cooling of disc is $\dfrac{dT}{dt}$, then the heat loss rate is given by:

$$Q = m_0 \times c \times \frac{dT}{dt}$$

where, m_0 is the mass of the brass disc, c is the specific heat of brass.

At steady state position, the amount of heat conducted by the bad conductor in one second will be equal to the amount of heat radiated in one second from the outer surface of the metallic disc. Hence,

$$\frac{KA(T_2 - T_1)}{l} = m_0 \times c \times \frac{dT}{dt}$$

$$K = \frac{m_0 \times c \times \dfrac{dT}{dt}}{A(T_2 - T_1)/l}$$

$$= \frac{m_0 \times c \times \frac{dT}{dt} \times l}{A(T_2 - T_1)} \tag{9.34}$$

where, K is the coefficient of thermal conductivity of the given sample, A is cross-sectional area of the sample, which is in contact with the used metallic disc, l is thickness of the sample, $(T_2 - T_1)$ is difference in temperature across the thickness of sample, m_0 is mass of the used metallic disc, and c is the specific heat of the metallic disc and dT/dt is the rate of cooling of the metallic disc at T_2.

9.19 LEE'S METHOD TO DETERMINE CONDUCTIVITY OF LIQUIDS

To calculate the thermal conductivity coefficient of liquid, convection current is important. To remove this problem, Lee designed an apparatus, in which heat flowed from the higher surface of the liquid to the lower surface.

Apparatus consists of four copper plates, C_1, C_2, C_3, C_4, of uniform thickness and a heating coil H (Figure 9.24). Four thermocouples are used to measure the temperatures. A glass plate is fitted between copper plates C_2, C_3. The sample liquid is taken in a cylindrical tube, which has an ebonite ring.

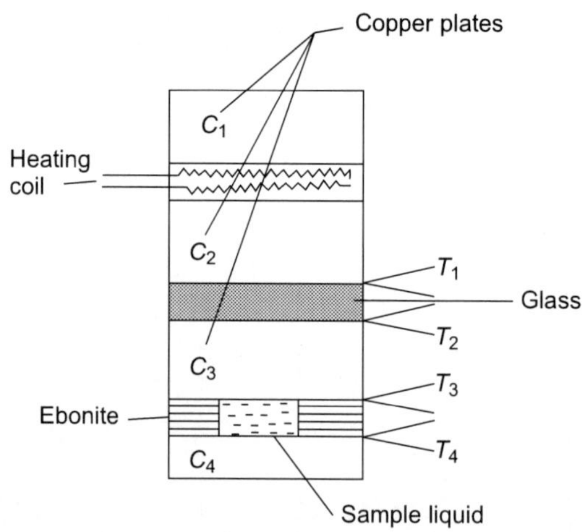

Figure 9.24

A current I is passed in liquid through the heating coil. At steady state, the temperatures T_1, T_2. T_3 and T_4 are recorded by the thermocouple.

Let us suppose that the cross-sectional area of glass plate is A, thickness of glass plate is l, cross-sectional area of liquid surface is A_1, cross-sectional area of the ebonite ring is A_2, thermal conductivity coefficient of glass is K, thermal conductivity coefficient of liquid is K_1, thermal conductivity coefficient of ebonite is K_2, thickness of the ebonite ring is l_1, and thickness of the liquid is l_1.

Amount of heat passed through the glass plate per second

$$= \frac{KA(T_2 - T_1)}{l}$$

Amount of heat passed through the liquid and ebonite ring per second

$$= \frac{K_1A_1(T_3 - T_4)}{l_1} + \frac{K_2A_2(T_3 - T_4)}{l_1}$$

Equating the above two equations, we get

$$\frac{KA(T_2 - T_1)}{l} = \frac{K_1A_1(T_3 - T_4)}{l_1} + \frac{K_2A_2(T_3 - T_4)}{l_1} \quad (9.35)$$

Equation (9.35) is used to determine the coefficient of thermal conductivity of liquid.

9.20 HEAT WAVES IN THE EARTH'S CRUST

Angstrom's method can be used to determine the thermal conductivity of the earth's crust. Earth rotates on its own axis. In day time and night time, earth's surface is heated and cooled. Due to day-night heating and cooling, a heat wave is propagated into the earth's interior surface. Periodic time of this heat wave is 24 hours. This earth wave is known as dirunal wave. These waves are created due to the earth's rotation about its own axis. As we know that the earth has another rotation around the sun also. That's the reason earth gets more heated in summer in comparison to winter. Hence, a second type of heat wave also travels into the interior surface of the earth. Due to their one year duration, these waves are known as annual waves.

Using Angstrom's method, one can calculate the coefficient of conductivity K of the earth's crust. The value of K for earth's crust is given by

$$K = \frac{\rho c \lambda^2}{4\pi T} \quad (9.36)$$

where, ρ is density, c is specific heat, T is periodic variation of temperature, which can be measured using sensitive thermometers at different depths of the earth. So, if we know the value of λ, we can calculate the value of thermal conductivity.

9.21 DAVY'S SAFETY LAMP

Davy's safety lamp works on the principle of conduction process of heat. To make this lamp, wire gauze is put over a Bunsen burner and then, the gas is lighted at the top of the wire gauze (Figure 9.25).

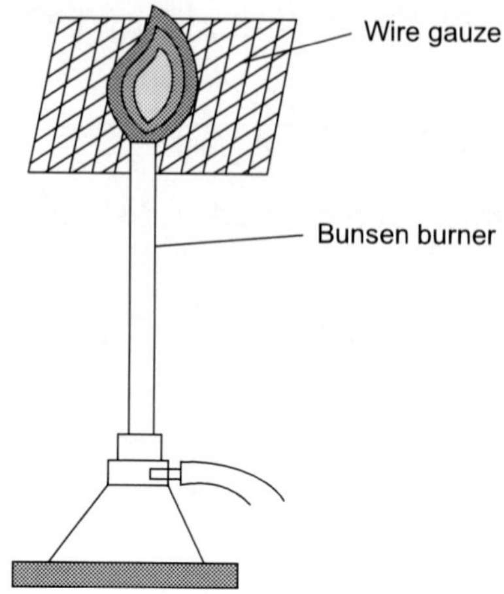

Figure 9.25

Gas passes from the wire gauze and a flame of it seems at the upper surface of the wire gauze. Heat flame appears due to the conduction process from the wire gauze at the top surface.

To cover the flame, high thermal conductivity metal gauze of cylindrical shape is used (Figure 9.26). The benefit of this lamp is that when it is taken inside a mine, it does not get ignited.

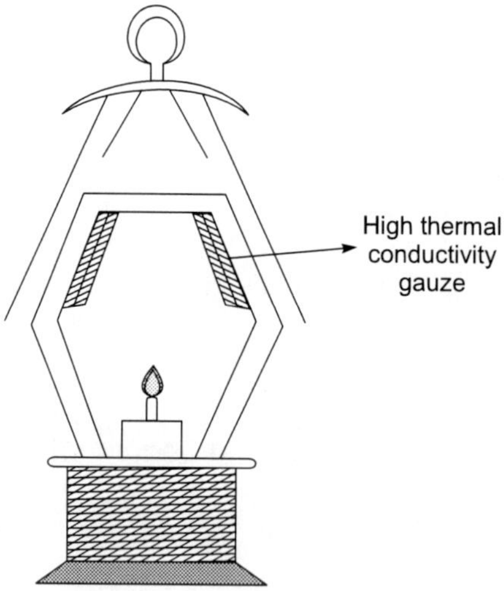

Figure 9.26 Davy's safety lamp

The ignition temperature of the gases is higher than the ignition temperature outside the wire gauze. If a wire gauze is not used in the safety lamp, then, explosion is possible due to outside gases. The colour of the flame gives signal of the presence of explosive gases in mines. In that case, the mine is ventilated.

9.22 APPLICATIONS OF CONDUCTION OF HEAT

As we know that metals are very good conductors of heat, and glass, cotton, wool, wood, etc., are bad conductors of heat. There are lot of applications of conduction of heat in our daily life:

(i) The hood of car becomes warm due to the conduction of heat through the engine to the hood of the car.

(ii) In an ice box, the spacing between two walls are filled with cork or left, due to the reason that both the materials are poor conductors of heat and can prevent conduction of heat from outside to inside the box.

(iii) A steel handle appears colder than a wooden handle in winter.

(iv) To construct cold storage, thick brick walls are used due to the reason of bad heat conductivity of bricks.

(v) Ironing a cloth is due to conduction of heat.

(vi) Ice melts in our hand due to conduction of heat from our hand to the ice.

(vii) When a bottle stopper is tightly fitted in bottle, then to remove it, the neck of the bottle is heated.

(viii) During winters, we turn on radiator to warm the room.

(ix) Saucepans, boilers, metals, cooking utensils, all are made of metal due to conduction.

(x) In hot countries, windows are made of double doors due to the reason that air in between two doors forms a non-conducting layer and heat is not transferred from outside to inside.

In cold countries, windows are also made for double door, so that non-conducting layer of air between the doors does not allow the passage of heat from inside to outside.

(xi) All woolen clothes have fine pores and air is filled in these pores. Wool and air, both are bad conductors of heat, so heat cannot flow outside from the body. That's the reason we wear woollen clothes during winter.

9.23 CONVECTION OF HEAT

The process of heat transfer by the transportation of large number of molecules within fluids, like gases and liquids, is known as convection.

When heat is supplied to a fluid from below, then thermal expansion starts. A fluid has different layers. The lower layers of the fluid become less dense because it is hotter than other layers. We know that the less dense fluid moves up, so hotter part of the fluid layer moves upside. When hotter layer moves upside, then the colder layer, i.e., denser fluid replaces it. This process is repeated again and again. This whole process is known as the transfer of heat through convection.

Convection process can either be forced or assisted and natural or free.

Forced or Assisted Convection

This type of convection occurs when a flow of fluid is caused by an external force, like a fan, a mixer or a pump.

Natural or Free Convection

This type of convection occurs due to the difference in densities by buoyancy forces. This difference is caused by variations in temperature of the fluid. When fluid is heated, the density of the fluid changes in the boundary layer and causes the fluid to move upside and be in place of that cooler fluid, which will also be heated and move upward. This continuous phenomenon is known as natural or free convection. Boiling or condensation processes of a fluid are also referred to as a convective process of transfer of heat.

Transformation of heat by movement of molecules from a hotter place to a colder place is known as convection. Sea breeze, trade wind, breeze and chimney are important examples of convection. Gases and liquid, both are bad conductors of heat but transformation of heat is possible through fluids by the method of convection.

When a liquid or gas is heated, they become lighter or less dense and expand. Hot gas or liquid starts to move upwards to the heated region and the cooler gas or liquid from the outer surroundings enter inside and fills that places which moved upward. Process goes on continuously and all the fluid of gases and liquid is heated up. Hence, the actual transformation or movement of heated molecules of gas or liquid, from hot region to cold region, is responsible for the transformation heat in the fluid.

Examples of Convection

As we know that convection process can either be forced or natural. There are number of examples of convection of heat:

1. **Chimneys:** We use chimney in our kitchen. In chimney, hot air of inside moves upward and fresh air of outside enters inside. This process continues and fresh oxygen is available for the combustion of fuel. Principle of chimney is also used in oil lamps or oil stove. When the wick of lamp or stove is lighted, heated air moves upside due to the burning of air and kerosene oil above the wick. When heated air moves up, fresh oxygen enters inside from the bottom holes of chimney. This process is continuous.
2. **Sea breeze:** Sea breeze phenomenon exists during the day time. Radiation from the sun heats up both, land and sea surfaces. The heating capacity of sea is higher than land, so it absorbs greater amount of the sun's radiation but gets heated up very slowly in comparison to the land. Due to this, the temperature above the land increases and warms up the air present in the surroundings top of it. As we know that hot air is less dense so, it expands and creates a low-pressure region over the land surface closer to the coast. In the meantime,

a relatively high pressure is created over the sea. Due to this difference in air pressure between these two surfaces, the air starts to flow from sea surface to land. Because of this, a sudden blow of high wind is felt which is known as the sea breeze.

3. **Land Breeze**: Land breeze exists during nights when the situation is just opposite. When the sun sets, the land and sea surfaces start cooling down. There is a very rapid loss of heat from land surface compared to water of sea because of different heat capacity of these two. As a result, the temperature of the water of sea surface is relatively much higher than land surface, which creates very low air pressure in the sea surface. This creates a flow of very cool breeze, called the land breeze.
4. **Trade wind**: The temperature of earth's surface at the equator is higher than that at the poles. Due to this, very warm wind moves upside, at the equator and cold wind moves towards the equator side from the pole side. In the southern hemisphere, the movement of air appears from south to west and in the northern hemisphere, it appears from west to east due to the rotation of the earth. This air flow is known as trade wind.

Applications of Convection

1. **Air conditioners:** In air conditioners, cool dry air moves outside of the room due to the force of rotary fan. The density of cool air becomes high in comparison to hot air, so hot air is drawn into the air conditioner and is cooled by air conditioner. The process continues and air is recirculated, which causes the cooling inside the room.
2. **The convection process is responsible for the household water system:** In the household water system, the water inside the boiler is heated by gas burners. The water becomes hot (less dense) and starts to expand. Expanded water moves upward and fills the upper half portion of the cylinder. In place of the hot water, cold water is the filled into the lower half portion of the cylinder. To take care the large expansion of the hot water, a overflow pipe is fitted to the cylinder. The position of hot-water tap in the water system must be lower than the tank to maintain the sufficient pressure difference between the tank and the tap, which is responsible for the flow of water out of the tap.
3. **Electric kettles:** In electric kettles, the heating coil is kept at the lower surface of the kettle. When we switch on the power plug, the water layer just near the heating coil gets heated and moves upwards. The region of hotter water is replaced by cooler water and the process goes on. Finally, all the water is heated due to convection.
4. **Centralized heating of buildings:** During winter season, hotels and office buildings are centrally heated, which is based on the principle of flow of convection currents.

5. **Refrigerators:** The functioning of refrigerators and an air conditioner is very much similar. The freezing unit of refrigerator is placed at the top portion to decrease the temperature of air (cool air), air being dense, sinks. At the mean time, the hot air from bottom surface rises upward. This process develops convection currents inside the cabinet of refrigerator due to which the contents inside it become cool.

9.24 CENTRAL HEATING SYSTEMS

During winters, the temperature of cold countries becomes very low and can drop below zero. To maintain the temperature, a central heating system is used, which depends on the principle of convection. There are different kinds of central heating systems.

Different types of energy sources, like water steam, liquids, petroleum, electricity, heat pumps, natural gas, etc., can be used in central heating system. The most common form of central heating systems is gas electricity due to easy maintenance. Due to the presence of natural gases in a gas heater, room becomes hotter, safe and less expensive.

A gas boiler is used in the water-based central heating systems. The main part of water-based central heating system is a boiler, which gets continuous supply of natural water. Boilers convert water in steam, which is circulated by pipes. In different water heating systems, different kinds of methods are used to develop heat, like base board units, floor system, radiators, etc. Most common method in such a central heating system is the use of set of radiators to develop the heat.

Hot water present in the boiler moves in upward direction and passes to different rooms by radiators. Due to the presence of this hot water, radiator gets heated and warms up the room (Figure 9.27). As heat is radiated to room, hot water converts in cold water and goes back to the boiler.

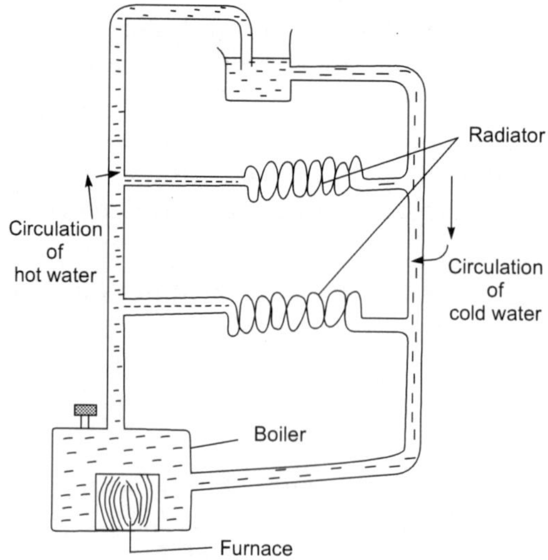

Figure 9.27 Processing of central heating system

Numerical Examples

1. Length of a copper rod is 20 cm, area of cross section is 0.784 cm^2 and coefficient of thermal conductivity is 380 W/m/K. One end of the rod is heated at 373°K and the other end is fixed at 303°K. Determine the quantity of heat which will pass for 10 minutes in the rod.

   Solution: Given that:

   $l = 0.20$ m

   $A = 0.784\ cm^2 = 7.84 \times 10^{-5}\ m^2$

   $T_1 = 303°K$

   $T_2 = 373°K$

   $T_2 - T_1 = 70°K$

   $K = 380$ W/m/K and time $t = 10$ min $= 10 \times 60$ sec $= 600$ sec

   We know that the quantity of heat flowing in a rod is given by

   $$Q = \frac{KA(T_2 - T_1) \times t}{l}$$

   $$= \frac{380 \times 7.84 \times 10^{-5} \times 70 \times 600}{0.20}$$

   $$= 6.25 \times 10^3\ J = 6.25\ kJ$$

2. A 30 cm long aluminum rod of cross-sectional area of 0.846 cm^2 is heated at one end at temperature 110°C. Coefficient of thermal conductivity of aluminum is 237 W/m/K. Other end of the rod is kept at 50°C. Calculate the quantity of heat which will pass for 20 minutes in the rod.

   Solution: Given that:

   $l = 0.30$ m

   $A = 0.846\ cm^2 = 8.48 \times 10^{-5}\ m^2$

   $T_1 = 323°K$

   $T_2 = 383°K$

   $T_2 - T_1 = 60°K$

   $K = 237$ W/m/K and time $t = 20$ min $= 20 \times 60$ sec $= 1200$ sec

   The amount of heat flowing in the aluminum rod is given by

   $$Q = \frac{KA(T_2 - T_1) \times t}{l}$$

$$= \frac{237 \times 8.48 \times 10^{-5} \times 60 \times 1200}{0.30}$$

$$= 4.82 \text{ kJ}$$

3. A nickel rod of cross-sectional area 6.24×10^{-5} m^2 is heated from one end at 127°C and other end is fixed at 50°C. The 5.24 kJ heat passes from nickel rod for 12 minutes. Determine the length of nickel rod used in the experiment. Given that coefficient of thermal conductivity of nickel is 80.1 W/m/K.

Solution: Given that:

$$A = 6.24 \times 10^{-5} \text{ m}^2$$

$$T_1 = 323°\text{K}$$

$$T_2 = 400°\text{K}$$

$$T_2 - T_1 = 77°\text{K}$$

$$Q = 5.24 \text{ kJ} = 5.24 \times 10^3 \text{ J}$$

$K = 80.1$ W/m/K and time $t = 12$ min $= 12 \times 60$ sec $= 720$ sec

The amount of heat flowing in aluminium rod is given by

$$Q = \frac{KA(T_2 - T_1) \times t}{l}$$

So, length of rod

$$l = \frac{KA(T_2 - T_1) \times t}{Q}$$

$$= \frac{80.1 \times 6.24 \times 10^{-5} \times 77 \times 720}{5.24 \times 10^3}$$

$$= 0.0525 \text{ m} = 5.25 \text{ cm}$$

4. In an experiment of Ingen-Hausz, some amount of wax melted over 20 cm of aluminum rod and 6 cm of copper rod. If the conductivity of copper is 413 W/m/K, determine the coefficient of thermal conductivity of aluminum.

Solution: Given that

$$l_1 = 20 \text{ cm}$$

$$l_2 = 6 \text{ cm}$$

$$K_1 = 413 \text{ W/m/K}$$

We know that, in Ingen-Hausz experiment,

$$\frac{K_1}{K_2} = \frac{l_1^2}{l_2^2}$$

or $$K_2 = \frac{l_2^2}{l_1^2} \times K_1$$

$$= \frac{6 \times 6}{20 \times 20} \times 413$$

$$= 37.17 \text{ W/m/K}$$

5. In an Ingen-Hausz experiment, some quantity of wax melted over l cm of iron rod and 14 cm or aluminum rod. The conductivity of iron is 392 W/m/K, and thermal conductivity of aluminum is 232 W/m/K. Determine the value of length of copper rod l.

Solution: Given that

$$l_1 = l \text{ cm}$$
$$l_2 = 14 \text{ cm}$$
$$K_1 = 83.5 \text{ W/m/K}$$
$$K_2 = 232 \text{ W/m/K}$$

We know that, in Ingen-Hausz experiment,

$$\frac{K_1}{K_2} = \frac{l_1^2}{l_2^2}$$

or $$l_1^2 = \frac{l_2^2}{K_2} \times K_1$$

$$= \frac{14 \times 14}{232} \times 83.5$$

$$l = 8.366 \text{ cm}$$

6. Two similar plates of equal thickness are combined together to make a single plate of just double thickness. The thermal conductivities of two plates are K_1 and K_2, respectively. Determine the coefficient of thermal conductivity of the new composite plate.

Solution: Given that,

$$l_1 = l_2 = l$$

So, the thickness of composite plate = $l_1 + l_2 = 2l$

Let us suppose that T_1 and T_2 are temperatures of hot and cold ends of plate, respectively, T is the temperature of composite plate and thermal conductivity of the composite plate is K.

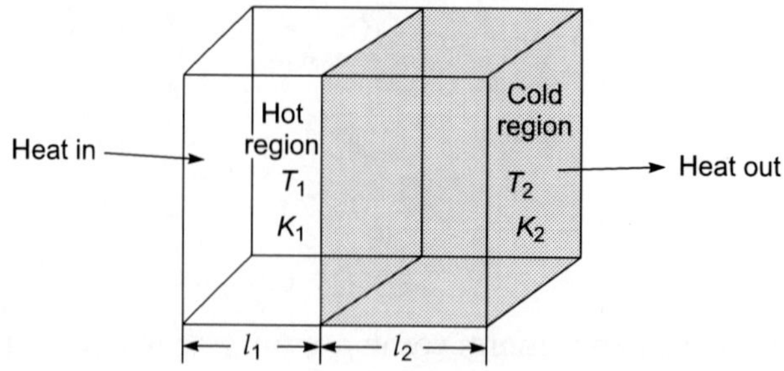

Figure 9.28

We know that, at steady state, the amount of heat flow in composite plate is given by

$$Q = \frac{K_1 A(T_1 - T)}{l} = \frac{K_2 A(T - T_2)}{l} \tag{1}$$

or

$$K_1 T_1 - K_1 T = K_2 T - K_2 T_2$$

$$K_1 T_1 + K_2 T_2 = K_2 T + K_1 T$$

$$(K_2 + K_1)T = K_1 T_1 + K_2 T_2$$

$$T = \frac{(K_1 T_1 + K_2 T_2)}{(K_2 + K_1)} \tag{2}$$

Quantity of heat flow in composite plate of thickness $2l$ and thermal conductivity K is given by

$$Q = \frac{KA(T_1 - T_2)}{2l} \tag{3}$$

From equations (1) and (3), we get

$$Q = \frac{K_1 A(T_1 - T)}{l} = \frac{KA(T_1 - T_2)}{2l}$$

or

$$K = \frac{2K_1(T_1 - T)}{(T_1 - T_2)} \tag{4}$$

Substituting the value of T from equation (2) to equation (4), we get

$$K = \frac{2K_1}{(T_1 - T_2)}\left[T_1 - \frac{(K_1 T_1 + K_2 T_2)}{(K_2 + K_1)}\right]$$

$$= \frac{2K_1}{(T_1 - T_2)}\left[\frac{T_1 K_2 + T_1 K_1 - K_1 T_1 - K_2 T_2}{(K_2 + K_1)}\right]$$

$$= \frac{2K_1}{(T_1 - T_2)}\left[\frac{T_1K_2 - K_2T_2}{(K_2 + K_1)}\right]$$

$$= \frac{2K_1K_2(T_1 - T_2)}{(T_1 - T_2)(K_2 + K_1)}$$

$$= \frac{2K_1K_2}{(K_2 + K_1)} \tag{5}$$

Equation (5) is the expression for required thermal conductivity of the composite plate.

7. The thermal conductivities of two metallic bars are 81 and 26 W/m/K. The cross-sectional area of bars is 49 cm and 61 cm, respectively. Two bars are brought together and joined at one end. The other end of the first and the second bar is kept at 99.8°C and 19.99°C, respectively. If there is no loss of heat outside the bar, determine the temperature of the composite bar at the joint.

Solution: Given that

$$l_1 = 49 \text{ cm}$$
$$l_2 = 61 \text{ cm}$$
$$K_1 = 81 \text{ W/m/K}$$
$$K_2 = 26 \text{ W/m/K}$$
$$T_1 = 99.8°\text{C}$$
$$T_2 = 19.99°\text{C}$$

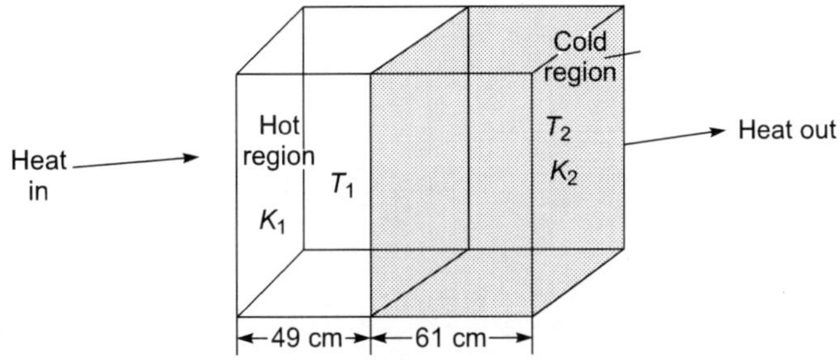

Figure 9.29

Let us suppose that the temperature at the common point is T, which is given by

$$T = \frac{\left(\dfrac{K_1T_1}{l_1}\right) + \left(\dfrac{K_2T_2}{l_2}\right)}{\left(\dfrac{K_1}{l_1} + \dfrac{K_2}{l_2}\right)}$$

$$T = \frac{\left(\frac{81 \times 99.8}{0.49}\right) + \left(\frac{50 \times 19.99}{0.61}\right)}{\left(\frac{81}{0.49} + \frac{26}{0.61}\right)}$$

$$T = \frac{16497.55 + 1638.52}{(165.30 + 42.62)}$$

$$T = \frac{16498.07}{207.92} = 79.86°\text{C}$$

8. The cross-sectional area of two metallic bars is 49 cm and 61 cm, respectively. These two bars are brought together to join at one end. The other end of the first and the second bar is kept at 99.8°C and 19.99°C, respectively. The temperature at the joint of the bar is 79.86°C. If the thermal conductivity of one bar is 81 W/m/K, determine the thermal conductivity of the other bar.

Solution: Given that

$$l_1 = 49 \text{ cm}$$

$$l_2 = 61 \text{ cm}$$

$$K_1 = 81 \text{ W/m/K}$$

$$T_1 = 99.8°\text{C}$$

$$T_2 = 19.99°\text{C}$$

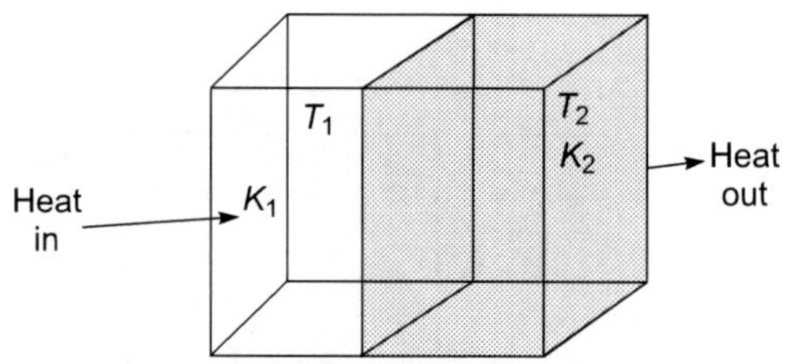

Figure 9.30

Let us suppose that the temperature at the common point is T, which is given by

$$T = \frac{\left(\frac{K_1 T_1}{l_1}\right) + \left(\frac{K_2 T_2}{l_2}\right)}{\left(\frac{K_1}{l_1} + \frac{K_2}{l_2}\right)}$$

$$T = \frac{\left(\frac{81 \times 99.8}{0.49}\right) + \left(\frac{K_2 \times 19.99}{0.61}\right)}{\left(\frac{81}{0.49} + \frac{K_2}{0.61}\right)}$$

$$K_2 = 26 \text{ W/m/K}$$

9. A layer of ice of thickness 0.03 m is floating on the top surface of a pond. If the thickness of layer of ice increases by 1 mm at atmospheric temperature of 20°C below zero, determine the time spent in this process. The thermal conductivity of ice is 0.0049 W/m/K, density of ice at 0°C is 0.90 g/cc and latent heat of ice = 80 cal/gm.

Solution: Given that

$$l_1 = 0.03 \text{ m} = 3 \text{ cm}$$

$$l_2 = 3 + 0.1 = 3.1 \text{ cm}$$

$$K = 0.0049 \text{ W/m/K}$$

$$L = 80 \text{ cal/gm}$$

$$\rho = 0.90 \text{ gm/cc}$$

We know that the time spent in increasing the thickness of layer of ice from l_1 to l_2 is given by

$$\int dt = \frac{\rho \times L}{KT} \int_{l_1}^{l_2} l\,dl$$

$$t = \frac{\rho \times L}{2KT} \times (l_2^2 - l_1^2)$$

$$= \frac{0.90 \times 80}{2 \times 0.0049 \times 20} \times (3.1 \times 3.1 - 3 \times 3)$$

$$= 221 \text{ sec}$$

10. An ice layer of thickness l cm on the top surface of a pond increases by 1 mm. The thermal conductivity of ice is 0.005 W/m/K, density at 0°C is 0.91 g/cc and latent heat of ice = 80 cal/gm. At atmospheric temperature of 20°C below zero, the time spent in this process is 222 sec. Determine the value of thickness, l.

Solution: Given that

$$l_2 = l + 0.1 \text{ cm}$$

$$K = 0.005 \text{ W/m/K}$$

$$L = 80 \text{ cal/gm}$$

$$\rho = 0.91 \text{ gm/cc}$$

We know that the time spent in increasing the thickness of layer of ice from l_1 to l_2 is given by

$$\int dt = \frac{\rho \times L}{KT}\int_{l_1}^{l_2} l\,dl$$

$$t = \frac{\rho \times L}{2KT} \times (l_2^2 - l_1^2)$$

$$l_1^2 = \left(l_2^2 - \frac{2KT \times t}{\rho \times L}\right)$$

$$l_1 = 0.03 \text{ m}$$

11. Charcoal powder is filled in two thin concentric circles of radii 6 cm and 12 cm, respectively. On supplying heat energy at the rate of 12.5 W at the center of a heater, 50°C temperature developed between the two shells. Determine the thermal conductivity of charcoal.

Solution: Given that

$$R_1 = 6 \text{ cm}$$

$$R_2 = 12 \text{ cm}$$

$$(T_1 - T_2) = 50°\text{C}$$

$$Q = 12.5 \text{ W} = 12.5/4.2 \text{ cal/sec}$$

We know that thermal conductivity of specimen enclosed in between two concentric spheres is given by

$$K = \frac{Q(R_2 - R_1)}{4\pi R_1 R_2 (T_1 - T_2)}$$

So, thermal conductivity of charcoal in the given experiment is

$$K = \frac{\left(\frac{12.5}{4.2}\right) \times (12 - 6)}{4 \times 3.14 \times 6 \times 12 \times 50}$$

$$= \frac{12.5 \times 6}{4.2 \times 4 \times 3.14 \times 6 \times 12 \times 50}$$

$$= \frac{75}{18907.2}$$

$$= 0.003966 = 3.96 \times 10^{-4} \text{ cal/cm/deg/sec}$$

12. Two thin concentric circles of radii 5 cm and 10 cm, respectively are filled with charcoal powder. The thermal conductivity of charcoal powder is 3.3×10^{-4}. If 10.5 watt heat energy is supplied at the center of a heater, determine the temperature developed between the two shells.

Solution: Given that

$R_1 = 5$ cm

$R_2 = 10$ cm

$Q = 10.5$ W $= 10.5$ J/sec $= 12.5/4.2$ cal/sec

Thermal conductivity of charcoal in the given experiment is given by

$$K = \frac{Q(R_2 - R_1)}{4\pi R_1 R_2 (T_1 - T_2)}$$

$$(T_1 - T_2) = \frac{Q(R_2 - R_1)}{4\pi R_1 R_2 K}$$

So,

$$= \frac{\left(\frac{10.5}{4.2}\right) \times (10 - 5)}{4 \times 3.14 \times 5 \times 10 \times 3.3 \times 10^{-4}}$$

$$= \frac{10.5 \times 5}{4.2 \times 4 \times 3.14 \times 5 \times 10 \times 3.3 \times 10^{-4}}$$

$$= 60°\text{C}$$

13. A metallic pipe is covered by a layer of 2.49 cm thickness insulating material. The external diameter of the pipe is 15 cm and steam carried by it is 200°C. If the outer surface of metallic pipe is kept at 100°C, determine the loss of per meter heat per minute. The thermal conductivity of insulating material is 0.21 W/m/K.

Solution: Given that

$l = 1$ m

$K = 0.21$ W/m/K

$R_1 = 7.5$ cm $= 7.5 \times 10^{-2}$ m

$R_2 = (10 + 2.49)$ cm $= 12.5$ cm $= 12.49 \times 10^{-2}$ m

$T_1 = 100°$C

$T_2 = 200°$C

We know that the amount of heat flow in a cylindrical shell is given by

$$Q = \frac{2\pi l\,(T_1 - T_2) \times K}{2.3026 \times \log_{10} \frac{R_2}{R_1}} \text{ J/min}$$

$$= \frac{2 \times 3.14 \times 1.5 \times 100 \times 0.21 \times 60}{2.3026 \times \log_{10} \frac{12.49}{7.5}} \text{ J}$$

$$= 27518.2 \text{ J} = 2.75 \times 10^3 \text{ kJ}$$

14. The following data were recorded in performing Lee's experiment:

Thickness of cardboard = 4.7 mm

Thickness measured for lower disc = 1 cm

Cooling time for lower disc from 94°C to 99°C = 4 min

Steady temperature recorded for copper disc = 99.4°C

Steady temperature recorded for lower disc = 83.4°C

The specific heat is 0.1 cal/g/°C and density of copper disc is 8.99 g/cc. Determine the coefficient of thermal conductivity of cardboard.

Solution: Given that

$$l_1 = 1 \text{ cm}$$

$$l = 0.47 \text{ cm}$$

$$\rho = 8.99 \text{ gm/cc}$$

$$c = 0.1 \text{ calg/°C}$$

$$T_1 - T_2 = 16\text{°C}$$

Rate of cooling of lower disc $= \frac{dT}{dt} = \frac{99 - 94}{4 \times 60} = \frac{1}{48}$ °C/sec

The amount of heat conducted per second from the upper disc to the lower disc by cardboard disc is given by

$$Q = \frac{K \times \pi R^2 (T_1 - T_2)}{l} \qquad (1)$$

If m is the mass of metallic disc and c is the specific heat of material and the rate of cooling is dT/dt,

$$Q = mc\frac{dT}{dt}$$

$$Q = (\pi \times R^2 l'\rho) \times c \frac{dT}{dt} \tag{2}$$

Hence, from equations (1) and (2), we get

$$\frac{K \times \pi R^2 (T_1 - T_2)}{l} = (\pi \times R^2 l'\rho) \times c \frac{dT}{dt}$$

$$K = \frac{(l'\rho) \times c \dfrac{dT}{dt}}{(T_1 - T_2)/l}$$

$$= \frac{(l'\rho) \times c \dfrac{dT}{dt} \times l}{(T_1 - T_2)}$$

$$= \frac{1 \times 8.99 \times 0.1 \times (1/48) \times 0.47}{16}$$

$$= 5.58 \times 10^{-4} \text{ cal/cm/s/°C}$$

15. Wiedemann-Franz law applies to copper and nickel at room temperature. Determine thermal conductivities of these materials using electronic contribution. Electrical resistivity of nickel and copper at room temperature are 1.39×10^{-8} Ωm and 1.64×10^{-8} Ωm, respectively. Lorentz's number is 2.45×10^{-8} WΩ/K^2.

Solution: Given that

$$L = 2.45 \times 10^{-8} \text{ W}\Omega/\text{K}^2$$

$$\rho_{Ni} = 1.39 \times 10^{-8} \ \Omega\text{m}$$

$$\rho_{Cu} = 1.64 \times 10^{-8} \ \Omega\text{m}$$

$$T = 300°\text{K}$$

We know that the relation between thermal and electrical conductivities using Wiedemann-Franz law

$$\frac{K}{\sigma} = L \times T$$

$$K = \sigma(L \times T)$$

So, thermal conductivity of copper

$$K_{Cu} = \sigma_{Cu}(L \times T)$$

$$= \frac{(L \times T)}{\rho_{Cu}}$$

$$= \frac{(2.45 \times 10^{-8} \times 300)}{1.64 \times 10^{-8}}$$

$$= \frac{735}{1.64} = 448.17 \text{ W/m/K}$$

Similarly, thermal conductivity of nickel

$$K_{Ni} = \frac{(L \times T)}{\rho_{Ni}}$$

$$= \frac{(2.45 \times 10^{-8} \times 300)}{1.39 \times 10^{-8}}$$

$$= \frac{735}{1.39} = 528.77 \text{ W/m/K}$$

16. During the performance of Ingen-Hausz experiment, wax melted over 15 cm length of copper rod and over 5 cm length of iron rod. If thermal conductivity of iron is 0.145, determine the thermal conductivity of copper.

Solution: Given that

$l_1 = 15$ cm

$l_2 = 5$ cm

$K_2 = 0.145$

We know that, in Ingen – Housz experiment,

$$\frac{K_1}{K_2} = \frac{l_1^2}{l_2^2}$$

or
$$K_1 = \frac{l_1^2}{l_2^2} \times K_2$$

$$= \frac{5 \times 5}{15 \times 15} \times 0.145$$

$$K_1 = \frac{3.625}{225} = 0.016 \text{ cal/cm/deg/sec}$$

17. Two opposite faces of a metallic plate of thickness 0.19 cm and area of cross section 200 square cm is kept at 100°C temperatures. Determine the quantity of heat that will flow through the metallic plate in one minute. Thermal conductivity of metal is given 0.2 in CGS units.

Solution: Given that:

$$A = 200 \text{ cm}^2$$

$$T_2 - T_1 = 100°\text{C}$$

$$Q = 5.24 \text{ kJ} = 5.24 \times 10^3 \text{ J}$$

$$l = 0.19 \text{ cm}$$

$K = 0.2$ in CGS and time $t = 1$ min $= 60$ sec

The amount of heat flowing in aluminum rod is given by

$$Q = \frac{KA(T_2 - T_1) \times t}{l}$$

$$= \frac{0.2 \times 200 \times 100 \times 60}{0.19}$$

$$= 1.98 \times 10^5 \text{ cal}$$

18. A 40 cm long bar of 6 cm^2 uniform cross-sectional area is made of two equal proportions of copper and iron. The initial and final end of composite rod is kept at temperature 225°C and 0°C, respectively. If the sides of bar are thermally insulated, determine the rate of flow of heat along the bar at steady state. Thermal conductivity of copper and iron are 0.9 and 0.12 in CGS units, respectively.

Solution: Let us suppose that T_1 and T_2 are temperatures of hot and cold ends of plate, respectively, T is temperature of composite plate and thermal conductivity of composite plate is K.

Given that,

$$l_1 = l_2 = 20 \text{ cm}$$

So, the thickness of composite plate $= l_1 + l_2 = 40$ cm

$$A = 6 \text{ cm}^2$$

$$T_1 = 225°\text{C}$$

$$T_2 = 0°\text{C}$$

$$K_1 = 0.9$$

$$K_2 = 0.12$$

We know that,

At steady state, the amount of heat flow in composite plate is given by

$$Q = \frac{K_1 A(T_1 - T)}{l} = \frac{K_2 A(T - T_2)}{l}$$

$$T = \frac{(K_1T_1 + K_2T_2)}{(K_2 + K_1)}$$

$$T = \frac{(0.9 \times 225 + 0.12 \times 0)}{(0.9 + 0.12)}$$

$$= \frac{202.5}{1.02} = 198.5°C$$

Quantity of heat flow in composite plate of thickness 40 cm and thermal conductivity K is given by

$$Q = \frac{K_1A(T_1 - T)}{2l}$$

$$= \frac{0.9 \times 6 \times (225 - 198.5)}{20}$$

$$= \frac{0.9 \times 6 \times 26.5}{20}$$

$$= \frac{143.1}{20} = 7.15 \text{ cal/sec}$$

19. Determine the time in which a layer of ice of thickness 10 cm over the pond surface increases by 5 cm. The temperature of outer surrounding is given as –10°C, coefficient of thermal conductivity of ice is 0.005, density of ice at 0°C is 0.91 gm/cm^3 and latent heat of ice is 80 cal/gm.

Solution: Given that

$$l_2 = 10 + 5 = 15 \text{ cm}$$

$$K = 0.005 \text{ W/m/K}$$

$$L = 80 \text{ cal/gm}$$

$$\rho = 0.91 \text{ gm/cc}$$

We know that the time spent in increasing the thickness of layer of ice from l_1 to l_2 is given by

$$\int dt = \frac{\rho \times L}{KT}\int_{l_1}^{l_2} l\,dl$$

$$t = \frac{\rho \times L}{2KT} \times (l_2^2 - l_1^2)$$

$$= \frac{0.91 \times 80}{2 \times 0.005 \times 10} \times [15 \times 15 - 10 \times 10]$$

$$= 8.99 \times 10^4 \text{ sec} = 24.97 \text{ hours}$$

Exercise

Short Answer Questions

1. What do you mean by the temperature gradient and thermal conductivity?
2. What is the dimension of coefficient of thermal conductivity? Define the thermal conductivity.
3. Define steady state in a metallic bar.
4. Explain heat conduction phenomenon.
5. Compare the electrical and thermal conductivities. What is the analogy between these two?
6. Write the fundamental law of conduction of heat.
7. Write the differences between steady state and variable state.
8. What do you mean by the periodic flow of heat?
9. State the Wiedemann-Franz law.
10. Explain any two applications of thermal radiation.
11. State Ingen-Hausz experiment and write its importance.
12. Compare electrical and thermal conductivities of a good conductor.
13. Compare series and parallel formations of a rod.
14. What do you mean by coefficient of thermal conductivity? Define thermal steady state.
15. What do you mean by the thermal conductivity of a substance. Obtain an expression for it.

Long Answer Questions

1. What do you mean by rectilinear flow of heat? Establish the Fourier differential equation for rectilinear flow of heat.
2. Obtain the relation to calculate the thermal conductivity of poor conductors using cylindrical shell method.
3. Describe the Lee's disc method to determine the thermal conductivity of poor conductors.
4. Determine the thermal conductivity of a rubber tube. Explain the theory used in the process.

5. What do you mean by thermal conductivity? Explain the Forbe's method to determine the thermal conductivity of a metallic rod.
6. Establish the differential equation for one-dimensional flow of heat energy in a metallic bar. Compare the results (i) there is no radiation loss from the sides, and (ii) radiation loss is taken into account.
7. Derive the Fourier equation for one-dimensional flow of heat. Obtain the expression for temperature T at a distance l from the hot end keeping zero radiation loss.
8. Describe the periodic flow of heat. Explain the Angstrom's method to determine the periodic flow of heat in a insulated bar.
9. Derive an equation of rectilinear flow of heat and on the basis of it, explain Ingen-Hausz's experiment.
10. Establish t relation for rectilinear flow of heat along a long bar, which is being heated at one end. If both the ends of rod are kept at constant temperatures T_1 and T_2, determine the steady state solution of it.
11. Discuss the flow of heat in composite wall and obtain the expression of heat flow through the wall.
12. Explain Lee's method to determine thermal conductivity of bad conductors which are taken in the form of disc.
13. Obtain the expression for rate of flow of heat along the wall of spherical shell.
14. What do you mean by the cylindrical flow of heat? Explain the experiment to determine the thermal conductivity of rubber or glass.
15. Explain the periodic flow of heat along a rod, which is periodically heated and cooled at one end.
16. Write the differences between thermal conductivity and thermal diffusivity of a substance. Explain the method to determine the thermal conductivity of bad conductors.
17. Explain the rectilinear flow of heat through a bar of uniform cross-sectional area. Prove that the steady state differential equation of metallic bar, which is heated at one end is given by

$$\frac{d^2T}{dl^2} = \mu^2 T$$

where, the symbols used in above equation have their usual meaning.
18. Explain the thermal conductivity of a metallic bar. Prove that the length up to which wax melts along a bar at steady state is proportional to the square root of thermal conductivity coefficient of metallic bar.
19. Define diffusivity of a substance. Obtain the Fourier equation for conduction of heat along one dimension conductor.
20. Determine the equivalent thermal conductivity of a composite metallic slab consisting of two slabs of different material.

Numerical Problems

1. Length of a copper rod is 30 cm, area of cross section is 0.682 cm^2 and coefficient of thermal conductivity is 270 W/m/K. One end of the rod is heated at 383°K and other end is fixed at 303°K. Determine the quantity of heat which will pass for 15 minutes in the rod.
2. A 40 cm long aluminum rod of cross-sectional area 0.747 cm^2 is heated at one end, at temperature 130°C. Coefficient of thermal conductivity of aluminum is 338 W/m/K. Other end of the rod is kept at 40°C. Calculate the quantity of heat which will pass for 25 minutes in the rod.
3. A nickel rod of cross-sectional area 7.25×10^{-5} m^2 is heated from one end at 137°C and other end is fixed at 40°C. The 6.27 kJ heat passes from nickel for 14 minutes. Determine the length of nickel rod used in the experiment. Given that the coefficient of thermal conductivity of nickel is 80.1 W/m/K.
4. Two thin concentric circles of radii 8 cm and 12 cm, respectively are filled with charcoal powder. The thermal conductivity of charcoal powder is 3.31×10^{-4}. If 12.6 watt heat energy is supplied at the center of a heater, determine the temperature developed between the two shells.
5. The area of cross section of two bars is 39 cm and 41 cm, respectively. These two bars are brought together to join at one end. The other end of the first and second bar is kept at 89.8°C and 29.99°C, respectively. The temperature at the joint of the bar is 69.86°C. If the thermal conductivity of one bar is 83 W/m/K, determine the thermal conductivity of the second bar.
6. Wax melted over 30 cm of nickel rod and 4 cm of aluminum rod. If the conductivity of nickel is 312 W/m/K. Using the experiment of Ingen-Hausz, determine the coefficient of thermal conductivity of aluminum.
7. An ice layer of thickness l cm on the top surface of a pond is increased by 2 mm. The thermal conductivity of ice is 0.0047 W/m/K, density at 0°C is 0.91 g/cc and latent heat of ice is 80 cal/gm. At atmospheric temperature of 20°C below zero, the time spent in this process is 222 sec. Determine the value of thickness l.
8. Two opposite faces of a metallic plate of thickness 0.21 cm and area of cross section 210 square cm are kept at 110°C temperature. Determine the quantity of heat that will flow through the metallic plate in two minutes. Thermal conductivity of metal is given 0.21 in CGS units.
9. Determine thermal conductivities of copper and nickel using Wiedemann-Franz law of electronic contribution. Electrical resistivities of nickel and copper at room temperature are 1.43×10^{-8} Ωm and 1.77×10^{-8} Ωm, respectively. Lorentz's number is 2.45×10^{-8} $W\Omega/K^2$.
10. A layer of ice of thickness 3 cm increases by 1 mm on the top surface of a pond at atmospheric temperature of 40°C below zero, determine the time spent in this process. The thermal conductivity of ice is 0.0039 W/m/K, density 0 at 0 C is 0.90 g/cc and latent heat of ice = 80 cal/gm.

Multiple-Choice Questions

1. Coefficient of thermal conductivity of a bad conductor is measured by
   (i) Searle's method (ii) Lee's disc method
   (iii) Forbe's method (iv) Ingen-Hausz method
2. The temperature of two ends of a rod are T_1 and T_2 respectively. The amount of heat flowing along the rod is Q cal/sec. If at constant temperature, all the linear dimensions are just doubled, the rate of flow of heat Q will be
   (i) $9Q$ (ii) $4Q$
   (iii) $2Q$ (iv) $Q/2$
3. If a plate is made by combining of two plates of equal thickness and conductivities K_1 and K_2, respectively. The composite plate has the coefficient of thermal conductivity
   (i) $\dfrac{K_2}{(K_1+K_2)}$ (ii) $\dfrac{2K_1K_2}{(K_1+K_2)}$
   (iii) $\dfrac{2K_1}{(K_1+K_2)}$ (iv) $\dfrac{2(K_1+K_2)}{K_1}$
4. A rod of conducting material is heated from both the ends. The flow of heat along the metallic rod is not affected by
   (i) Mass of the rod (ii) Cross-sectional area
   (iii) The temperature gradient (iv) Time
5. In the position of steady state of a substance, the temperature of substance
   (i) Does not change with time and remains constant at all the points of substance
   (ii) Decreases with time
   (iii) Increases with time
   (iv) None of the above
6. The temperature gradient in the earth's crust is given as 32°C per km. The mean thermal conductivity of the rocks is 0.008 in CGS units. If the radius of the earth is 6000 km, amount of heat loss daily by the earth is approximately
   (i) 100 cal (ii) 10^{20} cal
   (iii) 10^{60} cal (iv) 10^{40} cal
7. The rectilinear flow of heat along a spherical shell is given by
   (i) $K=\dfrac{Q(l_2-l_1)}{4\pi l_1 l_2(T_1-T_2)}$ (ii) $K=\dfrac{Q(T_1-T_2)}{4\pi l_1 l_2(l_2-l_1)}$
   (iii) $K=\dfrac{Q}{4\pi l_1 l_2(T_1-T_2)}$ (iv) $K=\dfrac{Q(l_2-l_1)}{(T_1-T_2)}$

8. Ingen-Hausz experiment is represented by

(i) $K_1K_2 = \frac{l_1^2}{l_2^2}$ (ii) $\frac{K_1}{K_2} = \frac{l_1^2}{l_2^2}$

(iii) $\frac{K_2}{K_1} = \frac{l_1^2}{l_2^2}$ (iv) $\frac{K_1}{K_2} = l_1^2 l_2^2$

9. An equal amount of ice melts in two identical pots in 180 and 120 seconds, respectively. The ratio of thermal conductivity of material of two pots is

(i) 1 : 4 (ii) 2 : 3

(iii) 4 : 1 (iv) 3 : 4

10. The two similar copper cubes are painted by two different paints. If both the cubes are heated up to same temperature and then kept in vacuum, the losss of heat will be most rapid in

(i) Painted shiny white (ii) Painted shiney black

(iii) Painted rough white (iv) Painted rough black

11. The thermal conductivity of specimen enclosed in between the two concentric spheres of radii R_1 and R_2 is given by

(i) $K = \frac{Q(R_2 - R_1)}{4\pi R_1 R_2 (T_1 - T_2)}$ (ii) $K = \frac{Q}{4\pi R_1 R_2 (T_1 - T_2)}$

(iii) $K = \frac{Q(R_2 - R_1)}{R_1 R_2 (T_1 - T_2)}$ (iv) $K = \frac{(R_2 - R_1)}{4\pi R_1 R_2 (T_1 - T_2)}$

12. The amount of heat flow in a composite plate of thickness l and thermal conductivity K is given by

(i) $Q = \frac{K_1(T_1 - T)}{2l}$ (ii) $Q = \frac{K_1 A(T_1 - T)}{2l}$

(iii) $Q = \frac{K_1 A T_1}{2l}$ (iv) $Q = \frac{K_1 A(T_1 - T)}{2lT_1}$

13. The Wiedemann-Franz law for electrical and thermal conductivities can be written as

(i) $\frac{K}{\sigma} = L \times T$ (ii) $\frac{K}{\sigma} = 2(L \times T)$

(iii) $\frac{K}{\sigma} = \frac{L}{2} \times T$ (iv) $\frac{K}{\sigma} = L \times T^2$

14. Heat conducted per second from the upper disc to the lower disc in Lee's disc method for conductivity of bad conductor is given by

(i) $Q = \frac{K(T_1 - T_2)}{l}$ (ii) $Q = \frac{\pi R^2 (T_1 - T_2)}{l}$

(iii) $Q = \frac{K \times \pi R^2 (T_1 - T_2)}{l}$ (iv) $Q = \frac{K \times \pi R^2}{l}$

15. The time spent in increasing the thickness of layer of ice from l_1 to l_2 at the top surface of a metal is given by

(i) $\int dt = \frac{\rho \times L}{KT} \int_{l_1}^{l_2} l dl$ (ii) $\int dt = \frac{L}{KT} \int_{l_1}^{l_2} l dl$

(iii) $\int dt = \frac{L}{\rho \times KT} \int_{l_1}^{l_2} l dl$ (iv) $\int dt = \frac{\rho}{KT \times L} \int_{l_1}^{l_2} l dl$

16. In the famous Ingen-Hausz experiment, the coefficient of thermal conductivity and length of the rod are related by

(i) $\frac{K}{L} = \text{constant}$ (ii) $\frac{2K}{L} = \text{constant}$

(iii) $KL = \text{constant}$ (iv) $\frac{K}{L^2} = \text{constant}$

17. Two plates of thermal conductivities K_1 and K_2 are joined to make a single plate of double thickness. The equivalent coefficient of thermal conductivity of the combined plate is

(i) $\frac{K_1 K_2}{K_1 + K_2}$ (ii) $\frac{K_1 K_2}{2(K_1 + K_2)}$

(iii) $\frac{2K_1 K_2}{K_1 + K_2}$ (iv) $\frac{4K_1 K_2}{K_1 + K_2}$

18. Two rods of thermal conductivities $2K$ and K are joined. The length of the rods are L and $2L$. The equivalent coefficient of thermal conductivity of the composite system is

(i) $\frac{4}{3}K$ (ii) $\frac{2}{3}K$

(iii) $\frac{8}{3}K$ (iv) $1.2K$

Answers

1. (i) 2. (iii) 3. (ii) 4. (i) 5. (i) 6. (ii) 7. (i) 8. (ii)
9. (ii) 10. (i) 11. (i) 12. (ii) 13. (i) 14. (iii) 15. (i) 16. (i)
17. (iii) 18. (iv)

Index

M

N

P

R

S

T